Springer Textbooks in Earth Sciences, Geography and Environment

The Springer Textbooks series publishes a broad portfolio of textbooks on Earth Sciences, Geography and Environmental Science. Springer textbooks provide comprehensive introductions as well as in-depth knowledge for advanced studies. A clear, reader-friendly layout and features such as end-of-chapter summaries, work examples, exercises, and glossaries help the reader to access the subject. Springer textbooks are essential for students, researchers and applied scientists.

Manuel Bustillo Revuelta

The Basics of Aggregates

Springer

Manuel Bustillo Revuelta
Faculty of Geology
Complutense University of Madrid
Madrid, Spain

Editorial Contact
Simon Shah-Rohlfs

ISSN 2510-1307 ISSN 2510-1315 (electronic)
Springer Textbooks in Earth Sciences, Geography and Environment
ISBN 978-3-031-42960-6 ISBN 978-3-031-42961-3 (eBook)
https://doi.org/10.1007/978-3-031-42961-3

This Springer imprint is published by the registered company Springer Nature Switzerland AG
The registered company address is: Gewerbestrasse 11, 6330 Cham, Switzerland

Paper in this product is recyclable.

This book is dedicated to my grandchildren Guillermo and Manuel. They are the future.

Foreword

Everything in our world is built on geology, and, with a few notable exceptions, the vast majority of that world is also built from geology. Despite the fact that geology is all around us and absolutely fundamental to where we live and how we live, the availability and supply of the vast majority of the mineral resources we rely upon in our daily lives are often taken for granted. On occasions, certain types of resources may appear as headlines in the media and attract public and political attention, normally as a consequence of their availability or cost. Hydrocarbon fuels are one example of this, and more recently the need for, and scarcity of, critical minerals has grown in profile. This reflects their importance in our transition to net zero and the geo-political influences on access, availability and supply, which in turn is influenced by their discrete geographical distribution around the world.

But of all the mineral resources that society requires, it is the incredibly diverse range of aggregates—sand, gravel and crushed rock—that provide the very foundation and fabric of the modern world we live in. From homes and hospitals, to roads and railways, to energy and water infrastructure, aggregates touch and influence every part of our life.

In this respect, they represent an essential resource that underpins global economic growth and development—this is particularly important in developing economies, where the priority is to help transition their populations out of poverty. For this reason, it is little wonder that after water, aggregates represent the largest natural resource that is consumed by global society—around 50 billion tonnes each year.

Looking to the future, aggregates will play a key role in helping society transition to a zero carbon economy, through the construction of new energy infrastructure and green transport infrastructure. Aggregate resources will also play an important role helping us adapt to a changing climate, helping to protect vulnerable coastlines from sea level rise and more extreme weather events as well as providing more resilient infrastructure.

Yet despite the strategic importance and value of this essential resource, their ubiquitousness means that too often the availability and supply of aggregates can be assumed. The growing influence of sustainable sourcing and particularly the transition towards a more circular economy for resource use means that it has never been more important for the essentiality of aggregate resources to be recognised. Sustainable supply means ensuring the right minerals are available to be used in the right place, in the right way and at the right time. This requires the availability and supply of aggregate resources to be properly planned for, monitored and managed.

Aggregates represent the largest volume of mineral resources flowing through our global economy, and this book shines a light on a fascinating but often overlooked subject area that is nevertheless absolutely vital for economic growth and development. It provides a comprehensive insight into the huge diversity of aggregate resources that exist around the world and explores their geological origins and how these influence their often unique properties, alongside the wide variety of different uses that these versatile minerals can be used for. Exploration, site investigation and extraction methods are considered, alongside the environmental consequences of extraction and how these can be effectively planned and managed to ensure the extraction and use of aggregate mineral resources contribute towards the objectives of sustainable development.

Mark Russell
Executive Director, Planning and Mineral Resources
Mineral Products Association
London, UK

Acknowledgements

As in my previous two books for Springer, this project would not have been possible without the assistance of my Editor, Alexis Vizcaíno; he trusted in me from the very beginning. I also take this opportunity to thank my students in the Faculty of Geology (Geologists and Engineering Geologists); they have, without doubt, been and are my guide for the last 40 years. Finally, I wish to express my grateful thanks to Mark Russell for the Foreword.

On the other hand, I would like to acknowledge the help given by many companies that have kindly provided images. However, it is essential to remember that many persons were behind these corporations. In particular, I would like to thank the following people:

Marcos Endrina (Conorsa), María Ángeles Vidal (Holcim), José Luis Corbacho (SAMCA), Pedro Rodríguez (Magnesitas de Rubián, S.A.), Octavio de Lera (Grupo Cementos Portland Valderrivas), Antonio Durán (Benito Arnó e Hijos, S.A.U.), Ángel Granda and Teresa Granda (International Geophysical Technology), Luis Fueyo (Fueyo Editores), Juan Luis and Ainhoa Etxebarria (Calcinor), José Alberto García Fuentes and Teresa Busqué (CEMEX), María Ángeles Bustillo, José Pedro Calvo, María José Huertas, Eduardo Revuelta, Miguel Ángel Sanz and Alwyn Annels.

Manuel Bustillo Revuelta

Contents

1	**Introduction**	1
1.1	Aggregate Types	5
1.2	World Aggregate Market	8
1.3	Mining Cycle of Aggregate Production	9
1.4	Aggregate Applications	12
1.5	Aggregates and Circular Economy	14
1.6	Criticality of Aggregate Resources	18
1.7	Environmental Concerns: Illegal Mining	19
1.8	Questions	20
	References	20
2	**Properties and Testing**	21
2.1	Introduction	22
2.2	General Properties and Tests	23
2.2.1	Petrographic Description	23
2.3	Geometrical Properties and Tests	26
2.3.1	Particle Size Distribution—Grading	26
2.3.2	Shape and Surface Texture	30
2.3.3	Fines	34
2.4	Mechanical and Physical Properties and Tests	38
2.4.1	Physical Properties and Tests	38
2.4.2	Mechanical Properties and Tests	40
2.5	Thermal and Weathering Properties and Tests	46
2.5.1	Thermal Properties and Tests	46
2.5.2	Weathering Properties and Tests	47
2.6	Chemical Properties and Tests	50
2.7	Questions	50
	References	51
3	**Geological Occurrence**	53
3.1	Introduction	54
3.2	Sand and Gravel	56
3.2.1	Fluvial Deposits	62
3.2.2	Glacial Deposits	69
3.2.3	Coastal Deposits	72
3.2.4	Marine Deposits	72
3.2.5	Eolian (Windblown) Deposits	76
3.3	Crushed Stone	76
3.3.1	Sedimentary Rocks	77
3.3.2	Igneous Rocks	87
3.3.3	Metamorphic Rocks	95
3.4	Questions	97
	References	99
4	**Exploration and Evaluation of Deposits**	101
4.1	Exploration Process	102
4.1.1	Introduction	102
4.1.2	Phases of Exploration	102
4.1.3	Methods of Exploration	107
4.1.4	Sampling	128
4.2	Evaluation Process	133
4.2.1	Introduction	133

4.2.2 Classical Methods ... 134
4.2.3 Geostatistical Methods ... 137
4.2.4 Economic Evaluation ... 138
4.3 Questions ... 141
References ... 141

5 **Extraction Methods** ... 143
5.1 Introduction ... 144
5.2 Crushed Rock ... 148
5.2.1 Surface Quarrying ... 148
5.2.2 Underground Quarrying ... 157
5.2.3 Super Quarries ... 161
5.3 Sand and Gravel ... 163
5.3.1 Dry Mining ... 164
5.3.2 Wet Mining ... 166
5.3.3 Marine Aggregate Extraction ... 167
5.4 Questions ... 169
References ... 170

6 **Processing Techniques** ... 171
6.1 Introduction ... 172
6.2 Sand and Gravel ... 175
6.2.1 Washing ... 178
6.2.2 Screening ... 187
6.2.3 Stockpiling and Loadout ... 197
6.2.4 Dewatering ... 197
6.2.5 Materials Handling ... 206
6.3 Hard Rock Aggregate ... 208
6.3.1 Crushing ... 208
6.3.2 Screening, Stockpiling, and Loadout ... 221
6.4 Questions ... 223
References ... 223

7 **Environment and Sustainability** ... 225
7.1 Mining, Environment and Sustainability ... 226
7.2 Potential Environmental Impacts and Their Management ... 229
7.2.1 Air Emission of Particulate Matter ... 230
7.2.2 Increased Noise Levels ... 236
7.2.3 Vibrations Associated with Blasting ... 240
7.2.4 Water Condition ... 244
7.2.5 Biodiversity Loss ... 246
7.2.6 Visually Disturbed Landscapes and Reclamation ... 251
7.3 Questions ... 257
References ... 257

8 **Applications** ... 259
8.1 Introduction ... 260
8.2 Concrete and Mortar ... 260
8.2.1 Concrete ... 260
8.2.2 Mortar ... 272
8.3 Roads ... 277
8.3.1 Unbound Materials ... 278
8.3.2 Bituminous Mixtures ... 281
8.4 Railway Ballast ... 286
8.5 Armourstone ... 291
8.6 Gabions and Matresses ... 295

8.7 Filter Media ... 298
8.8 Decoration ... 300
8.9 Questions ... 303
References ... 303

9 **Recycled Aggregates** ... 305
9.1 Introduction ... 306
9.2 Construction and Demolition Waste ... 306
9.2.1 CDW Composition ... 311
9.2.2 Management of CDW ... 317
9.2.3 Valorization Plants ... 326
9.3 Recycled Aggregates ... 335
9.3.1 Properties of Recycled Aggregates ... 336
9.3.2 Applications ... 343
9.4 Questions ... 346
References ... 347

Introduction

Contents

1.1 Aggregate Types – 5

1.2 World Aggregate Market – 8

1.3 Mining Cycle of Aggregate Production – 9

1.4 Aggregate Applications – 12

1.5 Aggregates and Circular Economy – 14

1.6 Criticality of Aggregate Resources – 18

1.7 Environmental Concerns: Illegal Mining – 19

1.8 Questions – 20

References – 20

M. Bustillo Revuelta, *The Basics of Aggregates*,
Springer Textbooks in Earth Sciences, Geography and Environment,
https://doi.org/10.1007/978-3-031-42961-3_1

1

Abstract

This chapter introduces the reader to the world of aggregates. Data on the worldwide consumption of aggregates are discussed as well as their main types and sources, including recycled aggregates. Aggregate applications are described and the mining cycle of aggregate production is briefly introduced. The importance of the influence of aggregate production in the circular economy is further discussed, including sustainability, and the criticality of aggregate resources is highlighted. Finally, illegal extraction of aggregates is cited as one of the most important environmental concerns of aggregate production.

Aggregates, made up of sediments and rock fragments, are the natural resource most used by humans after water. Demand worldwide is increasing exponentially due to the increase in urbanization, development, and population growth (◘ Fig. 1.1). "Developed countries cannot sustain their high level of productivity, and the economies of developing nations cannot be expanded, without the extensive use of aggregates" (Tost and Ammerer 2022).This is because they constitute the main component to carry out one of the human activities that best define, from an economic viewpoint, the degree of development of a country such as the construction of buildings, communication routes, and infrastructure works. In this sense, the aggregate industry follows economic cycles, reacting to the activity in the construction sector. Aggregates are often also referred to as "construction aggregates" because they are necessary for a broad range of construction purposes in buildings and civil engineering structures (◘ Fig. 1.2). These materials have enhanced the quality of life, and, although aggregates are often geologically abundant (natural aggregate resources are among the most abundant and broadly distributed in the Earth's crust), they can be considered critical material securing standards of our well-being.

Aggregates are used not only in the manufacture of concrete (the most widely used construction material since the beginning of the twentieth century) and mortar, but also in the preparation of the different layers that make up roads (◘ Fig. 1.3), in the materials that make up the base for railway structures (called, in this case, ballast—Fig. 1.4) and in a host of other applications. It means that the consumption of aggregates worldwide is several orders of magnitude higher than that of any other essential mineral resources. To get an idea of the importance of the consumption of this material, one kilometer of motorway needs 25,000 tons of aggregates, one cubic meter of concrete 2 tons of aggregates, and one kilometer of double rail approximately 10,000 tons of this material, to name just the three most important applications of aggregates.

Regarding the consumption of aggregates worldwide, The Global Aggregates Information Network (GAIN) offers the data shown in ◘ Table 1.1. As with most raw materials, China is the main producer, with an estimated production of 23.1 billion tons of aggregates for 2022, followed, far behind, by India, with an

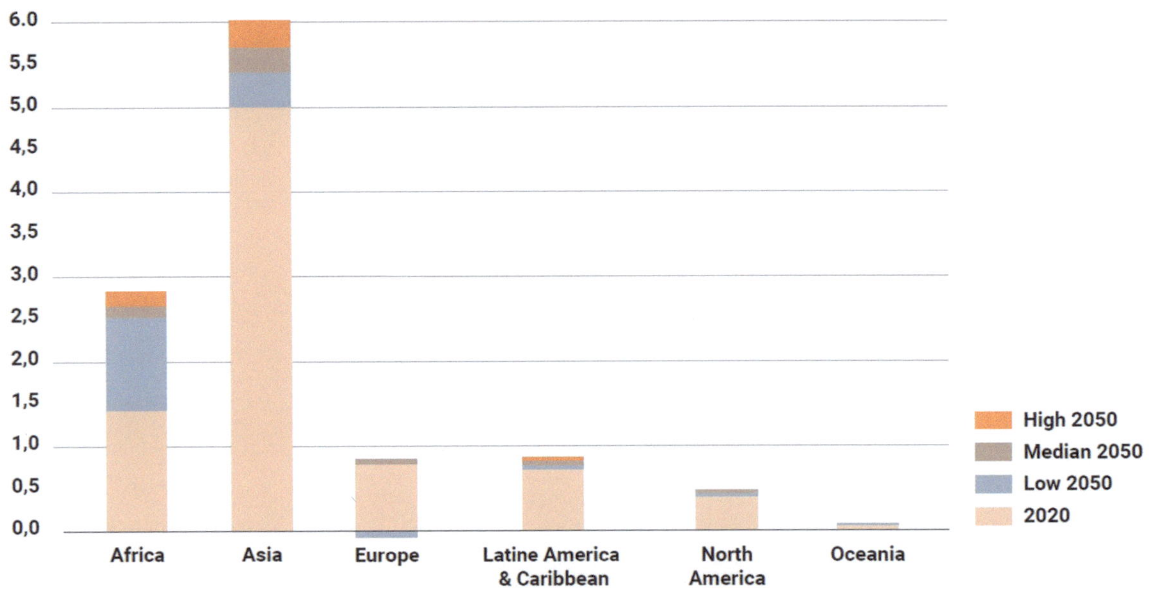

◘ **Fig. 1.1** Projected population increase and urban population trends between 2021 and 2050 (with low, median, and high scenarios) (UNEP 2022)

Fig. 1.2 Hoover dam (image courtesy of Luis Fueyo)

estimated production of 5330 million tons of aggregates. The United States, with an estimated aggregate production of 2540 million tons (960 million tons of sand and gravel and 1500 million tons of crushed stone—Willett 2023), and Europe, with an estimated aggregate production of 3015 million tons, are the group of the largest aggregate-consuming economies in the world. To serve as a comparison, around 20 million tons of copper, approximately 8000 million tons of coal, and 2500 million tons of iron mineralization were produced in the world for that same year, to name three of the most important solid raw materials consumed today. It is important to note that these amounts of aggregate consumption will increase enormously in the future due to the renovation of deteriorated roads, highways, bridges, airports, seaports, etc. According to the Organisation for Economic Co-Operation and Development (OECD), construction materials, specifically aggregates, dominate total materials use in 2011 and 2060 (Fig. 1.5).

Two of the basic characteristics that make aggregates so widely used are its low price (no more than 7 dollars per ton in many developed countries; the price of aggregates is stable compared to other minerals) and the ease, from a geological point of view, of finding a deposit that can be exploited. Nevertheless, the price depends on the specifications for specific end uses. The most typical example is the aggregates for railway ballast, which generate a very high price since their specifications are difficult to attain. On the other hand, lower quality aggregates can be suitable for applications of low intensity of utilization.

The production and consumption of aggregates are often locally or regionally controlled. Much of the value of the aggregates depends on their localization near the market and thus they have the so-called "high place value". Transporting aggregates long distances can greatly increase the final price and distant deposits can be uneconomic. As a rule, the international trade of aggregates is limited to local transactions across neighboring countries. It is estimated that "less than 5% of global aggregates production moves across borders, in particular to countries that have less geological availability of suitable materials for aggregates in combination with strong demand for large development projects (i.e., Singapore)" (UNEP 2019).

Since the places of production and consumption cannot be separated by more than 25 or 50 km at most,

Fig. 1.3 Unbound aggregate base layer (image courtesy of pavement interactive)

Fig. 1.4 Ballast (image courtesy of Plasser Australia)

Table 1.1 Estimated global aggregates tonnages 2019–2022 (*Source* GAIN)

GAIN country	Actual 2020 mt	Actual 2020 mt	% Change 2020 versus 2019	Estimated 2021 mt	% Change 2021 versus 2020	Estimated 2022 mt	% Change 2022 versus 2021
China	20,000	20,370	1.9	22,000	8.0	23,100	5.0
India	6035	4920	− 18.5	5125	4.2	5330	4.0
Europe	2986	2884	− 3.4	2955	2.5	3015	2.0
USA	2388	2331	− 2.4	2410	3.4	2540	5.4
Brazil	535	605	13.1	660	9.1	693	5.0
Canada	437	428	− 2.1	445	4.0	455	2.2
Mexico	400	325	− 18.9	345	6.3	366	6.0
Japan	375	355	− 5.3	360	1.4	345	− 4.2
South Korea	253	252	− 0.4	254	0.8	259	2.0
Australia	180	177	− 1.7	171	− 3.4	182	6.4
South Africa	180	160	− 11.1	155	− 3.1	165	6.5
UAE	158	135	− 14.5	145	7.4	150	3.4
Colombia	144	130	− 9.7	146	12.3	169	15.8
Argentina	137	69	− 49.5	76	10.1	69	− 9.2
Malaysia	129	75	− 41.6	69	9.3	75	10.3
New Zealand	42	40	− 4.8	42	5.0	45	7.1
GAIN totals	34,379	33,256	− 3.3	35,358	6.3	36,958	4.5
Rest of world	9883	8955	− 9.4	8942	− 0.1	9248	3.4
Global totals	44,262	42,211	− 4.6	44,300	4.9	46,206	4.3

efficient transportation of these materials is a significant factor to provide reasonable costs in construction projects. Nevertheless, although the potential sources of aggregates are widespread and large, the existence of factors such as conflicting land uses, economic considerations, local regulations, citizen opposition, and environmental concerns can drastically limit their availability. "The combination of the competition for land use and the social rejection of mining near urban areas is forcing the extraction points to be located away from the demand areas, increasing the transport distances with subsequent economic, social and environmental costs" (Escavy et al. 2022). Consequently, aggregate reserves mainly depend on land use, proximity to consumption centers, and local environmental concerns. In particular, the principal bottleneck for aggregate supply is land use competition. Social conflicts can also cause market supply shortages at the local level.

1.1 Aggregate Types

There are numerous definitions of the term aggregates depending on the source that is used. However, the simplest and most accurate definition is probably the one offered by the European Association of Aggregates (UEPG), which defines them as "granular materials used in construction". This granular material can be produced by crushing strong rocks such as granite and limestone or from particulate deposits such as sand and gravel. For this reason, primary aggregate materials are grouped into two main categories: the unconsolidated deposits of rock fragments that form part of surficial deposits and the hard rocks that constitute part of the Earth's crust. Figure 1.6 shows the main types of aggregates that can be used in the construction sector and Fig. 1.7 shows the split in terms of type, with crushed rock being the largest portion (45%) followed by sand and gravel (40%). Both types of primary or natural aggregates are utilized in construction, according to the specification standards and economic considerations. For instance, in the manufacturing of concrete sand and gravel are usually selected while crushed stone is preferred in asphaltic mixes since asphalt adheres better to rough and tough surfaces. Aggregate deposits, regardless of whether they are rocks for crushing or sand and gravel for direct utilization, are present in numerous geological environments, all with their unique characteristics (see ▶ Chap. 3).

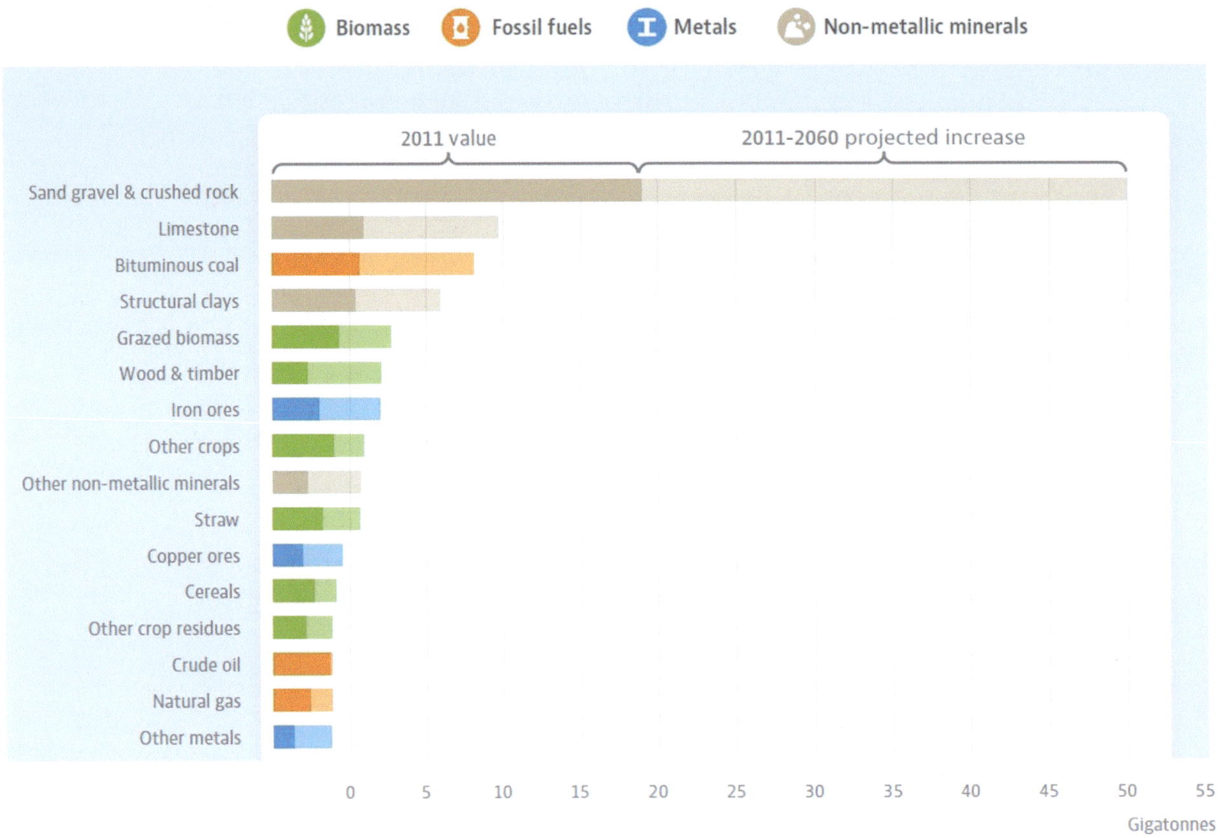

■ **Fig. 1.5** Materials use in 2011 and 2060 (OECD 2018)

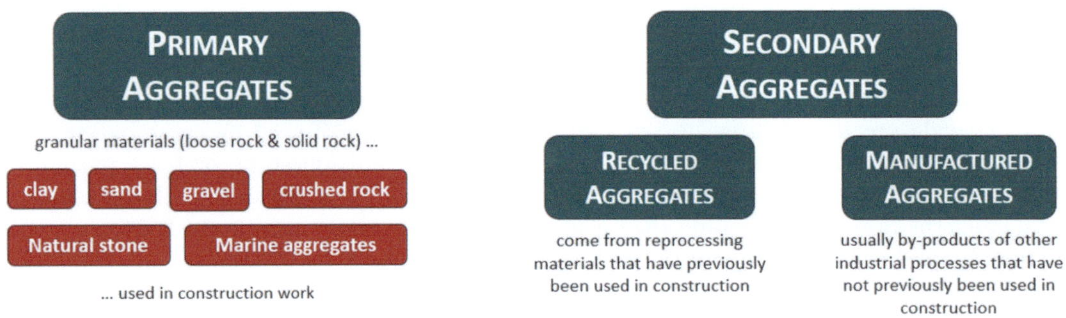

■ **Fig. 1.6** Basic classification of aggregates

From a geological viewpoint, glaciers, rivers and streams have contributed to the formation of most sand and gravel deposits (■ Fig. 1.8) (the most important transport medium is flowing water as in rivers). On the other hand, for every region of the world, the availability of rocks (i.e., granite or limestone—Fig. 1.9) for crushing relies upon the global geological history of the region. The relative proportions of crushed stone and sand and gravel production do not reflect the presence of sand and gravel deposits (including marine aggregates), which usually need the simplest treatments and are therefore cheaper to produce. This is because environmental regulations constrained the exploitation of this type of natural aggregate in many developed countries, producing a decline in its utilization.

Currently, another significant type of aggregate is the so-called recycled aggregate (see ▶ Chap. 9) (UEPG defines recycled aggregates (RA) as "reprocessed materials previously used in construction") (■ Fig. 1.10); they are becoming an important source of aggregates. In the last decades, the consumption of recycled aggregates has grown exponentially. Some examples include recycled concrete from construction and demolition waste (CDW) (■ Fig. 1.11) and reclaimed aggregates from asphalt plannings from road resurfacing and railway ballast. In the case of Europe, the European

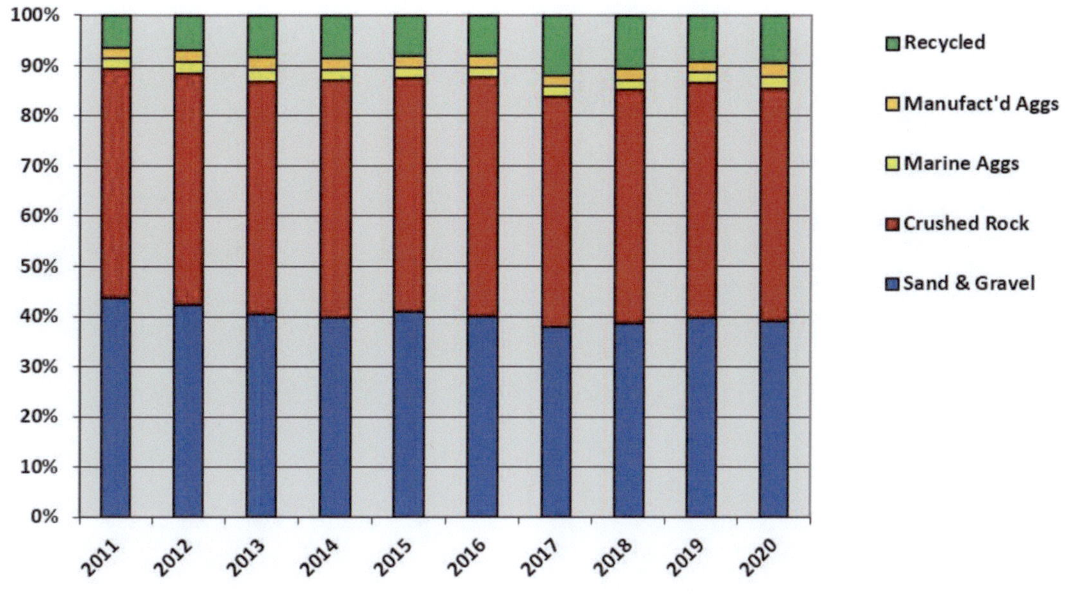

Fig. 1.7 Trend in aggregates type in (%) from total production (EU27 + UK + EFTA) (*Source* UEPG)

Fig. 1.8 Sand and gravel deposit

Commission has identified construction and demolition waste as a priority waste stream for reuse and recycling. The growth of the world population and economic conditions of developing countries has considerably augmented the development of the building and construction industry. However, while their

Fig. 1.9 Limestone aggregate quarry

Fig. 1.10 Recycled aggregates (image courtesy of Salmedina)

evolution has contributed to world economic growth, they also originate from multiple sources of waste. The utilization of recycled aggregates obtained from the treatment of this waste in the construction industry is the best way to solve this issue. Finally, manufactured aggregates, by-products of other industrial processes, can also be used in the construction industry.

1.2 World Aggregate Market

The construction aggregate market includes the supply and demand of construction aggregates. This market is global with many suppliers and consumers located in many different areas and countries worldwide. This is because, as noted above, transporting aggregates long distances greatly increases the price of the final product and thus there is little international trade.

Market segmentation is usually carried out based on the type of aggregate and type of application. Factors such as economic conditions, environmental legislation and government policies can seriously affect the market. Since the market is locally and regionally controlled, these factors can change from region to region and from country to country. The market can also vary if technological advancements (i.e., automation processes) replace the traditional methods of aggregate production.

The factors that drive the market are mainly the growing demand for infrastructure development, the increasing construction projects and the previously cited technological advancements. The growing demand for infrastructure development mainly affects developing countries although many developed countries also need to renew their transport networks (i.e., airports and highways). The increasing demand for construction aggregates is also controlled by new urbanization projects and the expansion of existing infrastructure. In this sense, the growing tourism industry and recreational projects can help to increase construction aggregate demand. Finally, the efficiency in the extraction and production processes has been clearly enhanced in recent years by technological advancements, which reduce costs and increase productivity.

Fig. 1.11 Construction and demolition waste

For example, new methods of processing can reduce the water consumption of aggregate plants, which is clearly excessive.

In contrast, the factors that restrict the market include environmental concerns, the existence of alternative products and the volatility of energy prices. The environmental impact of aggregate extraction (i.e., damage to biodiversity and water pollution) is probably the most important restriction to aggregate production worldwide. As a consequence, environmental legislation restricts aggregate exploitation. The development of new products such as recycled aggregates can also limit the demand for natural aggregates (sand and gravel and crushed rock). Finally, the volatility of energy prices can have a significant impact on the cost of aggregate production because the transportation cost to the construction site is commonly the most important cost in aggregate production, especially when the location is far.

From a geographical viewpoint, the Asia–Pacific region is the most important market for construction aggregates in terms of quantity and present and future growth. This is because there are many urbanization, industrialization, and infrastructure projects in countries such as China and India, which have a third of the world's population. Other traditional markets such as the USA and Europe markets are still important due to the great number of renovation and rehabilitation projects that demand construction aggregates.

1.3 Mining Cycle of Aggregate Production

As with all mineral resources, the mining cycle of aggregate production includes the following basic stages: (1) exploration and evaluation of deposits, (2) extraction, (3) processing, and (4) environmental restoration and/or rehabilitation. Aggregate production starts with exploration, continues through mining and processing and finishes with closure and postmining land use.

Mineral exploration (see ▶ Chap. 4), that is, "the process of analyzing an area of land to find mineral deposits" (Bustillo 2018), involves locating a suitable aggregate resource near where it is to be utilized. It covers the group of processes that gather information about the presence or absence of an aggregate deposit. The information collected during exploration is used to assess the size and quality of the aggregate deposit and to establish if there is the potential for it to be mined. The methods used in aggregate exploration can be organized in order of scale and stage, from photogeology to drilling, through remote sensing and geophysical surveys. Geophysical methods (Fig. 1.12) are essential in the exploration of aggregate resources, especially for sand and gravel deposits.

For its part, the estimation of aggregate reserves (see ▶ Chap. 4), that is, establishing the amount of aggregates present in the deposit, acquires a critical character. Moreover, the results of the evaluation process will later influence the selection of the exploitation method

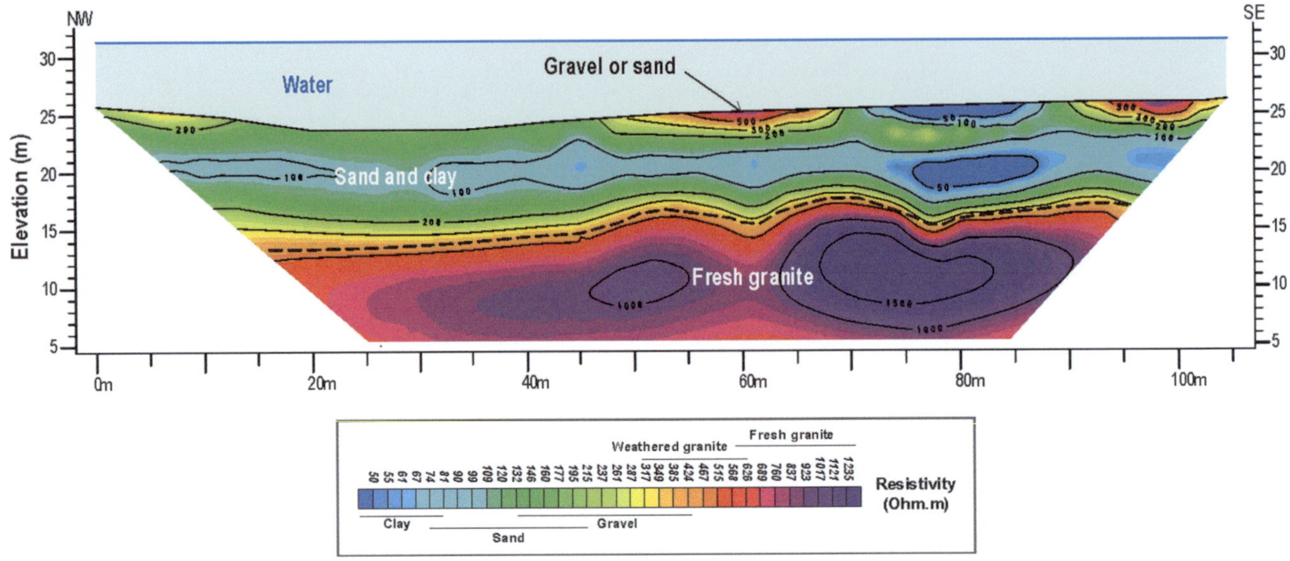

■ **Fig. 1.12** Graphic display of an IP survey (illustration courtesy of International Geophysical Technology)

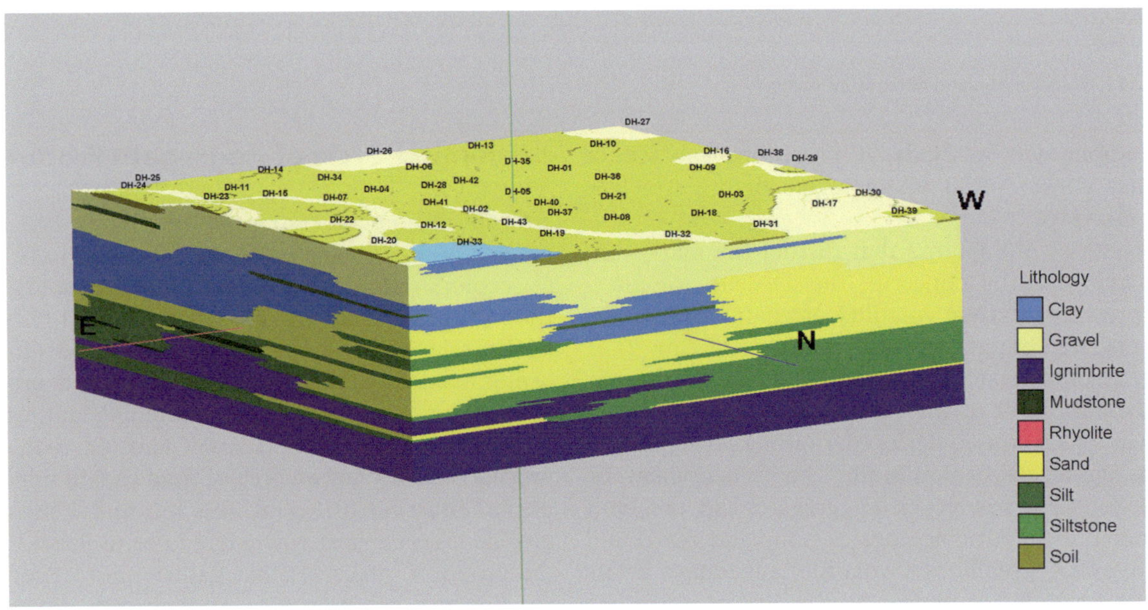

■ **Fig. 1.13** Modelling the deposit for aggregate reserves estimation using Rockworks software

(■ Fig. 1.13). The estimation of the reserves can be carried out using geometric methods or geostatistical methods. The latter appeared at the beginning of the sixties and reached its maximum splendor with the development of computers in the early eighties. Nevertheless, the complexity of these methods makes them rarely used in the evaluation of aggregate deposits.

Extraction (see ▶ Chap. 5) refers to all methods of removal of aggregates from their natural position. This term is used interchangeably with the terms mining and quarrying. The methods to extract aggregates mainly depend on whether the material is sand and gravel or rock, the natural conditions at the site, and the final product. The economic success of aggregate extraction is mainly controlled by the quality of the aggregates and their quantity in the deposit as well as the political, legal, administrative, social, and economic environments in which the extraction occurs.

Although aggregate sources can be widely variable, the methods of extraction are very similar throughout the world, with crushed rocks mainly extracted from quarries (■ Fig. 1.14) and sand and gravel from gravel pits. In gravel pits, the aggregate can be extracted above or below the waterline. In several countries (i.e., the

■ **Fig. 1.14** The greatest aggregate quarry in the world, which is located in United Arab Emirates (image courtesy of Stevin Rock)

United Kingdom), aggregates are also obtained from sea-dredged materials. The mining system and excavating equipment are different depending on the type of aggregate. A large proportion of world aggregate production is obtained using quarrying methods because sand and gravel are not always readily available. Quarrying is a term used to describe the production of construction and building materials such as crushed rocks for aggregate.

The next stage in the mining cycle of aggregate production, aggregate processing (see ► Chap. 6), consists of transporting material extracted in the mine (quarry rock or sand and gravel) to the processing plant (■ Fig. 1.15) for generating the final product. Therefore, the aim of the processing plant is to prepare the materials in an adequate form for their use as aggregates. It is established in key parameters such as particle size and distribution, particle shape, physical and mechanical properties, and lack of contaminants. In this sense, some properties (i.e., particle shape and rock type) will influence the type and range of plant utilized (aggregates are processed using either a mobile device or at a fixed plant). Processing techniques basically involve crushing and screening in crushed stone aggregates whereas sand and gravel materials are processed using screening and washing methods. Both types of aggregates finally need to be stockpiled (■ Fig. 1.16) and loaded out for transport to the market. The transport is carried out on trucks, by trains or barges. Truck is the preferred method of transport at close distances while other lower cost means of transportation such as rail or ship are selected for long distances. Aggregates mostly travel by road for distances up to tens of km while transportation of hundreds of kilometers is carried out preferably by railway.

Many are the interactions between mines and the environment. The environmental impact of mining is perhaps part of the price that humankind must pay for the benefits of mineral consumption because some environmental degradation due to mining is unavoidable. In fact, the rejection of mining activities is not something new, "Roman naturalist Pliny the Elder wrote in the first century AD that mining activities were inappropriate in the homeland and should be developed only in conquered lands" (Escavy et al. 2022).

Environmental considerations are an important part of the modern mining industry and they must be included in all project planning. In summary, the exploitation of aggregate resources is not the problem, but it must be developed in a green and modern

Fig. 1.15 Aggregate processing plant

execution. Aggregate mining operations affect flora and fauna as well as land and water (Fig. 1.17), similar to other major industrial operations. For this reason, aggregate mining companies are carrying out considerable efforts to decrease the environmental impact of mine works and diminish the footprint of their operations throughout the mining cycle, including working to reclaim ecosystems post mining. Old mine sites can be converted to wildlife habitat and refuge recreational areas, shopping malls, golf courses, airports, lakes, underground storage facilities, solid waste disposal areas, mining and power plant waste storages, museums, sites of special scientific interest and regionally important geological sites, industrial land, pisciculture ponds, and many other economically or ecologically productive land utilizations that can benefit society.

1.4 Aggregate Applications

Aggregates are used in many applications (see ► Chap. 8), some of which are essential for modern society. They can be used as a construction material in two large types of applications: unbound applications (the aggregates are not bound) and bound applications (mixes containing binding agents such as cement or bitumen). In order of importance, the main applications of aggregates are the production of concrete (and mortar), unbound and bound (bituminous) pavements, and railway ballast. Other important uses of aggregates are for armourstone and gabions and as filter media; finally, decorative/ornamental uses can be an example of the infinite applications of aggregates. It is necessary to note that the physical, chemical and mineralogical properties of aggregates (see ► Chap. 2) must be known prior to their utilization in different applications.

The largest proportion of aggregates is utilized in the production of concrete (Fig. 1.2), which is the most widely used building material in the construction industry. Aggregates constitute approximately 80% of the concrete weight and 70% of the concrete volume. The aim is to utilize as much aggregate as possible to bind the elements together with the hardened cement paste: aggregates form the skeleton of concrete. Fine aggregate (sand) and coarse aggregate (gravel) are used in concrete production. Sand and gravel aggregates

Fig. 1.16 Stockpiled sand and gravel aggregates

for concrete are defined based on particle size rather than composition. According to the European legislation (EN standards), the term gravel (or more correctly coarse aggregate) is utilized for defining "particles between 4 and 80 mm", and the term sand (or fine aggregate) is used for fragments that are "finer than 4 mm, but coarser than 0.063 mm". As in the case of concrete, plastering and rendering mortars are made with aggregates. The main differences between concrete and mortar components are the types of cementitious materials and the aggregate size.

Aggregates are the principal material in pavement construction (Fig. 1.3) since they constitute, in percentages greater than 90%, most of the layers of road surfaces. This implies a large consumption of these materials in road construction, not only if the road is new but also in its rehabilitation. In fact, the construction of roads is the second most important application of aggregates after concrete production. Unbound aggregates have formed the basis of all pavements since engineering began. They are utilized in road pavements as base and subbase layers. Asphalt is a mixture of approximately 90–95% well-graded aggregates together with filler and, in some cases, additives. Bitumen makes up the remaining less than 5% of the mixture, where it constitutes the black, adhesive component that binds the aggregate together to form a cohesive mixture. The aggregates utilized in a bituminous mixture are formed by either coarse or fine aggregates or, more commonly, a mixture of both sizes.

One of the most demanding applications for crushed aggregate is railway ballast (Fig. 1.4), also known as railroad ballast. Rail tracks are built on compacted granular structures termed ballast, which are laid on natural materials. Railway ballast is a selected crushed and graded aggregate placed upon the railroad roadbed to provide drainage, track stability, flexibility, and uniform support for the rail and ties. This type of aggregate also provides distribution of the track loadings to the subgrade as well as facilitating maintenance. In addition, it deters the growth of vegetation that might interfere with the track structure. The quality of the rock utilized is an essential characteristic in railway track construction. Substantial amounts of crushed coarse aggregate are used as railway track ballast since it is typically made of crushed rock with sizes ranging

Fig. 1.17 Pond related to aggregate extraction

between approximately 20 and 60 mm. The rocks for track ballast are required to be strong, clean and angular with a high resistance to abrasion. Most railway ballast is sourced from siliceous (i.e., igneous) rocks and many countries prohibit the utilization of sedimentary rocks (i.e., limestones or dolostones).

Armourstone consists of coarse aggregates utilized in hydraulic structures and other engineering works (Fig. 1.18). In the context of coastal engineering, armouring means permanent protection to a structure in water, providing safety for both the environment and people. It is principally selected in breakwaters, which are permanent structures built in coastal areas (harbors and shores) for protecting against tides, currents, waves, storms, etc. Armourstone is classically formed by large equant blocks of rocks with masses generally greater than 0.25 tons and, according to data from the European Aggregates Association, it represents approximately 3% of aggregate production in Europe.

To finish with the applications of aggregates, gabions (Fig. 1.19) are robust structures comprising a double-twist wire-mesh basket filled with hard, durable stone. They perform a great variety of functions within coastal, estuarial and fluvial environments. Drainage and filter aggregates are sand, gravel or crushed stone or mixtures thereof. They play significant roles in many engineering projects. Decorative/ornamental aggregates can be used in a wide variety of spaces such as commercial and residential spaces, domestic pathways, driveways, car parks, architectural landscaping, gardening (patios and rockeries), and sports grounds.

1.5 Aggregates and Circular Economy

Although the circular economy (CE) has gained ground in recent years, the exact concept is still unclear and there is no consensus on its definition. The circular economy was first defined by Pearce and Turner (1990) "to explain the feasibility of considering the natural environment in the economic flows through the closing of the industrial cycles". Within this framework, development is approached from a sustainable perspective. The circular economy is a requirement to reach sustainability. Because the accelerated growth of global consumption has resulted in the overexploitation of natural

Fig. 1.18 Aggregate used as armourstone (Santander, Spain)

Fig. 1.19 Gabions

1

resources, the circular economy has emerged as a paradigm to promote more responsible patterns of production and consumption. Ideally, it means moving from everything being disposable to everything being recoverable. Consequently, CE is profoundly changing the pillars of traditional industry. In summary, to achieve sustainable development it is essential to change drastically the way to produce and consume products.

The circular economy is a response to the need to clearly separate economic growth from environmental pressure according to a system based on the three R's, that is, reduction, reuse and recycling, of materials in the production, distribution, and consumption processes. In this sense, to prevent waste generation, minimizing the waste generated and promoting recycling are the concepts underpinning the circular economy (Fig. 1.20). In summary, CE reduces material utilization, redesigns materials to be less resource-intensive and uses waste to produce new materials and products. The circular approach means re-evaluating and modifying the production cycles. It is an innovative full and radical approach compared to the classic production model. The circular economy "is in contrast to the linear, traditional and classical economy, that does not consider neither the origin of resources nor the destination of waste" (Migliore et al. 2020).

Obviously, the circular economy principles also find application in the construction sector, which is specially typified by an extreme consumption of natural raw materials for producing huge amounts of construction and demolition waste. In this sector, it is essential to develop the circular economy with the aim of making progress toward the main sustainability targets whereas conserving natural resources. In order to achieve these goals, the construction industry needs to consider all the aspects related to the circular economy at the early stages of planning. The drive to adopt a circular economy requires changes in the utilization of resources, including nonrenewable aggregates.

One of the pillars of the circular economy must be the construction sector, with the aim of recovering of materials obtained through the replacement of natural raw materials (primary aggregates) with secondary raw materials (recycled aggregates—Fig. 1.21). Recycled aggregates mean a change from a linear to a circular model and sustainability (▶ Box 1.1: Aggregates and Sustainability). It must not be forgotten that the traditional linear approach to aggregates in the construction industry involves extracting, transporting, and utilizing them. Increasing the utilization of recycled aggregates means that the reserves of natural, nonrenewable aggregates will be better maintained and that less waste will be landfilled. Within this scheme, the circular economy of the construction sector, based on a renewed balance between the three R's, aims to seek the sustainability of the building construction and demolition processes. By applying the circular economy principles, it is possible to situate the construction industry on the path of Agenda 2030.

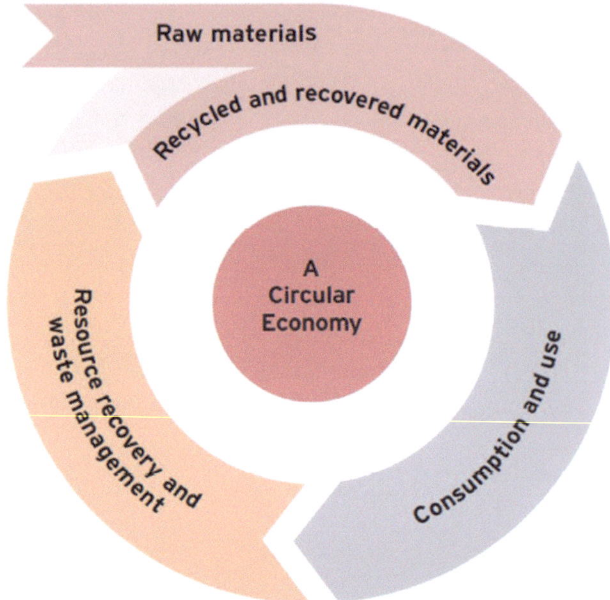

Fig. 1.20 The circular economy for sustainable development (Perkins et al. 2021)

Box 1.1—Aggregates and Sustainability

In recent decades, the terms "sustainable development" and "sustainability" have been used by governments and policy makers worldwide. It is agreed that sustainable development was defined early in 1987 by the Brundtland Commission (Our Common Future, World Commission on Environment and Development, United Nations) as "a system of development that meets the basic needs of all people without compromising the ability of future generations to meet their own life-sustaining needs". Since then, a rich discussion has ensued about what this means in practical terms. Although many other sets of words have been suggested for defining sustainable development, the Brundtland Commission definition has stood the test of time. In the mining world, these words were first utilized in the early 1990s in the Rio Summit (1992). Sustainability is a more general term that captures the idea that we need to maintain certain important aspects of the world over the long term. The UNEP (2019) *report entitled "Sand and Sustainability: Finding New Solutions For Environmental Governance of Global Sand Resources" defines sustainability as "transforming our ways of living to*

Fig. 1.21 Parameters influencing the aggregates recycling market (Tost and Ammerer 2022)

maximize the chances that environmental and social conditions will indefinitely support human security, well-being and health" (in this and other reports of United Nations Environment Programme sand also means aggregates).

When the United Nations Environment Programme (UNEP) released its first report on sand in 2014, the issue of sand and sustainability had not received considerable scientific, or policy, attention at the intergovernmental level. Subsequently, in the UNEP (2019) *report, it is highlighted that "we have been exceeding easily available sand resources at a growing rate for decades" and that "the needs and expectations of our societies cannot be met without improved governance of global sand resources". In other words, sustainability of aggregate production is imperative because the environmental and social impacts of aggregate mining are an issue of global significance. After this report, the sand and sustainability challenge was already being considered by the international community.*

Recently, the United Nations Environment Programme (UNEP) published its latest report (UNEP 2022) *on sand and sustainability, entitled "Sand and Sustainability: 10 strategic recommendations to avert a crisis". In this report, after accepting that "the sand and sustainability challenge is undoubtedly complex" and that "real solutions will need to be cross-sectoral, and in some cases cross-border, where immediate actions across all scales of governance are needed to avert a global crisis", a group of world experts propose 10 solution-oriented recommendations that include "adopting integrated policy and legal frameworks, mapping sand resources, promoting resource efficiency and circularity, sourcing responsibly and restoring our ecosystems. The overarching purpose of this report is to encourage policy makers at all governance levels to adopt relevant policies and standards, and promote best practices". Consequently, this report aims to "(a) raise awareness around the world on sand extraction and use, and its related impacts, (b) urge policymakers to explore and adopt policies on sand extraction and use that are appropriate to their contexts and jurisdictions; (c) shape common goals across sectors that will help achieve just and responsible sand governance and management everywhere, and (d) propose solutions for finding pathways toward a more sustainable use of sand". The ultimate goal of this report is to inform decision-making and support actions at the intergovernmental level, all of which based on the responsible and just management of aggregate resources. In summary, if aggregate production can be managed, it is possible to avoid a crisis and move toward a circular economy.* Figure 1.22 *shows the suggested hierarchy of available solutions to improve sustainable consumption and production of aggregates* (UNEP 2019).

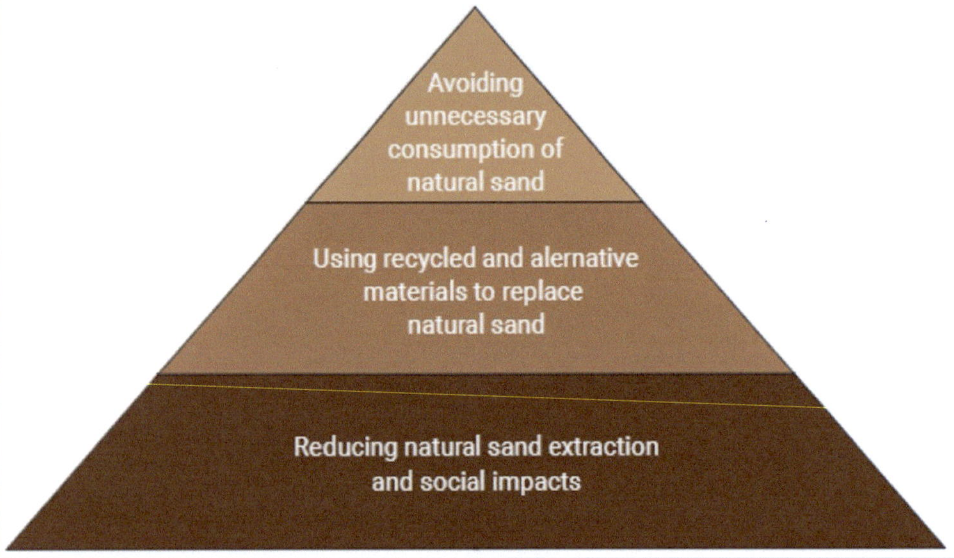

Fig. 1.22 Suggested hierarchy of available solutions to improve sustainable consumption and production of aggregates

For the purpose of a circular economy in the aggregate sector, the use of life cycle assessments of aggregates can be a very useful tool. The idea of life cycle assessment (LCA) "was conceived in the 1960s when environmental degradation and in particular the limited access to resources started becoming a concern" (Bjørn et al. 2018). Although these studies were primarily done for companies, the 1980s and 1990s saw an increase in methodological development and international collaboration and coordination in the scientific field. Today, LCA can be defined as "a tool to assess the potential environmental impacts and resources used throughout a product's life cycle, i.e., from raw material acquisition, via production and use stages, to waste management" (ISO 14044 standard). The application of LCA in the construction sector can be essential for reducing the environmental impacts originating from the sector. Nevertheless, "there is little specific and marginal inventory data relevant for the LCA of aggregates for construction" (Korre and Durucan 2009). Consequently, this methodology can be clearly enhanced in the coming years to obtain a definitive circular economy in the aggregate industry.

1.6 Criticality of Aggregate Resources

The criticality of a raw material has become essential in recent decades, given its notable importance in supplying resources for a country. The case of supply and demand for rare earth elements is undoubtedly the most recent example. The criticality of raw materials includes environmental, socioeconomic and geopolitical aspects related to the availability and utilization of raw materials. Although there is no currently agreed definition of criticality or assessment methodologies, critical materials can be defined as raw materials that are essential to the economy or security of a country or region, and have a supply chain vulnerable to disruption. This definition combines raw materials of high importance to the economy and of high risk associated with their supply. For this reason, for instance, both Europe and the USA have published their lists of critical minerals or raw materials. These lists are not permanent but will be dynamic lists updated periodically (i.e., every three years in Europe) and represent current data on supply, demand, concentration of production, and current policy priorities. Critical materials are not static, but change over time. It is important to note that due to many different causes, demand for one raw material can increase significantly in a short time and far exceed the supply available at that moment, converting it into a critical raw material.

Aggregate resources could become critical although they are geologically abundant. The regional focus to evaluate the criticality of aggregates is the basic differentiation from other raw materials because these resources can be at a risk of depletion at a regional scale whereas their world reserves can be defined as almost infinite. According to a recent study published by the OECD, "the use of construction materials is projected to almost double between 2017 and 2060 with the largest growth in aggregates (sand and gravel, and crushed

Fig. 1.23 Illegal river sand mining (image courtesy of Surendar Solanki—Gaon Connection)

rock)" while construction materials use per capita is projected to rise in most countries (OECD 2018). At the moment, aggregates do not belong to the so-called critical raw materials group in Europe and the USA but sustainable extraction and handling of aggregates is essential to avoid a future scarcity of resources. Due to the irreplaceable role of aggregates in the construction industry, their future availability deserves special attention. Consequently, the strategic value of aggregates as a critical resource needs to be quickly recognized.

For instance, Ioannidou et al. (2017) show an evaluation of the local criticality of construction aggregates applied to the cantons of Switzerland. The evaluation is based on three dimensions: supply risk (SR), environmental implications (EI) and vulnerability to supply restriction (VSR). The first dimension includes three components: (1) geological, technological, and economic, (2) social and regulatory, and (3) geopolitical aspects. The environmental impact of natural aggregates is calculated with a life cycle assessment study and the third dimension comprizes three components: importance, substitutability, and susceptibility. The results show that inside the country, the criticality of aggregates is driven by the supply risk, which depends on the surface and number of quarries and their distribution in the region. The regions that present the highest risk in the supply of construction resources are the cantons where the large Swiss cities are located, such as Basel and Zurich.

1.7 Environmental Concerns: Illegal Mining

By its very nature, mining aggregates have an impact on the land upon which they operate (see ► Chap. 7). It is possible that the environmental impact is the most pressing issue of aggregate production. The impacts occur from the initial exploration and present operation of the site to its closure and rehabilitation (with biodiversity in mind). Aggregate extraction and processing are the main causes of environmental issues associated with aggregate production. Aggregate quarries and pits are faced with NIMBY (not-in-my-backyard) syndrome, which describes the low public acceptance of people who live nearest to the extraction location.

Although the environmental impacts produced by the exploitation of aggregates are of a very varied nature (see ► Chap. 7), illegal aggregate mining (Fig. 1.23) is considered one of the most important globally (UNEP 2014). If no managed correctly, aggregate mining in places with fragile ecosystems produces a huge environmental impact. On the other hand, the great demand of aggregates worldwide facilitates the existence of aggregate illegal exploitation. Sand mafias in India is an example of this type of problem (UNEP 2014). Moreover, as the aggregate price increases, so does the traffic of aggregate by local mafias since aggregate trading is a lucrative business. It is really a great problem in some parts of the world among emerging and developing countries, "especially

1

if weak governance and corruption is present" (Saviour 2012). For instance, in Morocco, half of the sand comes from illegal coastal sand extraction (UNEP 2019).

1.8 Questions

Short Questions

- Explain the basic characteristics that make aggregates so widely used in the world as a construction material
- Comment the main types of aggregates
- What is the difference between sustainable development and sustainability?
- Explain the mining cycle of aggregate production
- Define critical material.

Long Question

- Explain in detail the concept of circular economy and its application to aggregates.

References

Bjørn A, Owsianiak M, Molin C, Laurent A (2018) Main characteristics of LCA. In: Hauschild MZ, Rosenbaum RK, Olsen SI (eds) Life cycle assessment. Springer International Publishing AG, Cham, Switzerland, pp 9–16

Bustillo M (2018) Mineral resources—from exploration to sustainability assessment. In: Springer textbooks in earth sciences, geography and environmental sciences. Springer International Publishing AG, Cham, 653 p

Escavy JI, Herrero MJ, Lopez-Acevedo F, Trigos L (2022) The progressive distancing of aggregate quarries from the demand areas: magnitude, causes, and impact on CO_2 emissions in Madrid Region (1995–2018). Resour Policy 75:102506

Ioannidou D, Meylanb G, Sonnemannc G, Haberte G (2017) Is gravel becoming scarce? Evaluating the local criticality of construction Aggregates. Resour Conserv Recycl 126:25–33

Korre A, Durucan (2009) Life cycle assessment of aggregates. Waste & Resources Action Programme, 41 p

Migliore M, Talamo C and Paganin G (2020) Construction and demolition waste. Strategies for circular economy and cross-sectoral exchanges for sustainable building products. Springer, Cham, pp 45–76

OECD (2018) Global material resources outlook to 2066—economic drivers and environmental consequences. OECD Highlights, OECD Publishing, Paris, 24 p

Pearce DW, Turner RK (1990) Economics of natural resources and environment. John Hopkins University Press, Harvester Wheatsheaf, p 392

Perkins L, Royal ACD, Jefferson I, Hills CD (2021) The use of recycled and secondary aggregates to achieve a circular economy within geotechnical engineering. Geotechnics 1:416–438

Saviour N (2012) Environmental impact of soil and sand mining: a review. Int J Sci Environ Technol 1(3):125–134

Tost M, Ammerer BA (2022) Sustainable supply of aggregates in Europe. Montanuniversität Leoben, Chair of Mining Engineering and Mineral Economics, Leoben, Austria, 123 p

UNEP (2014) Sand, rarer than one thinks. UNEP Global Environmental Alert Service (GEAS), 15 p

UNEP (2019) Sand and sustainability: finding new solutions for environmental governance of global sand resource: synthesis for policy-makers. GRID-Geneva, United Nations Environment Programme, Geneva, Switzerland, 56 p

UNEP (2022) Sand and sustainability: 10 strategic recommendations to avert a crisis. GRID-Geneva, United Nations Environment Programme, Geneva, Switzerland, 90 p

Willet JC (2023) Sand and Gravel (Construction) and Stone (Crushed). U.S. Department of the Interior, U.S. Geological Survey, Mineral Commodity Summaries, pp 150–152 and 165

Standard

ISO 14044: 2006. Environmental management—life cycle assessment—requirements and guidelines

Properties and Testing

Contents

2.1 Introduction – 22

2.2 General Properties and Tests – 23
2.2.1 Petrographic Description – 23

2.3 Geometrical Properties and Tests – 26
2.3.1 Particle Size Distribution—Grading – 26
2.3.2 Shape and Surface Texture – 30
2.3.3 Fines – 34

2.4 Mechanical and Physical Properties and Tests – 38
2.4.1 Physical Properties and Tests – 38
2.4.2 Mechanical Properties and Tests – 40

2.5 Thermal and Weathering Properties and Tests – 46
2.5.1 Thermal Properties and Tests – 46
2.5.2 Weathering Properties and Tests – 47

2.6 Chemical Properties and Tests – 50

2.7 Questions – 50

References – 51

M. Bustillo Revuelta, *The Basics of Aggregates*,
Springer Textbooks in Earth Sciences, Geography and Environment,
https://doi.org/10.1007/978-3-031-42961-3_2

2

Abstract

This chapter discusses the basic properties of aggregates and the associated test standards. Aggregate properties are defined, and the significance of each property is briefly discussed. At the same time, commonly used tests for determining the properties of aggregates are described. General properties of aggregates mainly include the petrographic description of the aggregate, which provides valuable information about the mineralogy, chemical composition, grain size, and texture of the aggregate particles. Geometrical properties are those related to particle size distribution (gradation), shape (i.e., roundness) and surface texture (i.e., broken surfaces) as well as the content and quality of the fines. All these properties are key features that control the utilization of aggregates in many end-uses. The mechanical and physical properties of aggregates include physical properties such as density, water absorption, and porosity as well as mechanical properties such as resistance to abrasion and resistance to polishing. The importance of the thermal (thermal volume change, thermal conductivity and integrity during heating) and weathering (freezing and thawing and wetting and drying) properties of aggregates depends on their end-use, with some of the properties (i.e., thermal properties) being rarely needed. Chemical properties are very important in applications such as concrete or asphalt because some types of aggregate contain minerals that are chemically reactive and can affect the final behavior of the mixture.

2.1 Introduction

The behavior of most of the construction materials manufactured with aggregates depends both on the proportion or dosage of the different components and on the properties of each of them. Examples are numerous of failures traceable directly to inadequate aggregate selection and use. For instance, the correct determination of the properties of the aggregates used in the production of concrete will allow establishing not only their most appropriate dosage but also predicting, as far as possible, their future behavior in service. This is because the properties of aggregates highly influence the durability and structural performance of concrete. The properties of the aggregates utilized in concrete (◘ Fig. 2.1) must be well understood. In summary, recognition of aggregate properties is essential to evaluate the suitability of the different aggregates used in each application.

It is important to note that the properties of aggregates depend on the intrinsic or inherent features of the material (i.e., mineralogy, density, texture, alteration, porosity, and chemical composition) as well as their extrinsic characteristics such as particle size fraction, shape, and fracture faces. Inherent properties depend on the characteristics of the parent rock whereas extrinsic features are the result of the production process (i.e., crushing and sieving in crushed stone or sieving and cleaning in sand and gravel). In other words, extrinsic properties are absent in the parent rock. These two types of characteristics must always be taken into account because the quality of a rock as an aggregate may not be adequate after a crushing process. However, aggregate characteristics such as geometrical features or the content of fines can be clearly improved by adopting suitable extraction and processing methods. The suitability of an aggregate for a particular use depends principally on its physical and mechanical properties although chemical and mineralogical properties can also be important in some applications.

Aggregates used in asphalt do not necessarily require the same characteristics as those utilized in concrete. For this reason, distinct properties of aggregate are needed for different applications. For instance, aggregates including certain siliceous components can be a very significant problem in concrete but are not a problem in asphalt. Consequently, for a specific end-use, exclusively certain aggregate properties are necessary, that is, aggregates do not indispensably must to show a high quality in each property.

There is a whole series of laboratory tests that allow the characterization of the aggregate properties with the objective of estimating the suitability of the aggregate as a component for the production of construction materials such as concrete and bituminous mixtures (the two most important applications of aggregate) (◘ Fig. 2.2). This is because laboratory testing is the best method of scientifically evaluating the properties and suitability of aggregates. The acceptance of a certain type of aggregate for a particular use must be based upon specific information obtained from tests utilized for measuring the quality of the aggregate, that is, aggregate testing is of value in assessing its suitability for each application. For this purpose, a wide group of tests has been devised over the years. A general overview of the most important properties of aggregates and their corresponding tests is described in this chapter while more specific considerations are taken into account in each specific aggregate application (i.e., concrete, road pavement, and rail ballast).

The most used standards for testing procedures for aggregates are those described by European legislation

Fig. 2.1 Aggregates used for manufacturing concrete

(EN standards) and by the American Society for Testing and Materials (ASTM standards). As a method of organization, the description of properties and tests in the present chapter is carried out in accordance with European legislation and standards although that both types of standards, EN and ASTM, have similar guiding principles. The properties of aggregates and the associated tests can be grouped into the following categories: (a) general properties and tests, (b) geometrical properties and tests, (c) mechanical and physical properties and tests, (d) thermal and alteration properties and tests, and (e) chemical properties and tests.

2.2 General Properties and Tests

The general properties group of European legislation does not include aggregate properties in itself but rather certain aspects such as sampling and petrographic description of the material. Since sampling is closely related to the investigation and evaluation of aggregate mineral deposits, it will be considered in the next chapter, which is specifically devoted to those topics, including the correct sampling methods and adequate reduction of the weight of the samples.

2.2.1 Petrographic Description

The petrographic description is carried out in conjunction with a handlens, stereoscopic microscope, and polarizing microscope (Fig. 2.3). It provides valuable information about the mineralogy, chemical composition, grain size, and texture of the aggregate particles, which will determine the future behavior of the aggregate in the selected application. For example, a detailed mineralogical analysis allows the detection of the presence of dangerous components when the material is used as aggregate (i.e., silica minerals—Fig. 2.4—that can be reactive with the alkalis in concrete—the alkali-silica reaction). A correct description of aggregate composition must include not only rock composition and mineral types but also the mineral characteristics in each grain.

The grain size and texture of the aggregate help to determine if the material is compact, the laminar character of the components, and many other factors that have a notable impact on the homogeneity, porosity, and resistance of the rock. For instance, the surface texture of the aggregate affects the quality of concrete in the fresh and hardened state, having a remarkable influence on the mechanical strength of concrete,

Fig. 2.2 Bituminous material in road construction (image courtesy of Nynas)

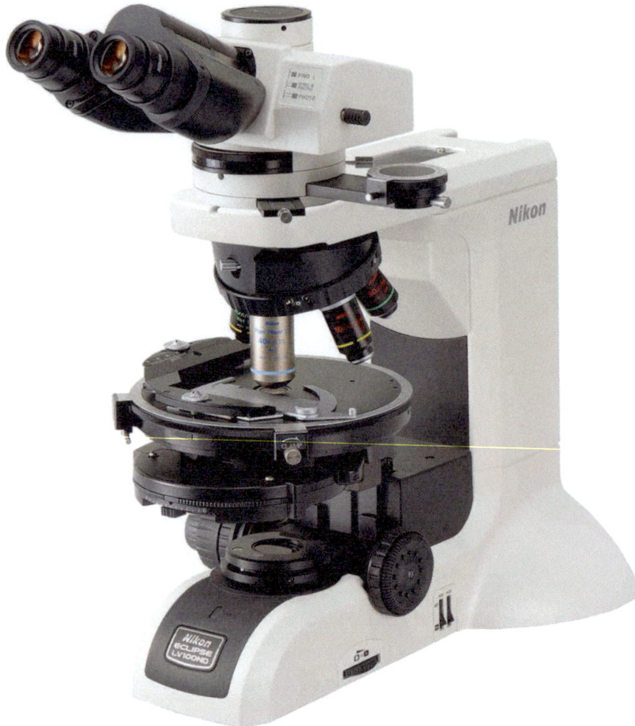

Fig. 2.3 Polarizing microscope (image courtesy of Nikon)

Fig. 2.4 Silica mineral (chalcedony) that reacts with the alkalis in concrete (image courtesy of María Ángeles Bustillo)

particularly the flexural strength. In summary, petrographic examination of aggregates is a useful aid in assessing their quality.

The petrographic description of aggregates is covered in the European legislation by the EN 932-3 standard (▶ Box 2.1: Petrographic Description) whereas the ASTM C295/C295 M standard covers this topic in the USA. The American standard guide, entitled "Standard Guide for Petrographic Examination of Aggregates for Concrete", states the principal objectives of petrographic examination: "to determine the physical and chemical characteristics of the material that may be observed by petrographic methods and that have a bearing on the performance of the material in its intended use, to describe and classify the constituents of the sample, to determine the relative amounts of the constituents of the sample that are essential for proper evaluation of the sample when the constituents differ significantly in properties that have a bearing on the performance of the material in its intended use, and to compare samples of aggregate from new sources with samples of aggregate from one or more sources, for which test data or performance records are available".

Box 2.1—Petrographic Description

*EN 932-3 standard, entitled "Tests for general properties of aggregates. Part 3: Procedure and terminology for simplified petrographic description", states that "the sample shall be first subjected to a visual examination to determine the constituent rock or mineral types. It may be appropriate to wash the sample. Each rock type shall be carefully inspected using a hand lens or a stereoscopic microscope and other appropriate means. If necessary, where appropriate, thin sections (*Fig. 2.5*) should be examined using a polarizing microscope".*

The standard is defined for aggregates obtained from natural deposits, which are formed by mineral particles and rock fragments. The method of description and nomenclature shall be used only for particle sizes between 0.1 and 63 mm. "For an aggregate sample, the description of the sample (or grain size fraction) shall include: (a) brief information about the shape, surface conditions (roughness, etc.) and roundness of particles, (b) a petrographic identification based on counting a sufficiently representative number of particles". This standard also defines the concepts of igneous, sedimentary, and metamorphic rocks, which is essential to correctly classify a material that can be used as aggregate. Annex A of the standard, entitled "Nomenclature", provides a list of simple petrographic terms applicable to most types of rock utilized for aggregates. For example, granite is defined in the annex as "a light coloured rock containing alkali feldspars and quartz, together with mica (biotite and/or muscovite)" and limestone as "a rock consisting predominantly of calcium carbonate ($CaCO_3$*)".*

The EN 932-3 standard establishes a final report with the following headings: "(1) essential data needed to identify the sample, (2) petrographic description of the different rock types or of the different aggregate size fractions, including the results of any particle counting, and (3) geological information on source, i.e. on sample origin".

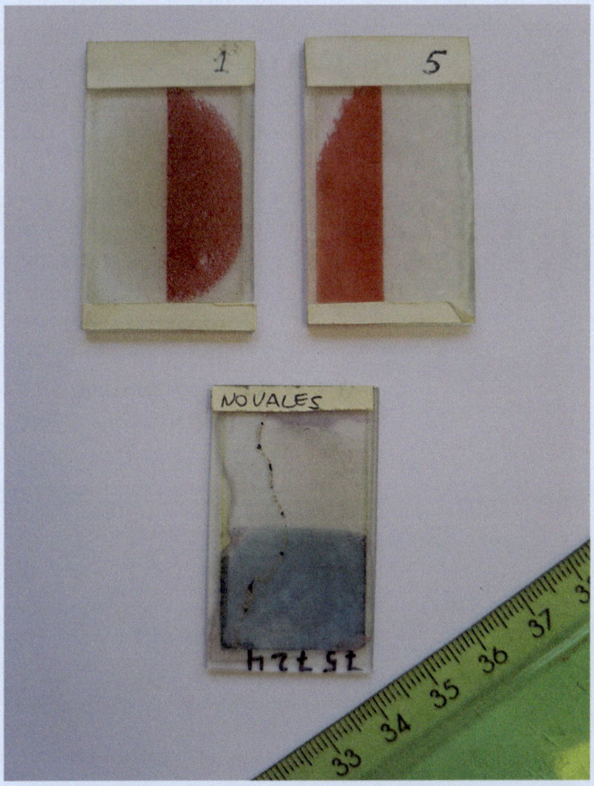

Fig. 2.5 Thin sections used for petrographic description

2

It is important to note that the ASTM C295/C295 M standard is devoted specifically for aggregates for concrete, not for aggregates in a general sense. For this reason, the standard includes numerous objectives related to the interaction between aggregates and Portland cement. For instance, it states that "petrographic examinations may also be used to determine the proportions of cubic, spherical, ellipsoidal, pyramidal, tabular, flat, and elongated particles (◘ Fig. 2.6) in an aggregate sample or samples". This is because these types of particles augment the mixing water requirement and reduce the concrete strength. Obviously "petrographic examination should identify and call attention to potentially alkali-silica reactive and alkali-carbonate reactive constituents, determine such constituents quantitatively, and recommend additional tests to confirm or refute the presence in significant amounts of aggregate constituents capable of alkali reaction in concrete".

◘ **Fig. 2.6** Flat and elongated particles

2.3 Geometrical Properties and Tests

Geometrical properties are those related to particle size distribution (gradation), shape (i.e., roundness), and surface texture (i.e., broken surfaces) as well as the content and quality of the fines. All these properties are key features that control the utilization of aggregates in many end-uses.

2.3.1 Particle Size Distribution—Grading

Particle size distribution or grading refers to the distribution of particle sizes in an aggregate sample. The term particle size distribution is often utilized instead of grading. The ISO 19595 standard defines grading as "particle size distribution expressed as the percentages by mass passing a specified set of sieves (◘ Fig. 2.7)". Particle size distribution serves to designate the size of the aggregates. Aggregates are designated (i.e., EN 12620 standard—aggregates for concrete) by their minimum size d and maximum size D in mm, (size of the lower (d) and upper (D) sieves of the fraction considered) according to the following expression: aggregate d/D (for example, an aggregate that has a minimum size of 2 mm and a maximum size of 4 mm will be reported as 2/4). For this purpose, the maximum size D of an aggregate is called the minimum opening of the EN 933-2 sieve through which more than 90% by weight passes, and the minimum size d of an aggregate is referred to as the maximum opening of the EN 933-2 sieve through which less than 10% by weight passes.

Grading is a property of great significance in almost all construction aggregate applications and often defines the product. In the first stage, sand and gravel and crushed stone aggregates are characterized by the

◘ **Fig. 2.7** Sieves used for determining particle size distribution

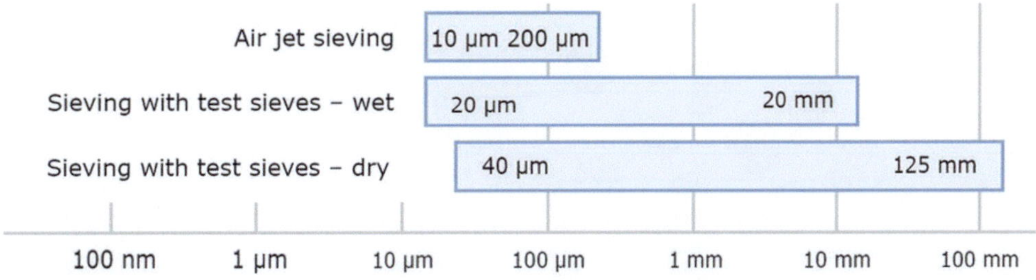

Fig. 2.8 Measuring range for sieving

size of their particles (the minimum size analyzed is 0.063 mm and sedimentation methods are used to determine the grading of particles with lower sizes). Sedimentation methods involve dispersion of the fines and determination of the suspended material. Sieving is a typical test performed on aggregates for use in concrete. This is because aggregate grading is essential in relation to the properties of concrete. A well-graded aggregate gives workable mixtures that are readily transported, placed, and compacted. The individual grading of each aggregate in concrete, as well as their combinations, is essential to establish the plastic and hardened properties.

Particle size distribution depends on the size distribution present in the raw material for sand and gravel aggregates or the crushing process for crushed aggregates. In sand and gravel aggregates, the particle size distribution is an intrinsic feature of the deposit, reflecting the environment of formation. As a consequence, the material cannot form a regularly graded assemblage conforming to a particular grading class. It is common the existence of deficiencies or excesses in certain size ranges. A general rule is that the larger the gravel-to-sand ratio is, the better the deposit. This is because aggregates, in most construction applications, require a wide spectrum of particle sizes although silt-sized or smaller fine particles are undesirable. For this purpose, grain size analysis is one of the most important test procedures, principally utilizing the mechanical sieving method. The range of gradings acceptable for a certain end-use is termed a grading envelope. In some cases, grading in aggregates is improved by processing to meet specifications.

During sieving, the sample is subjected to vertical movement (vibratory sieving) and horizontal motion, with both movements superimposed. The choice of the most appropriate sieving method is based on the degree of fineness of the sample (Fig. 2.8). In this sense, dry sieving is the principal method for particle sizes ranging from 40 µm to 125 mm (dry sieving is sieving in the absence of a liquid and wet sieving is sieving with the help of a liquid—ISO 2395). Nevertheless, the measurement range is controlled by material properties such as density, electrostatic charging, or a tendency to agglomerate.

Particle size distribution in aggregates is obtained in accordance with ASTM C136/C136 M and EN 933-1 standards. Grading is carried out by sieve analysis in which the sample is passed through a column of sieves (Fig. 2.9) and the weight retained (Fig. 2.10) on each sieve is calculated. The ASTM C136/C136 M and EN 933-1 standards define a method using sieves to determine the particle size distribution of aggregates (▶ Box 2.2: Determination of Particle Size Distribution: Sieving Method). Both tests consist of separating, using a series of sieves, a sample in several particle size fractions of decreasing size; as a rule, the number

Fig. 2.9 Column of sieves

2

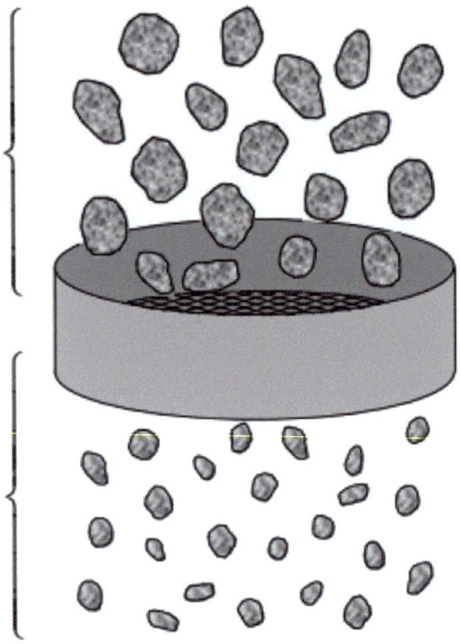

Fig. 2.10 Particles retained or passing the sieve

of sieves and the aperture sizes are selected based on the type of sample and the precision needed. The sieves cover sizes between 0.063 and 32 mm or more and have square openings.

Table 2.1 Weight of sample in sieving analysis (EN 933-1 standard)

Maximum particle size (mm)	Minimum permissible sample weight (g)
1	100
2	200
4	500
8	800
16	1000
32	2000
63	10000

The amount of sample required for the test is obviously a function of the maximum size of the aggregate, so that the values in Table 2.1 (EN 933-1 standard) show the amount of initial sample, needless to say that the amount of initial sample must be greater than the value indicated in the table. Nevertheless, the amount of material used in the analysis must be limited to avoid clogging of the sieves to permit the particles to move down to the sieve efficiently.

Box 2.2—Determination of Particle Size Distribution: Sieving Method

*In sieve analysis, a sample of dry aggregate of known weight is separated through a series of sieves with progressively smaller openings. The operative procedure of the sieving method is very similar in all standards. It consists of washing the sample and later sieving it using a column of sieves; the column is prepared with a set of sieves arranged, from top to bottom, in decreasing order of aperture sizes (*Fig. 2.11*). Agitation of the column can be carried out manually or mechanically (*Fig. 2.12*).*

*In particular, The American Concrete Institute states that "a sample of the aggregate is shaken through a series of wire-cloth sieves with square openings (*Fig. 2.13*), nested one above the other in order of size, with the sieve having the largest openings on top, the one having the smallest openings at the bottom, and a pan underneath to catch material passing the finest sieve". The materials retained on the different sieves are weighed after removing the sieves one by one. For an accurate determination of the quantity of material that is finer than 62 μm, the sample must be washed, and the amount of fine material must be calculated. This can be done on the sieve analysis sample before sieving (in this case including the results in the sieve analysis) or it can be done on another sample.*

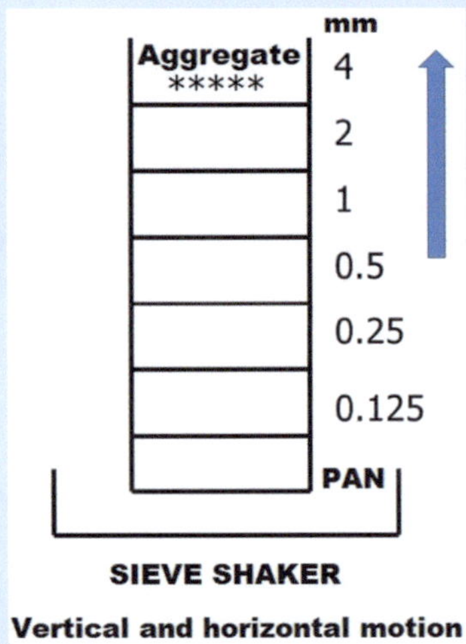

Fig. 2.11 Position of the sieves

Particle size distribution can be expressed as a percent retained by weight on each sieve size (individual percent retained). It is the percentage of material contained between

Fig. 2.12 Apparatus for mechanical sieving

Fig. 2.13 Square openings of the sieve

successive sieves, recorded to the nearest whole percent. The weight retained on each sieve is divided by the total weight of the sample (the sum of the masses retained on each sieve and the pan) and multiplied by 100, obtaining the retained weight in percent for each fraction. Subsequently, to construct the particle size curve, the accumulated retained weight is calculated (cumulative percentage). It is obtained by adding to each fraction the total weight of the previous fractions.

It is important to ensure that proof sieving is done correctly for material retained on each sieve and that no sieves are overloaded during the grading operation as this can skew the results. The results obtained are accurate if the total mass of the sample after sieving is compared with the original mass of the sample placed on the sieves. If quantities are different by more than a certain percentage (i.e., 0.5–1%), the results should not be utilized to accept the test and it shall be repeated.

To prevent overloading in the sieves, the fraction retained on each sieve shall not exceed the following:

$$\frac{A \times \sqrt{d}}{200}$$

where

A – *is the area of the sieve in mm²*

d – *is the size of the sieve openings in mm.*

The sum of the accumulated percentages retained on the sieves, divided by 100, is termed the fineness modulus (FM). It is a dimensionless parameter used as a single number to characterize a particle size distribution, being a measure of the average particle size. The higher the fineness modulus is, the coarser the aggregate is. According to the EN 12620 standard (aggregates for concrete), FM is calculated using the following sieves: 4 mm—2 mm—1 mm—0.5 mm—0.25 mm—0.125 mm—0.063 mm.

Although FM is an approximation of the average size of the aggregate, there can be infinite granulometries with the same fineness modulus. This parameter is not a unique measure of grading. In concrete, the importance of this value must be remarkable since all mixtures that have a similar FM will need the same amount of water to produce a mixture of the same consistency. The fineness modulus is calculated for fine aggregate rather than for coarse aggregate. In concrete, the utility of FM lies in detecting slight variations in the aggregate from the same source because these variations can affect the workability of the fresh concrete.

As mentioned above, the aggregate particle size and grading requirements control the number of sieves and shape of the openings selected for sieve analysis. Regarding the size of the sieves and the shape of the openings, they are established in EN 933-2, ASTM E11 and ISO 3310-1 and -2 standards. For example, the EN 933-2 standard specifies that test sieves with openings larger than 4 mm must be manufactured of perforated sheet metal while sieves with openings smaller than 4 mm must be made of braided wire mesh; in

2

both cases, the shape of the openings must be square (Fig. 2.13). With respect to the nominal size of the openings, the EN 933-2 standard establishes, in addition to another required sieve, the following basic series of sieves: 0.063 mm—0.125 mm—0.250 mm—0.500 mm—1 mm—2 mm—4 mm—8 mm—16 mm—31.5 mm—63 mm—125 mm.

In sieving an aggregate sample, the results are also influenced by the shape of the particles. The volume and size of the particles retained on a certain sieve are constrained by shape (see next heading); for instance, particles of elongate shape in any one size approximate in size to the particles of flaky shape corresponding to the next coarser size. Thus, sieving does not exactly classify particles according to their sizes when the shapes are very different. Consequently, the particle size in aggregates with a high proportion of elongate particles would be coarser than in one rich in flake shapes.

The results of a sieving analysis are expressed in tabular or graphical format. The cumulative percentage on each sieve is the value used in the plotting of the grading curve. Grading charts or grading curves are utilized to graphically show the results of a sieve analysis. In this type of data representation, the cumulative percentage on each sieve is plotted on the vertical axis whereas the size of the sieves is plotted on the horizontal axis. Figure 2.14 shows a typical particle size curve. It is constructed normally on semilogarithmic paper in which the scale corresponding to the sizes is presented with a geometrical progression of ratio 2. If aggregate grading curves are adequately plotted on identical graph types, it is possible to easily check the conformity of an aggregate with the specification requirements for an application. These requirements are expressed as grading limits (upper and lower) and are termed grading envelopes when plotted in graphical format.

In some applications, the utilization of graded aggregates, that is, those graded continuously from coarse to fine sizes is mandatory. However, gap-graded aggregates are sometimes utilized in certain uses. Gap-grading occurs when one or more intermediate-size fractions are absent. This means that certain particle sizes in the aggregate are omitted, particularly in the 5–10 mm range. Gap-grading is represented in the particle size distribution curve by a horizontal line in the sizes eliminated (Fig. 2.15). Although gap-grading has some advantages in some aggregate applications, continuous grading is commonly used.

2.3.2 Shape and Surface Texture

- Shape

Another geometrical characteristic associated with the size of the particles, and key in aggregate characterization, is the shape and surface texture of the particles. It affects the mechanical properties of the aggregates. According to the American Concrete Institute, particle shape can be defined in terms of compactness, which is "a measure of whether the particle is compact in shape, that is, if it is close to being spherical or cubical as opposed to being flat (disk-like) or elongated (needle-like)". Shape also determines the void content and packing density of aggregates. Poorly shaped particles have lower compacted densities and higher void contents. This is because rounded particles create less particle-to-particle interlocking than angular particles.

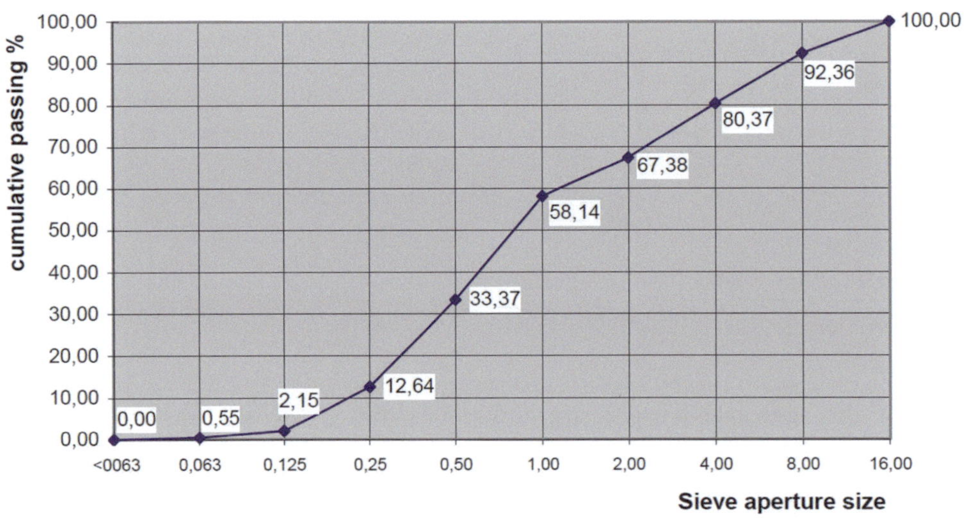

Fig. 2.14 Grading curve of an aggregate

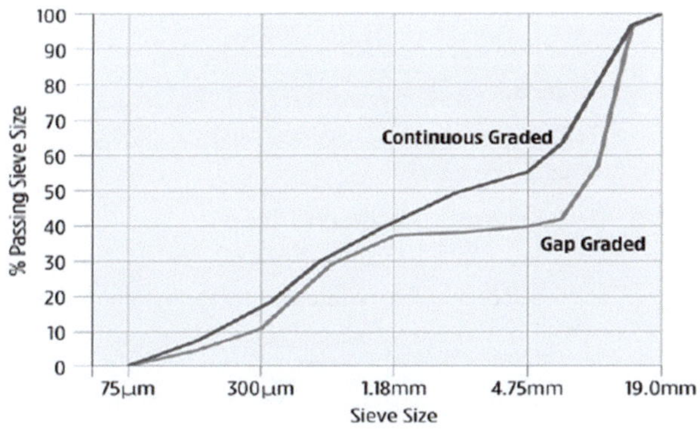

Fig. 2.15 Gap-grading curve

Fig. 2.16 Particle shapes (CCA 2020)

As a rule, the most suitable aggregates are those with a high proportion of roughly equidimensional particles.

In both types of natural aggregates, sand and gravel and crushed stone, the particles within a specific size fraction may show a broad spectrum of shapes (Fig. 2.16 and Table 2.2). The former reflects different petrological characteristics and environmental factors whereas the efficiency of the crushing equipment influences the shape of the particles in crushed stone. Thus, upon crushing rocks break into fragments that are elongated and flattened rather than cuboidal. For example, rocks with natural bedding planes (i.e., certain shales and sandstones) tend to generate flaky particles and strong and hard rocks generate a higher proportion of flakes than weak rocks but the latter produce more fines. For its part, sand and gravel sediments tend to have more spherical grains. The ASTM C125 standard defines flatness as "a flat particle has width/thickness ≥ 3" and elongation as "an elongated particle has width/thickness ≤ 3". Figure 2.17 shows the main dimensions of a particle.

Particle shape refers not only to the basic shape of aggregates but also includes other features such as roundness/angularity and sphericity (Fig. 2.18). Sphericity is defined by Waddell (1933) as "the ratio of the surface area of an equal-volume sphere to the actual surface area of the particle", that is, how closely the particle approaches a spherical shape. This index has a maximum value of 1, corresponding to a grain with a perfectly spherical shape. Roundness indicates the sharpness of the edges and corners of the particle (degree of angularity of a grain). Sphericity is a feature closely linked to petrogenetic factors. Certain physical processes, mainly freeze–thaw cycles, are sometimes responsible for the low sphericity of detrital materials originating in glacial environments. In summary, the distribution of sphericity will depend on the petrophysical characteristics of the original rock mass and on the sediment-genetic environment. Angularity is a measure of the lack of rounding of aggregate particles. It mainly depends on the strength and abrasion resistance of the rock as well as the degree of wear suffered by the particles. Angularity adversely affects the workability of concrete although it can improve other concrete properties such as the stability of interlocking particles. The simplest method to estimate sphericity and roundness is visual comparison to a calibrated set of standards (Fig. 2.19).

In European legislation, the quantification of shape is carried out by measuring the three dimensions of a number of particles and estimating the so-called flakiness index (EN 933-3 standard) and shape index (EN 933-4 standard). This is easier to do for coarse than fine particles. In the USA, the ASTM D4791 standard describes a test method that "covers the determination of the percentages of flat particles, elongated particles, or flat and elongated particles in coarse aggregates". The flakiness index is estimated as the mass of particles that pass the bar sieves with parallel slots. It refers to the proportion of flaky particles in a sample and indicates the aggregate interlocking properties. The shape index can be defined as the ratio thickness/length in several particles, utilizing for example a caliper (Fig. 2.20).

2

Table 2.2 Particle shape classification (Alexander and Mindex 2005)

Classification	Description	Examples
Rounded	Fully water-worn or completely shaped by attrition	Gravels and sands derived from marine, alluvial, or windblown sources
Irregular	Naturally irregular, or partly shaped by attrition and having rounded edges	Other gravels, typically dug from pits
Angular	Possessing well-defined edges formed at the intersection of roughly planar faces	Crushed rocks of natural or artificial origin; talus rocks
Flaky	Material in which the thickness is small relative to the other two dimensions	Poorly crushed rocks, particularly if derived from laminated or bedded rocks; other laminated rock
Elongated	Material, usually angular, in which the length is considerably larger than the other two dimensions	
Flaky and elongated	Material having the length considerably larger than the width, and the width considerably larger than the thickness	Poorly crushed rocks, as above. Poor processing techniques can exacerbate the undesirable shape, and vice versa

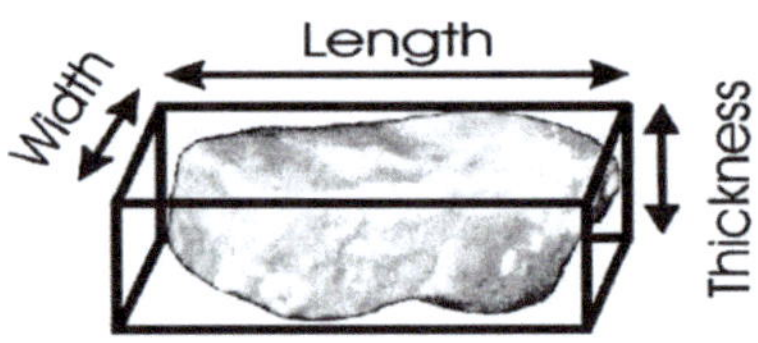

Fig. 2.17 Main dimensions of a particle

The EN 933-3 standard can be applied to "aggregates of natural or artificial origin, including light aggregates, but not to aggregates with sizes smaller than 4 mm or greater than 80 mm". To determine the test in a sample, it is first reduced into several size fractions, and two of them are selected (i.e., passing 10 mm and retained on 8 mm or passing 6.3 mm and retained on 5 mm). These size fractions are further sieved using special sieves (Fig. 2.21). The apertures of these special sieves are the larger of the two size fractions divided by two; for instance, aperture of the flakiness sieve is 5.0 mm for the size 8–10 mm). Flakiness is a measure of particles that are more or less half the nominal size in thickness. In this way, the flakiness index (FI) is calculated from the total mass of particles passing through the bar sieves and it is expressed as a percentage of the total dry mass of the particles tested, that is:

$$FI = (M_2/M_1) \times 100$$

where

M_1 – is the sum of the masses of the particle size fractions in grams

M_2 – is the sum of the masses of the particles passing through the sieve bar width for each size fraction in grams.

As in a previous procedure, there is an apparatus (gauges) (Fig. 2.22) to visually estimate the possible presence of flaky particles in the aggregate.

The second test to estimate the shape of the particles (shape index) describes a method to calculate the shape of the particles in aggregates with particle sizes between 63 and 4 mm. The method is very simple and consists of measuring, using a caliper, the length (L) and thickness (E) of a number of particles

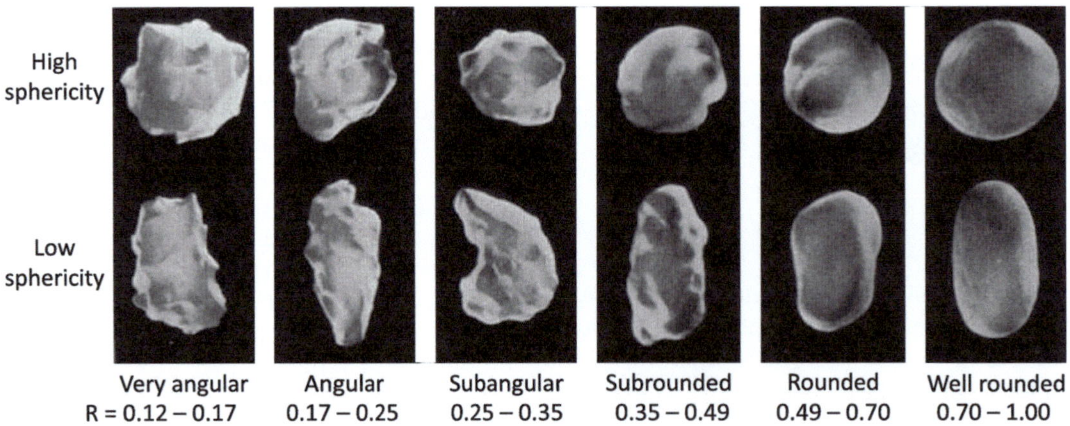

Fig. 2.18 Examples of roundness/angularity and sphericity

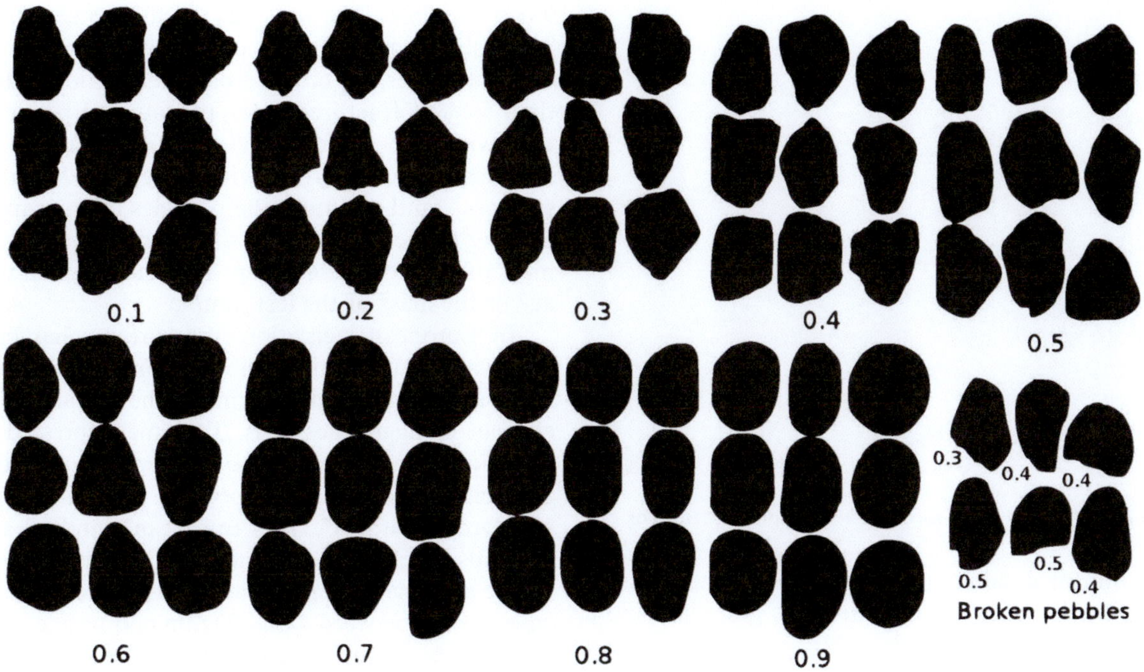

Fig. 2.19 Krumbein's chart for visual determination of roundness (Krumbein 1941)

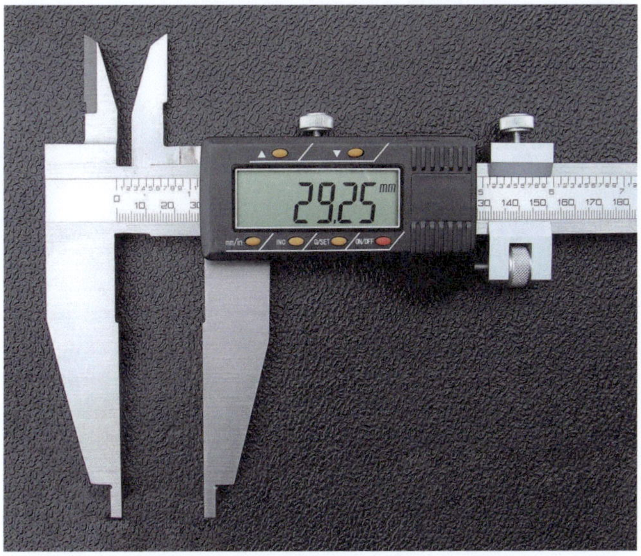

Fig. 2.20 Caliper for determining the shape index

Fig. 2.21 Sieves used for the determination of flakiness index (EN 933-3 standard)

and calculating the shape index as the mass of particles with an L/E value greater than 3 (these particles are considered noncubic) and expressing the index as a percentage of the total dry mass of the tested particles.

- Surface Texture

Surface texture is the relative irregularity of the aggregate particle surface. It is outlined in a qualitative form utilizing terms such as glassy, rough, or honeycomb (Table 2.3) rather than being described quantitatively. Surface texture plays an essential role in developing the bond between the aggregates and the cementing materials, affecting frictional properties and intergranular slip in unbound aggregates as well as the adhesion of Portland cement to the particles. To manufacture concrete, smooth particles need less mixing water and less amount of cement for a fixed water/cement ratio to produce concrete with a certain workability while rougher textures generate better bonding between aggregates and paste, enhancing the mechanical properties of concrete. Surface texture also influences the

2

Fig. 2.22 Device (gauges) used to detect the presence of slabs

workability of concrete and hot mix asphalt. The surface texture depends on the hardness, grain size, pore structure, and structure of the rock.

Some aggregates can originally have adequate surface texture but can polish smoothly later under traffic. They are inadmissible for final wearing surfaces. This is the case for carbonate rocks (i.e., limestones). Similar to particle shape, the surface texture is significantly affected by the source and production of aggregates. Sand and gravel suffering attrition tend to be fairly smooth while crushed stone will fracture with surface texture characteristics of their mineralogy and composition.

Related to surface texture, the term fractured faces express the surfaces of an aggregate particle produced by crushing processes and limited by sharp edges (Fig. 2.23). This feature plays a significant role in aggregate applications because the greater their angularity, the greater the strength of the agglomerate (i.e., concrete or hot mix asphalt).

The determination of this property is very simple and is carried out with the test described in the EN 933-5 standard. The test uses the following definitions: "(a) fully crushed particle: particle with more than 90% fractured faces, (b) crushed particle: particle with more than 50% of fractured faces, (c) rounded particle: particle with 50% or less fractured faces and (d) totally rounded particle: particle with more than 90% rounded surfaces", and separates the particles of the sample according to the previous classification. Subsequently, the mass of each of these groups is calculated as a percentage of the mass of the test sample.

In European legislation, the surface characteristics of the aggregate particles can be expressed using the EN 933-6 standard, which determines the so-called flow coefficient of aggregates. It is "the time, expressed in seconds, for a specified volume of aggregate to flow through a given opening, under specified conditions using a standard apparatus". This European standard specifies methods "for the determination of the flow coefficient of coarse and fine aggregates; it applies to the coarse aggregate of sizes between 4 and 20 mm and to fine aggregate of size up to 4 mm". The flow coefficient in coarse aggregates is related to the percentage of crushed or broken surfaces of an aggregate commented on above, so it may be utilized in association with the method specified in the EN 933–5 standard. The flow coefficient results are also influenced by the shape of the aggregate particles.

2.3.3 Fines

The fines, the particle size fraction of an aggregate that passes through the 0.063 mm sieve, produce a decrease in the quality of aggregates and negatively affect the properties of the material. For instance, in the case of concrete, the presence of fines in the aggregates produces an increase in the amount of mixing water, for the same workability of the concrete, impairing the strength and durability of the concrete. In the case of aggregates for bituminous mixtures, the presence of

Table 2.3 Surface texture of aggregates (Alexander and Mindex 2005)

Surface texture	Characteristics	Examples
Glassy	Conchoidal (i.e. curved) fracture	Glassy or vitreous materials such as slag or certain volcanics
Smooth	Water-worn or smooth due to fracture of laminated or fine-grained rock	Alluvial, glacial or windblown gravels and sands; fine-grained crushed rocks such as quartzite. dolomite, etc.
Granular	Fracture showing more or less uniform size rounded grains	Sandstone, coarse grained rocks such as certain granites etc.
Rough	Rough fracture of fine- or medium-grained rock containing no easily visible crystalline constituents	Andesite. basalt, dolerite, felsite, greywacke
Crystalline	Containing easily visible crystalline constituents	Granite, gabbro. gneiss
Honeycombed	With visible pores and cavities	Brick, pumice, foamed slag, clinker, expanded clay

Fig. 2.23 Fractured faces produced by crushing processes

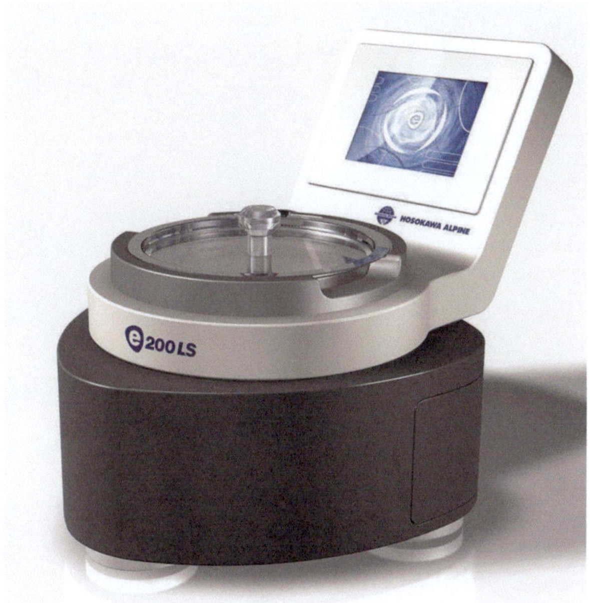

Fig. 2.24 Air-jet sieving machine (image courtesy of Hosokawa Micron Ltd)

fines is detrimental because it reduces the adhesion between the particles and the bituminous binders. In European legislation, fines content can be calculated through the EN 933-10 standard, entitled "… Part 10: Assessment of fines. Grading of filler aggregates (air-jet sieving)". This standard describes a method for determining the mass distribution of the particle size of fillers by sieving a sample in an airstream using an air jet sieve machine (Fig. 2.24). It is utilized to determine the particle size distribution of dry, powdery materials ranging in size between 25 and 2000 μm.

The fines portion of sand can be constituted by several types of materials. In the case of natural sands from sand and gravel deposits (Fig. 2.25), the fines would be mainly fine sand, silts, and clays while rock dust would be dominant in crushed sands. Some types of clay materials (i.e., montmorillonite and illite) must be avoided because they impart dimensional instability (i.e., shrinkage and swelling) to the final product. Clay materials are identified petrographically or by tests such as the sand equivalent and the methylene blue indicator tests. A further undesirable feature of clays is that they can also adversely affect setting times and the hardening of concrete. The ASTM D2419, ASTM C1777 as well as EN 933-8 and EN 933-9 standards cover the estimation of the amount and type of finest grain size in aggregates. The finest content, for instance, shall be declared in accordance with the relevant category specified in the EN 12620 standard table for concrete aggregates.

The EN 933-9 standard defines a method to determine the sand equivalent value in a sample with particle sizes between 0 and 2 mm. In this test, borrowed from the Americans approximately 1950, the plastic fines are forced to remain in suspension whereas the more desirable fine material is allowed to settle. This test indicates the relative amount of undesirable plastic fines in an amount of fine aggregate. According to this standard, "a test portion of sand and a small quantity of flocculating solution are poured into a graduated cylinder and are agitated to loosen the clay coatings from sand particles in the test portion (Fig. 2.26). The sand is irrigated using an additional flocculating solution forcing the fine particles into suspension above the sand. After 20 min, the sand equivalent value (SE) is calculated as the height of sediment expressed as a percentage of the total height of flocculated material in the cylinder (Fig. 2.27)". Higher percentages indicate low clay contents. A mechanical sand equivalent shaker can be used to shake the sample (Fig. 2.28). The clay content in the fines fraction of the aggregates may also be estimated with the so-called methylene blue test. It is a well-established procedure for determining the presence of clay minerals in an aggregate (▶ Box 2.3: Methylene Blue Test).

Fig. 2.25 Blocks of granular aggregates separated by their high fines content

Fig. 2.26 Settling in a cylinder for calculation of the sand equivalent value (image courtesy of Holcim)

Fig. 2.27 Measuring the height of the sediment (image courtesy of Holcim)

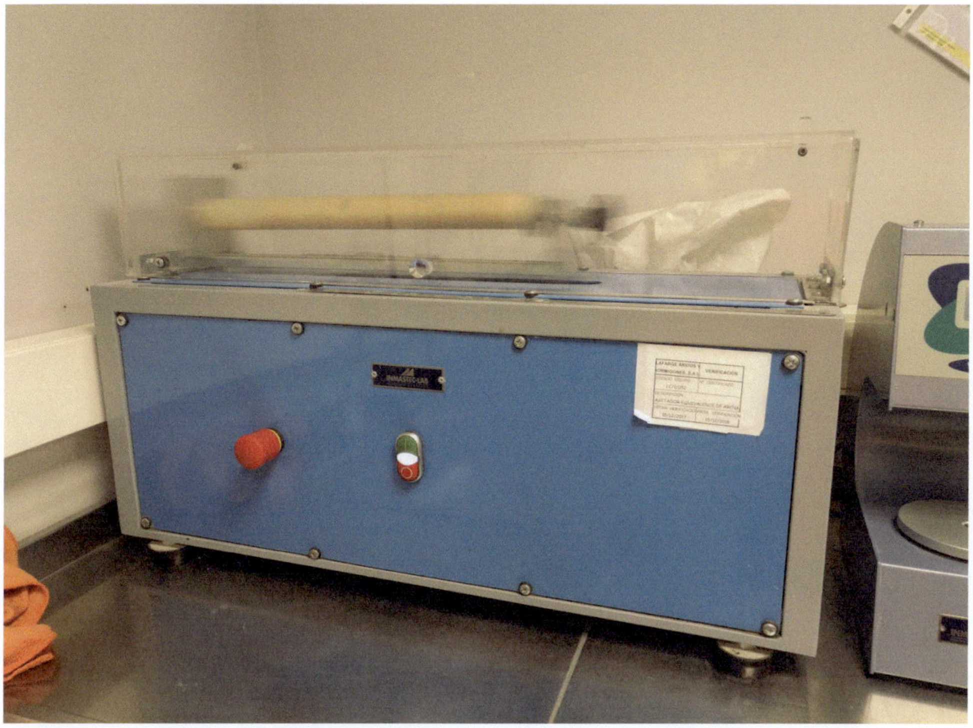

Fig. 2.28 Mechanical sand equivalent shaker (image courtesy of Holcim)

Box 2.3—Methylene Blue Test

Since the organic molecule methylene blue is absorbed by all minerals wettable by water, "the quantity of blue absorbed by nonclay inert fines is negligible compared to the amount absorbed by clays" (Tourenq and Denis 2000). *The methylene blue test specifies a method to determine the value of methylene blue (MB) of the particle size fraction 0/2 mm of fine aggregates or of the total mixture of aggregates as well as a procedure to calculate the same parameter in the particle size fraction 0/0.125 mm. This method includes the reaction of deleterious clay fines with a blue dye. The measure of the dye uptake, as a color change, estimates clay contamination. Different versions of the test based on an end-point titration method may be found in country legislations worldwide. This test is required when the sand equivalent value is below the specification; otherwise, MB is not necessary and the tested sand is accepted.*

In the EN 933-8 standard, "increments of a solution of methylene blue are added successively to a suspension of the test portion in water. The adsorption of dye solution by the test portion is checked after each addition of solution by carrying out a stain test on filter paper (Fig. 2.29) *to detect the presence of free dye. When the presence of free dye is confirmed, the methylene blue value (MB) is calculated and expressed as grams of dye adsorbed per kilogram of the size fraction tested". Previously, a suitable amount of the aggregate was dried and placed in a beaker with 500 ml of demineralized water, and the mixture was stirred for 5 min* (Fig. 2.30). *When this period is finished, an amount of 5 ml of the methylene blue dye solution was added. The new mixture was stirred again for 1 min and the staining method was carried out.*

After each injection of dye, the stain test consists of taking a drop of suspension by means of the glass rod and depositing it on the filter paper. The stain that originates is formed by a central deposit of material, usually solid blue in color, surrounded by a colorless wet zone. According to the EN standard, "the quantity of drop taken shall be such that the diameter of the deposit is between 8 and 12 mm". The test is considered positive if, in the wet zone, a halo consisting of a persistent light blue ring of around 1 mm is developed around the central deposit. The halo must remain visible for 5 min for considering that the test has been finished.

The methylene blue (mb) value is calculated with the following expression:

$$MB = (V_1/M_1) \times 100$$

where

MB – *is the methylene blue value*
M_1 – *is the mass of the test sample, expressed in grams*
V_1 – *is the total volume of the added dye solution, expressed in milliliters.*

The size of the test sample does not have to have a certain weight, but it does have to be greater than 200 g (the process to be followed in the particle size fraction between 0 and 0.125 mm is similar to the previous one, but using a sample of only 30 g).

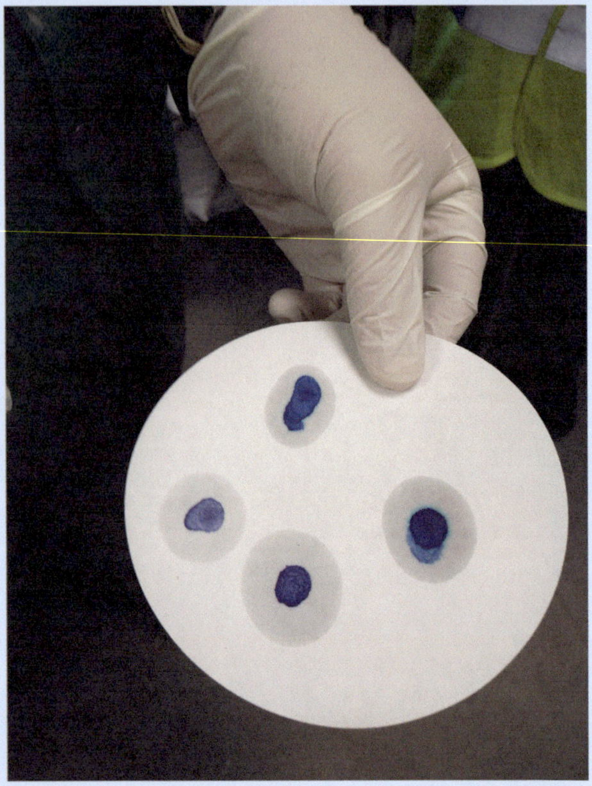

Fig. 2.29 Staining test on filter paper in methylene blue test (image courtesy of Holcim)

Fig. 2.30 Stirring the sample in the methylene blue test (image courtesy of Holcim)

2.4 Mechanical and Physical Properties and Tests

This large group of properties includes physical properties such as density, water absorption and porosity as well as mechanical properties such as resistance to abrasion and resistance to polishing.

2.4.1 Physical Properties and Tests

- Density

The density of a solid is defined as the ratio of its mass to the volume it occupies. There are several measures of density. Absolute density is the density of a solid excluding the internal enclosed pores and it is not commonly used in the aggregate market because its calculation involves pulverizing the material to eliminate the enclosed impermeable pores. Apparent density refers to the density of a solid including impermeable pores but excluding permeably or capillary pores. It is the most frequently and easily determined value and is important in concrete technology.

Specific gravity (also referred to as relative density) is the "ratio of the density of a material to the density of distilled water at a stated temperature" (ASTM C127 standard); the specific ratio is useful since it is independent of the units utilized for absolute density. Low values of this physical property indicate aggregates that are porous, weak or absorptive; high-specific gravity values frequently indicate high-quality aggregates. For most applications, specific gravity itself is not a critical property. As an example, an aggregate with a density of 3.00 is not essentially preferable than other with a density of 2.55. Nevertheless, the density of the aggregate is very important if structural considerations require that the concrete has a maximum or minimum weight.

Bulk-specific gravity can be defined as the ratio of the weight of a given volume of aggregate, including all voids, to the weight of an equal volume of water. It is an important characteristic of a fine aggregate since the closeness by which the grains are packed is the most important feature to be measured; a sand that has no voids would have a bulk density equal to the specific gravity of the mineral. The specific gravity of the aggregate is of significance when designing or structural considerations require concrete with a maximum or minimum weight. Finally, particle density is defined as the mass of a number of aggregate particles divided by their saturated surface-dried volume, for example the mass of a solid cubic meter of aggregate.

Density is one of the parameters used to classify aggregates. In terms of oven-dried particle density, the aggregates are classified as (a) lightweight aggregate (aggregate with oven-dried particle density ≤ 2000 kg/m^3), (b) normal weight aggregate (aggregate with oven-dried particle density higher than 2000 kg/m^3 and lower than 3000 kg/m^3), and (c) heavyweight aggregate (aggregate with oven-dried particle density ≥ 3000 kg/m^3).

In European legislation, the most common standard for determining the density of aggregates is the EN 1097-6 standard. It includes different methods to determine the particle density and water absorption of an aggregate. The main tests are the wire basket method for aggregates passing the 63 mm test sieve and retained on the 31.5 mm test sieve and the pyknometer method (◘ Fig. 2.31) for aggregates with sizes between 31.5 mm and 0.063 mm. Regarding the basis of the calculation method, particle density is calculated from the relationship between its mass and volume. The mass is calculated by weighing the test sample in the saturated and surface-dried condition and again in the oven-dried condition. The volume is calculated from the mass of water displaced, either by a mass reduction in the wire basket method or by weightings in the pyknometer method.

- Water Absorption

The internal structure of aggregate particles is made up of solid matter and voids that can contain water. As a rule, most aggregate particles are capable of absorbing water. Therefore, water absorption can be defined as the penetration of water into aggregate particles with a resulting increase in particle weight. For the pores of an aggregate particle to fill with water, the pores must be interconnected and open to the surface so that water from the exterior can penetrate the solid. This is not always the case, and what is measured is an apparent porosity that does not account for the impermeable pores. Absorption is expressed as the ratio of the increase in mass of an oven-dried sample after saturation to the mass of the saturated-surface-dry sample, in percent.

◘ **Fig. 2.31** Pyknometer apparatus for determining the density of particles

Absorption by various types of aggregates ranges from virtually zero to over 30% of the weight of the dry aggregate. This amount can be very small (i.e., in dense, fine-grained rocks) or very large (i.e., in lightweight and other porous materials).

The total water content in an aggregate is equal to the sum of absorption and surface moisture content. Surface moisture is an essential parameter in concrete production. This is because it influences the quantity of water that should be used in a concrete mixture to obtain a designed water/cement ratio. It is important to note that aggregates exposed to rain accumulate a remarkable quantity of moisture on the surface of the particles and many times keep it over long periods. Moisture content must be estimated constantly since it can change with weather and varies from one stockpile to another. ACI (2007) defines "four moisture conditions for aggregates depending on the amount of water held in the pores or on the surface of the particle: (1) damp or wet: aggregate in which the pores connected to the surface are filled with water and with free water also on the surface, (2) saturated surface-dry: aggregate in which the pores connected to the surface are filled with water but with no free water on the surface, (3) air-dry: aggregate that has a dry surface but contains some water in the pores, and (4) oven-dry: aggregate that contains no water in the pores or on the surface" (Fig. 2.32).

In European legislation, water absorption is calculated together with density using the EN 1097-6 standard or independently through the EN 1097-5 standard procedure whereas the ASTM C566 standard determines the percentage of evaporable moisture in a sample of aggregate. It calculates this value by drying both surface moisture and moisture in the pores of the aggregate. The EN 1097-5 standard provides a measure of the total free water present in a sample of aggregate, including the water on the surface of the aggregate and the water in the accessible pores within the aggregate particles. With this method (oven-drying), the "water content is determined as the difference in mass between the wet and the dry mass and is expressed as a percentage of the dry mass of the test portion".

- Porosity and Permeability

Absorption, porosity, and permeability of aggregates influence the behavior of the aggregates in applications such as concrete and bituminous mixtures. For instance, porosity and permeability affect the bond between the aggregates and the cement paste in concrete, its resistance to freezing and thawing, and its chemical stability.

Most conventional dense mineral aggregate particles have a measurable porosity. Porosity is the percentage of the total volume of aggregate particles occupied by pore spaces; the pores in the aggregate vary in size over a broad range. While the pore volume includes all

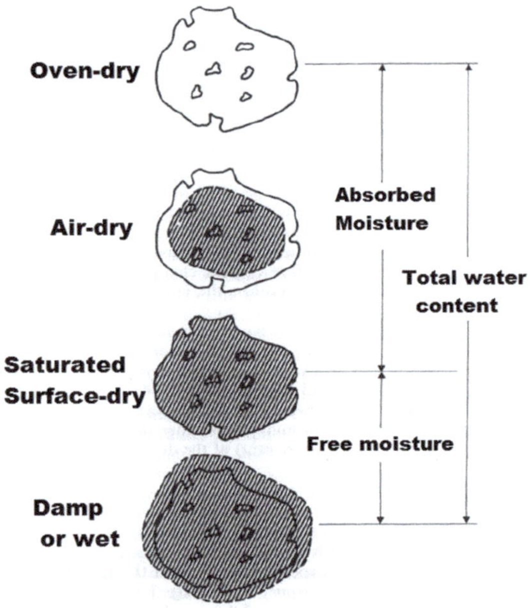

Fig. 2.32 Moisture conditions for aggregates (Neville and Brooks 2010)

pores, it is possible to measure only the interconnected porosity in typical laboratory tests. Porosity is measured by drying a sample at 100–110 °C to constant mass and then saturating the sample in water. The porosity of an aggregate is indicated by the quantity of water that it absorbs.

In general, good-quality aggregates should be dense and have low porosity. Porosity "affects the strength and elastic characteristics of aggregates particles and may affect their absorption, permeability and durability characteristics" (NSSGA 2013). In general, igneous and metamorphic rocks have very low porosity whereas the majority of sedimentary rocks tend to have a higher porosity. As a rule, aggregates with particles of lower porosity are preferred in the majority of construction applications. Both EN 1097-3 and ASTM C29/29 M standards cover the calculation of voids between particles in fine, coarse, and mixed aggregates using the calculation of the aggregate bulk density.

For example, the EN 1097-3 standard states that the voids in the test portion are expressed as a volume percentage of a cylinder, as in the following calculation:

$$\text{Percentage Voids} = 100[a - ((b/1000)/a)]$$

where

a – is the relative density of the aggregate on an oven-dried basis

b – is the bulk density, oven-dried, compacted or uncompacted.

It is very important to distinguish between permeability and porosity. The former is the capacity of an aggregate particle/s to transmit a fluid (Fig. 2.33). Most of igneous, metamorphic, and fine-textured sedimentary rocks are relatively impermeable to water. Nevertheless, some materials can possess a high permeability and may be undesirable for construction uses. On the other hand, the particles of an aggregate can have a high porosity but are impermeable to the flow of water. This is because the openings are not connected or because the openings are of small size.

2.4.2 Mechanical Properties and Tests

According to Langer (2006), "when crushed, the stone particles should be strong, which means they should resist abrasion; hard, which means they should resist loads; tough, which means they should resist impact; and sound, which means they should be able to withstand stresses caused by repeated freezing and thawing or wetting and drying". Compressive strength, resistance to impact, and soundness are essential features in crushed stone aggregates. For instance, materials that

Fig. 2.33 Permeability: fluid moving through the aggregate particles

contain weak, cleavable, absorptive, or swelling particles such as some shales, clayey rocks, and very coarse crystalline rocks are low-quality aggregates. The hardness and strength characteristics of aggregates determine their ability to resist mechanical breakdown. Abrasion resistance, abrasiveness, and polishing are also aggregate properties of great importance.

- Abrasion

The abrasion resistance of an aggregate is "its ability to resist being worn away by friction with other materials" (NSSGA 2013), which is utilized as a general index of the aggregate quality. For instance, mechanical abrasion tests are essential to evaluate the suitability of a rock for use as railroad ballast. An aggregate with a low abrasion resistance can increase the quantity of fines in the concrete during mixing, which increases the water requirement of the mix. Abrasion testing measures the durability of aggregates when subjected to abrasion and impact. The most common test for abrasion resistance in coarse aggregate is the Los Angeles abrasion test (▶ Box 2.4: Los Angeles Abrasion Test).

Box 2.4—Los Angeles Abrasion Test

*The Los Angeles Abrasion test is widely used as a marker of the related quality of the aggregate and is often described erroneously as a hardness test. It is a dry abrasion test and is carried out by introducing a certain amount of aggregate into a rolled steel drum (*Fig. 2.34*), mounted horizontally, which is loaded with an abrasive charge formed by standard-size steel balls (*Fig. 2.35*) (the machine can be fitted inside a noise reduction and safety cabinet). The drum is rotated, generating finer particles by the interaction of the steel balls and the aggregate particles, and the percentage of aggregate worn away is measured. The difference between the original weight and the final weight (sieved through a specified size) is expressed as a percentage of the original weight of the sample aggregate; this value is termed the Los Angeles (LA) abrasion value and many specifications worldwide set an upper limit on this mass loss, that is, a high LA value means that the aggregate has poor abrasion resistance.*

Los Angeles abrasion test represents the degradation of the aggregate caused by abrasion and impact processes (with the latter possibly more significant); although it is entitled as an abrasion test, it is truly an estimation of aggregates impact and crushing strengths. Thus, the proper strength of the particles as well as the presence of cracks or even the shape of the particles can influence the results. The main advantage of this test over other impact tests is that a larger and more representative amount of aggregate can be tested.

2

Fig. 2.34 Steel drum in the Los Angeles abrasion test

Fig. 2.35 Steel balls in the Los Angeles abrasion test

In the EN 1097-2 standard, the "testing machine consists of a hollow steel cylinder, closed at both ends, with an inside diameter of 711 mm ± 5 mm and an inside length of 508 mm ± 5 mm. The steel cylinder must be mounted on stub shafts attached to the ends of the cylinder but not entering it and must be mounted in such a manner that it may be rotated about its axis in a horizontal position. An opening in the cylinder (150 mm ± 3 mm) must be provided for the introduction of the test sample. The opening must be closed with a dust-tight cover that is easily removed". The abrasive charge consists of 11 solid, steel spheres with weights between 400 and 445 g and diameters of 45–49 mm. The final weight of the abrasive charge ranges between 4690 and 4860 g. The test sample is formed by at least 15 kg of sample sizing between 10 and 14 mm and a final sample weight of 5000 g. In the ASTM C131/C131 M standard, used for aggregates in concrete production, the aggregate sample is subdivided into different size fractions by sieving and the obtained fractions are then recombined to the grading shown in Table 2.4.

The test procedure includes the following main steps (EN 1097-2 standard): "(1) place the test specimen and abrasive charge in the Los Angeles abrasive testing machine; (2) start the testing machine and run it for 500 revolutions at a rate of 31–33 rpm, (3) when the testing machine has completed the required number of revolutions, empty the entire contents into a pan and remove the abrasive charge from the pan, (4) separate the test specimen on the 1.6 mm sieve and weigh and record this value, and (5) calculate the Los Angeles abrasion value using the following equation":

$$\text{Percent Wear} = \frac{(\text{A} - \text{B})}{\text{A}} \times 100$$

where

A – is the weight of the original test specimen to the nearest 1 g (5000 g)
B – is the weight retained on the 1.6 mm sieve to the nearest 1 g.

For instance, a Los Angeles abrasion value of 30 means that 30% of the original sample weight passed through the specified size (1.6 mm in European standard—EN 1097-2). For instance, the resistance to fragmentation of coarse aggregates in European regulations for aggregates in concrete (EN 12620 standard) shall be determined in terms of the Los Angeles coefficient and this value shall be declared in accordance with the relevant category specified in Table 2.5.

The test can be performed on aggregates of different sizes. For example, ASTM C131/C131M for aggregates between 2.36 and 37.5 mm and ASTM C535 for aggregates between 19 and 75 mm (these two standards are utilized as specification quantification tests for concrete aggregates).

Table 2.4 Sample size and gradings for the Los Angeles test (ASTM C131/C131 M standard)

Sieve size (square openings)		Mass of indicated sizes, g			
Passing	Retained on	Grading			
		A	B	C	D
37.5 mm (1½ in.)	25.0 mm (1 in.)	1250 ± 25	–	–	–
25.0 mm (1 in.)	19.0 mm (¾ in.)	1250 ± 25	–	–	–
9.0 mm (¾ in.)	12.5 mm (1/2 in.)	1250 ± 10	2500 ± 10	–	–
12.5 mm (1/2 in.)	9.5 mm (3/8 in.)	1250 ± 10	2500 ± 10	–	–
9.5 mm (3/8 in.)	6.3 mm (¼ in.)	–	–	2500 ± 10	–
6.3 mm (¼ in.)	4.75 mm (No. 4)	–	–	2500 ± 10	–
4.75 mm (No. 4)	2.36 mm (No. 8)	–	–	–	5000 ± 10
Total		5000 ± 10	5000 ± 10	5000 ± 10	5000 ± 10

Table 2.5 Categories for maximum values of Los Angeles coefficients in coarse aggregates in concrete (EN 12620 standard)

Los Angeles coefficient	Category LA
≤ 15	LA_{15}
≤ 20	LA_{20}
≤ 25	LA_{25}
≤ 30	LA_{30}
≤ 35	LA_{35}
≤ 40	LA_{40}
≤ 50	LA_{50}
> 50	$LA_{Declared}$
No requirement	LA_{NR}

Sometimes, it is necessary to test an aggregate under wet abrasion conditions to identify aggregates that may breakdown into fines, generated for example in handling, batching and mixing processes. In these cases, an alternative to the Los Angeles abrasion test is the Micro-Deval test, which is utilized for fine aggregates. The Micro-Deval test is similar in principle to the Los Angeles abrasion test. Both tests will provide specific information about the relative quality and durability of aggregates. The main advantages of the Micro-Deval test are the smaller equipment size and lower sample amount.

The main goal of the Micro-Deval test is to determine the abrasion loss in the presence of water and an abrasive charge. Many aggregates are more susceptible to abrasion when wet than dry, and the use of water in this test checks the reduction in resistance to degradation. In European regulation, the EN 1097-1 standard specifies a procedure for measuring the resistance to wear of a sample of aggregate. The test determines the Micro-Deval coefficient, which is the percentage of the original sample reduced to a size smaller than 1.6 mm during rolling. The test consists of "measuring the wear produced by friction between the aggregates and an abrasive charge in a rotating drum (Fig. 2.36) under defined conditions; when rolling is complete, the percentage retained on a 1.6 mm sieve is used to calculate the Micro-Deval coefficient" (EN 1097-1 standard). A lower value of the Micro-Deval coefficient indicates a better resistance to wear. The sample is normally tested under wet conditions although it can also be tested under dry conditions.

Fig. 2.36 Machine for Micro-Deval test (image courtesy of Holcim)

Similar to the Los Angeles abrasion test, the Micro-Deval coefficient is calculated using the following expression:

$$M_{DE} = (500 - m)/5$$

where

- M_{DE} – is the Micro-Deval coefficient (under wet conditions)
- m – is the mass (in grams) of the oversize fraction retained on a 1.6 mm sieve.

Another test that is also intended for coarse aggregates used in the road is the EN 1097-9 standard, which determines the resistance of aggregates to abrasive wear by studded tire using the so-called Nordic abrasion machine. The test is performed on aggregates with sizes ranging from 11.2 to 16.0 mm and consists of a rotating aggregate in a drum containing steel abrasive balls and water. The abrasion loss rate of aggregates is calculated after the specified number of revolutions stated in the standard.

- Polishing

All aggregates polish under traffic depending on the aggregate type, traffic characteristics, and road geometry. The action of road vehicle tires on road surfaces polishes the exposed aggregate surface, and the state of polish is a very important factor that affects the resistance to skidding. This is related to the ability of the aggregates to lose their initial, rough surface, and develop a polish. Although the correlation between tests of polishing on aggregates and the skid resistance of the corresponding pavement is still not clear, it allows establishing the ranking of polishing characteristics.

The hardness of the minerals in an aggregate controls the polishing resistance of the aggregate. Since the rate of polishing is mainly related to the type of minerals in the aggregate, "a high percentage of hard, well-bonded mineral grains in a softer matrix should be present in the aggregates if it is to resist the abrasive smoothing action of tires" (NSSGA 2013). Aggregates obtained from hard, fine-grained rocks tend to polish quickly but this process depends on the composition and state of weathering of the rocks. Igneous rocks with a small proportion of soft minerals have a high polishing resistance whereas limestone has a low resistance to polishing and hence is frequently limited for use in pavement surface courses.

Polishing is determined through EN 1097-8, ASTM E303, and ASTM D3319 standards. The latter "covers a laboratory procedure by which an estimate may be made of the extent to which different coarse aggregates may polish". A polishing value is calculated that can be utilized to establish a classification of coarse aggregates based on their ability to resist polishing under traffic. ASTM E303 determines "the relative effects of various polishing processes on materials or material combinations using the British pendulum skid resistance tester". For its part, the EN 1097–8 standard estimates the so-called polishing stone value (PSV) of an aggregate, which "provides a measure of the resistance of road stone to the polishing action of vehicle tires on a road surface under conditions similar to those occurring on a road". This standard is the most suitable test for aggregates that may be used in a road surface course (▸ Box 2.5: Polishing Stone Value).

Box 2.5—Polishing Stone Value

The polishing stone value is calculated in an aggregate passing through the 10 mm sieve and retained on the 7.2 mm grid sieve. The sample must be taken from a normal run of production from the plant. The standard advises that "chippings that have been freshly crushed in the laboratory or recovered from bituminous materials may give misleading results". The test is carried out in two steps: (a) the test specimen is subjected to the action of polishing produced by an accelerated polishing machine (▪ Fig. 2.37*) and (b) measurement of the polishing state of the sample using a friction test (i.e., British pendulum friction test—*▪ Fig. 2.38*). With the results of the second step, PSV is determined.*

The British pendulum tester is a dynamic pendulum impact-type machine that is utilized for measuring the energy loss when a rubber slider edge is propelled over a test surface. This device is employed to study problems in the design and maintenance of public highways, and to test the frictional resistance of new roads, road markings, and iron works. It provides a routine method of checking the resistance of wet and dry surfaces to slipping and skidding.

As a rule, "it has been found that rock types consisting of a variety of mineral grains of different hardness or size, or of harder grains in a softer cementing matrix, give higher PSV values compared to rocks composed of uniform grains of uniform hardness in a similarly hard matrix" (Bustillo 2021). *Finally, several annexes in the standard control the different steps in the process. Annex B controls the materials, Annex C controls the calibration of the accelerated polishing machine, and Annex D controls the calibration of the friction tester and slider.*

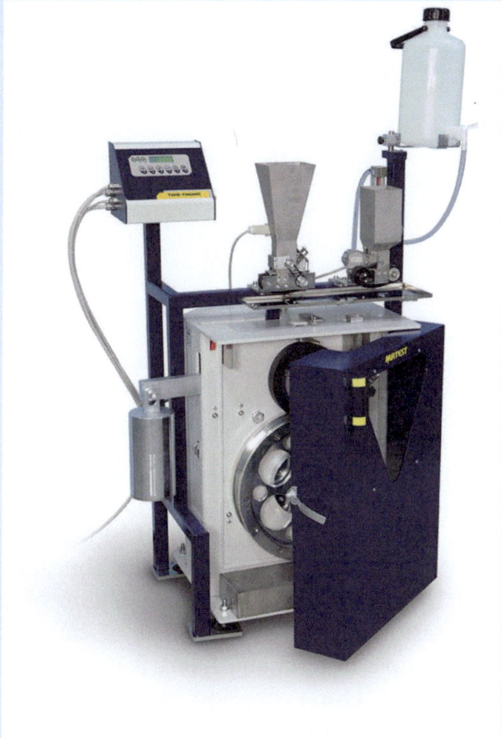

▪ Fig. 2.37 Accelerated polishing machine (image courtesy of Matest)

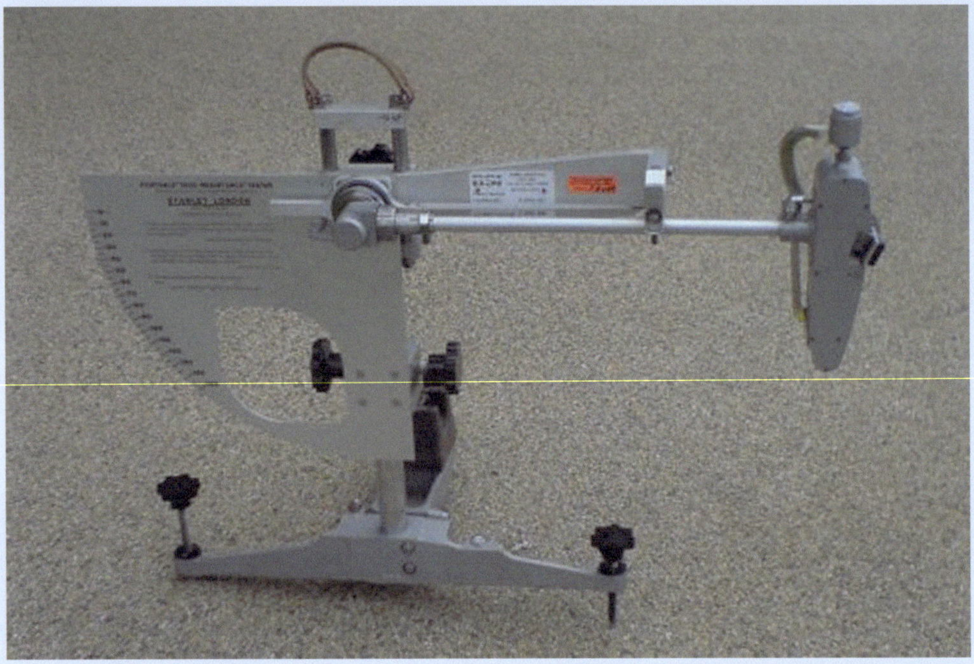

Fig. 2.38 Friction pendulum for PSV calculation

In addition, annex A of the EN 1097-8 standard includes an optional method for determining the aggregate abrasion value (AAV). This method should be used for aggregates that may suffer the effects of the abrasion action produced by road traffic. The determination of the aggregate abrasion value (AAV) consists of embedding an aggregate sample in resin and fixing it in contact with a polishing wheel (Dorry abrasion machine—Fig. 2.39) that rotates in a horizontal plane, continuously feeding with abrasive sand. The CAA coefficient is determined from the difference between the mass of the specimen before and after abrasion. The aggregate for testing must pass the 14 mm sieve and be retained in the 10.2 mm sieve.

2.5 Thermal and Weathering Properties and Tests

The importance of the thermal and weathering properties of aggregates depends on their end-use, with some of these properties (i.e., thermal properties) being rarely needed. In European regulations, thermal and weathering properties are determined by the following parts of the EN 1367 test methods series: determination of resistance to freezing and thawing (EN 1367-1 standard), magnesium sulfate test (EN 1367-2 standard), boiling test for Sonnenbrand basalt (EN 1367-3 standard), determination of drying shrinkage (EN 1367-4 standard), determination of resistance to thermal shock (EN 1367-5 standard), determination of resistance to freezing and thawing in the presence of salt (NaCl) (EN 1367-6 standard), determination of resistance to freezing and thawing of lightweight aggregates (EN 1367-7 standard), and determination of resistance to the disintegration of lightweight aggregates (EN 1367-8 standard). Similar or comparable ASTM standards for these European tests are C666/C666 M for the resistance of concrete to rapid freezing and thawing and C88 for soundness of aggregates by utilization of sodium sulfate or magnesium sulfate.

2.5.1 Thermal Properties and Tests

Three thermal properties can be significant in some aggregate applications (i.e., in the performance of concrete): thermal volume change, thermal conductivity, and integrity during heating. Thermal volume change is the volume modification of aggregates generated by temperature variations. This property is expressed through the coefficient of thermal expansion. Aggregates should have a coefficient of thermal expansion that is approximately "the same in all directions and at all exposure temperatures; furthermore, all minerals present in the aggregates, ideally, should have the same coefficient of thermal expansion to prevent internal fracture of the aggregates" (NSSGA 2013).

When heated, concrete expands uniformly with temperature rise over the temperature range of 0–60 °C. Because aggregate is the most important component

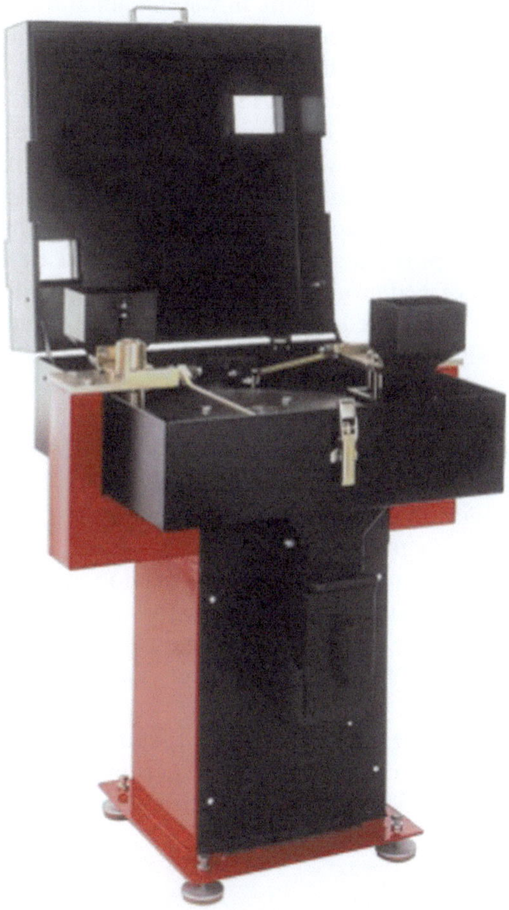

Fig. 2.39 Dorry abrasion machine

in concrete, the coefficient of thermal expansion of the aggregate used in the manufacture of concrete dominates in determining the expansion of the concrete, although the aggregate content and mix proportions markedly influence its final value. If a significant difference exists between the coefficient of thermal expansion of the aggregate and the cement paste, the durability of the concrete subjected to freezing and thawing can be negatively affected.

Concretes produced with different aggregates can perform very differently when subjected to high or low temperatures. For instance, concrete made with siliceous rocks, if exposed to fire, is likely to spall and crack to a greater extent than concrete manufactured using carbonate rocks. The thermal expansion coefficient of the most common rocks used as aggregate ranges from 5 to 12×10^{-6}/°C. It increases with increases in the silica content of the rock but wide variations can occur within a rock group. Table 2.6 shows that rocks containing nearly 100% silica have the highest coefficients whereas limestone has the lowest value since it has very low silica content.

Table 2.6 Thermal expansion coefficient of the most common rocks used as aggregate (Smith and Collis 2001)

Rock	Silica content (%)	Coefficient of thermal expansión 1×10^{-6}/°C
Chert	94	11.8
Quartzite	94	10.3
Sandstone	84	9.3
Granite	66	6.8
Basalt	51	6.4
Limestone	Traces	5.5

Thermal conductivity is the ability of aggregates to transmit heat. Aggregates "with low thermal conductivity are desired to decrease the depth of frost penetration through a pavement; conversely, aggregates that transmit heat rapidly minimize the development of large differential temperatures between the top and bottom of a rigid pavement slab" (NSSGA 2013). Integrity during heating is the capacity of aggregates to maintain their main characteristics during high temperature changes. Aggregates must not be adversely affected by high temperatures so that the aggregate particles fracture, become weak, or have other characteristics modified by heat.

The thermal behavior of aggregates is controlled in European legislation by EN 1367-5 standard. It "specifies methods for the determination of resistance to thermal shock of aggregates, subject to heating and drying in the production of hot bituminous mixtures". Regarding the principle of the test, it involves two test samples; one sample is soaked heated to 700 °C for 3 min and the resistance to fragmentation is calculated using EN 1097-2 (see mechanical tests). The strength loss is calculated by comparison with the result of the second sample that has not been heated. The increase in undersize passing the 5 mm sieve after the thermal shock is also calculated.

2.5.2 Weathering Properties and Tests

The weathering properties of aggregates are decisive in some applications, especially in concrete, road, and track ballast uses. In regions with severe winters, a major cause of aggregate deterioration in exposed concrete is freezing and thawing. Therefore, the climate has a major influence on weathering and thus on the decomposition (soundness) of rocks. Although soundness strictly refers to a physical property of aggregates, it is a term defined as "the ability of an aggregate to resist excessive changes in volume as a result of changes in physical conditions such as freezing and thawing,

2

thermal changes at temperatures above 0 °C and alternate wetting and drying" (Alexander and Mindex 2005). It relates to the physical competence or physical durability of the material. Consequently, this property must clearly be distinguished from volume changes caused by chemical reactions between aggregates and their environment. Soundness is complicated to measure, a fact hindered by the long-term nature of most durability processes. In this sense, petrological examination of aggregates can offer a valuable estimation of expected soundness by determining the mineralogy as well as the nature and distribution of pores and fissures.

Aggregates can be susceptible to freeze–thaw damage depending on their pore structure and absorption and permeability properties. An aggregate particle can absorb so much water that it cannot accommodate the expansion that takes place during water freezing; this process is the primary mechanism of freeze–thaw damage. This is because any porous solid that can imbibe water and freeze will suffer hydraulic, expansive pressure. The main controlling factor is the degree of saturation because partially saturated pores may permit the relief of this type of pressure. Consequently, the most interesting conclusion is that aggregate freeze–thaw issues will take place only if the aggregate becomes critically saturated. Regardless of the freezing rate, there can be a critical size of particle above which a particle will fail if frozen. This particle size depends on the rate of freezing and the porosity, permeability, and tensile strength of the aggregate particle.

The performance of aggregates subjected to freezing and thawing can be studied with two methods (a) past performance in the field and (2) laboratory tests. In the first method, if aggregate from the same source has previously given adequate results, it can be considered suitable. With respect to laboratory tests, ASTM C666/C666 M and EN 1367-1 standards control the resistance of aggregates to freezing and thawing, with the former devoted to concrete applications.

The EN 1367-1 standard "specifies a test method that provides information on how an aggregate behaves when it is subjected to the cyclic action of freezing and thawing". The principle of the test is the following: "test portions of single-sized aggregates (having a particle size between 4 and 63 mm), having been soaked in water at atmospheric pressure, are subjected to 10 freeze–thaw cycles. This involves cooling to – 17.5 °C under water and thawing in a water bath at approximately 20 °C; after completion of the cycles, the aggregates are examined for any changes" (i.e., loss in mass). Alternatively, the ASTM C666/C666 M standard evaluates the freeze–thaw performance of aggregates in concrete through quick cycles of freezing and thawing between 4.4 and – 17.8 °C. In particular, two different procedures are established in this standard (a) rapid freezing and thawing in water and (b) rapid freezing in air and thawing in water. Climatic chambers (Fig. 2.40) are used in these tests for controlling temperature and humidity.

On the other hand, ASTM C88 and EN 1367-2 standards measure the resistance of aggregates to accelerated physical weathering by salt crystallization. The principle of the tests is that the internal expansive force generated by the salt crystal growth in the aggregate pores is similar to that produced by the expansion of water on freezing. In the ASTM C88 standard, the aggregate is immersed in prepared solutions of magnesium or sodium sulfate of specified density at a temperature of 20–22 °C for a period of 16–18 h. The immersion cycle is repeated five times and the sample is sieved, recording the weight retained; the loss of mass in the sample is expressed as a percentage of the initial weight (sulfate soundness loss). In the EN 1367-2 standard, aggregates in the size range of 10–14 mm are subjected to five cycles of immersion in a saturated solution of magnesium sulfate followed by oven drying at 110 °C. The degradation process is measured by the extent to which the material finer than 10 mm is produced.

Weathering caused by wetting and drying can also influence aggregate durability. The expansion and contraction coefficients of rocks can change according to

Fig. 2.40 Climatic chamber

temperature and moisture content variations. Concrete surfaces may be exposed to wetting and drying cycles due to rainfall, rise and fall of water, etc. and severe strain can develop in some aggregates when alternate wetting and drying takes place. With a certain type of rock, this may create a permanent volume increment in the concrete and an eventual breakdown. In general, the majority of dense aggregates have a low shrinking and swelling potential and aggregates impart dimensional stability to concrete. Nevertheless, certain aggregates are exceptions to this general rule. Moreover, clay lumps and other friable particles can degrade quickly with repeated wetting and drying.

The EN 1367-4 standard "specifies a method for determining the effect of aggregates on the drying shrinkage of concrete based on the testing of concretes of fixed mix proportions and aggregates of 20 mm maximum size". In this European test, the aggregate "is mixed with cement and water and cast into prisms of specified dimensions; the prisms are subjected to wetting followed by drying at 110 °C and the change in length from the wet to the dry state is determined. The excess drying shrinkage of the concrete is attributed to the aggregate, and is expressed as the average change in length of the prisms, as a percentage of their final dry lengths".

In European legislation, the last standard devoted to the weathering properties of aggregates is related to the Sonnenbrand effect in basalt (▶ Box 2.6: Sonnenbrand in Basalt).

Box 2.6—Sonnenbrand in Basalt

Sonnenbrand (sunburn in English) is a German word and finds its origin in the old German mining industry. Originally "Sonnenbrenner", basalts or not, were those rocks that had lost their capability to be processed in the quarry. It was caused by their loss of their internal moisture, "for example by drying up in the sunshine" (Tooren 1992). *Although the Sonnenbrand problem seems to be a modern problem, it was first described in 1786 by Faujas de St. Fond. This author outlined the gray spherical spots present in a basalt that tended to disintegrate caused by the existence of cracks from spot to spot. From 1875, the term "became increasingly reserved for basalts with gray spherical spots and cracks between the spots, that is, basalts that tend to deteriorate very quickly"* (Pukall 1939).

Since the historical background of the Sonnenbrand phenomenon lies mainly in Germany, the term Sonnenbrand has been preferentially used in European standards (i.e., EN 1367-3: Boiling test for "Sonnenbrand basalt"). In this standard, Sonnenbrand is defined as a "type of rock decay that may be present in some basalts and which manifests itself under the influence of atmospheric conditions". This process "starts with the appearance of gray/white star-shaped spots; usually hairline cracks are generated radiating out from the spots and interconnecting them; this reduces the strength of the mineral fabric, and as a result the rock decays to small particles". The last paragraph of the definition in the European standard states that "depending on the source this process may take place within months of extraction or extend over several decades". ◘ Figure 2.41 *shows the typical star-shaped spots in a basalt sample.*

◘ **Fig. 2.41** Typical star-shaped spots in a basalt sample (Gisbert et al. 2016)

Sonnenbrand has been reviewed for over 200 years but most of its characteristics (i.e., extent, formation of sunburn areas, and origin) remain unexplained today. It is an unusual type of disintegration of basaltic rocks and should not be confused with any of the miscellaneous symptoms of basalt weathering. In the first stage of Sonnenbrand development, spots appear (three-dimensional, isometric fair discolorations) with diameters ranging from tenths of a millimeter to 20 mm. A system of irregular capillary cracks ("Haarrise" in German literature) is further developed, usually between the spots (◘ Fig. 2.42). *As a consequence, the formation of these disjunctions leads to the breaking up of the sunburn basalt into debris.*

Regarding the genesis of the Sonnenbrand phenomenon, the connection of the Sonnenbrand origin with late magmatic processes and nonuniform crystallization of analcime and nepheline from the residual melt seems clear, that is, Sonnenbrand "occurs exclusively in basalts and is bound to the existence of nepheline, analcime and/or rock glass" (Weiher et al. 2007). Gisbert et al. (2016) *stated that "the combination of rock temperature and water supply strongly influence the conditions of the development and type of Sonnenbrand".*

Basalts with a tendency to sunburn are extremely susceptible to weathering, disintegrate easily and are not suitable for many technical applications (i.e., track ballast).

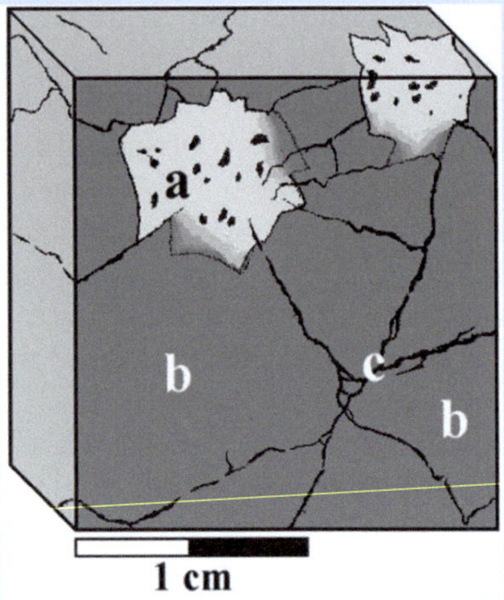

Fig. 2.42 Sketch of the sunburn rock: **a** spot, **b** healthy parts of rock and **c** capillary cracks (Zagozdzon 2003)

Petrographic examination provides a first evaluation of the process. However, if there is a suspicion of the Sonnenbrand process in basalts, the stability must be defined according to the EN 1367-3 standard. It "specifies methods for the determination of the presence of signs of "Sonnenbrand" in basalt and the disintegration of aggregate produced from basalt showing such signs; the test applies to pieces of rocks, and graded basalt coarse aggregates".

In the test procedure, aggregate samples are tested to determine, after boiling, the loss of mass and the loss of strength in percent. for the loss of mass, the following formula calculates this loss:

$$M_1 = [(m_0 - m_1)/m_0] \times 100$$

where

M_1 *is the loss of mass as a percentage*
M_0 *is the mass of the sample before boiling*
M_1 *is the mass of the retained fraction on the half-size sieve after boiling.*

For the loss of strength, the formula is very similar to the previous one:

$$S_{LA} = [(LA_1 - LA_0)/LA_0] \times 100$$

where

S_{LA} *is the loss of strength in percentage*
LA_0 *is the Los Angeles coefficient of the unboiled sample*
LA_1 *is the Los Angeles coefficient of the boiled sample.*

2.6 Chemical Properties and Tests

The chemical properties of aggregates are very important in applications such as concrete or asphalt. This is because some types of aggregate contain minerals that are chemically reactive and can affect the final behavior of the mixture. In other words, some components found in aggregates can have a detrimental effect on the cementing and overall performance characteristics of concrete and asphalt. In some cases, serious deterioration is produced by chemical action initiated by binder compounds. Consequently, the aggregate must be compatible with the binder to avoid this problem. The alkali-silica reaction in concrete is the most outstanding chemical reaction related to aggregate mineral composition. Since the influence of the chemical composition of the aggregates is closely related to their final application, it will be taken into account in each aggregate application (see ▶ Chap. 7). The main test methods for the chemical properties of aggregates are described in the different parts of the European EN 1744 standard series, with EN 1744-1 (chemical analysis) being the most interesting from a general viewpoint.

2.7 Questions

Short Questions

- What do aggregate properties depend on?
- What does grading mean? Explain the term
- Explain in short the sieving method for a sieve analysis
- Define particle shape and sphericity
- What is the objective of the methylene blue test?
- Define density, absolute density, apparent density and relative density
- What is the difference among the following terms: lightweight aggregate, normal weight aggregate and heavyweight aggregate?
- Explain the differences between porosity and permeability
- What does the term soundness mean?
- Explain the significance of the freeze–thaw damage

Long Questions

- Explain in detail the Los Angeles abrasion test
- Explain the significance of the Sonnenbrand phenomenon

References

ACI (2007) Aggregates for concrete. American Concrete Institute, ACI Education Bulletin E1-07, 29 p

Alexander M, Mindex S (2005) Aggregates in concrete. In: Modern concrete technology, vol. 13. Taylor & Francis, Oxford, England, 435 p

Bustillo M (2021) Construction materials—geology, production and applications. Springer, Cham, Switzerland, 603 p

CCA (2020) Guide to concrete construction—Part II: constituents of concrete—Section 3: aggregates. Cement Concrete & Aggregates Australia, Sydney, Australia, p 21

Gisbert G, Aulinas M, Garcia-Valles M, Fernandez D, Gimeno D, Zagozdzon P (2016) Caracterización petrológica y geoquímica del moteado leucocrático en rocas basálticas alcalinas. IX Congreso Geológico De España, Geo-Temas 16:1576–5172

Krumbein WC (1941) Measurement and geological significance of shape and roundness of sedimentary particles. J Sediment Res 11:64–72

Langer W (2006) Crushed stone. In: Kogel JE, Trivedi NC, Barker JM, Krukowski ST (eds) Industrial minerals & rocks—commodities, markets and uses. Society for Mining, Metallurgy and Exploration, Inc., Colorado, USA, pp 171–180

Neville AM, Brooks JJ (2010) Concrete technology, 2nd ed. Pearson Education Limited, Essex, England, 442 p

NSSGA (2013) The aggregates handbook, 2nd ed. The National Stone, Sand and Gravel Association, Alexandria, Virginia, USA, 906 p

Pukall K (1939) Beitrage zur frage des sonnenbrandes der Basalte. I. Zeitschrift Für Angewandte Mineralogie 1:195–222

Smith MR, Collis L (eds) (2001) Aggregates: sand, gravel and crushed rock. Geological Society, London, Engineering Geology Special Publication, vol 17, London, UK, 360 p

Tourenq, Denis (2000) Properties of aggregates—tests and specifications. In: Primel L, Tourenq C (eds) Aggregates. A.A. Balkema, Rotterdam, Netherlands, pp 109–142

van Tooren MM (1992) Sonnenbrand in basalts. In: Pieters WE, van Tooren MM, Verhoef PNW (eds) Geomaterials characterisation and testing—evaluation report. Memoirs of the Centre of Engineering Geology in the Netherlands, vol 104, pp A6-1–A6-9

Waddell H (1933) Sphericity and roundness of rock particles. J Geol 41:310–331

Weiher B, Lehrberger G, Thuro K (2007) Petrophysical properties of sunburn basalt from the upper palatinate in north-eastern Bavaria. In: Otto F (ed) Tagung für Ingenieurgeologie und Forum "Junge Ingenieurgeologen." DGGT, Bochum, pp 77–87

Zagozdzon P (2003) Sunburn in the tertiary basalts of Silesia (SW Poland). Geolines 15:188–193

Standards

ASTM C117—2017. Standard test method for materials finer than 75-μm (no. 200) sieve in mineral aggregates by washing

ASTM C125—2018. Standard terminology relating to concrete and concrete aggregates

ASTM C1252—2017. Standard test methods for uncompacted void content of fine aggregate (as influenced by particle shape, surface texture, and grading)

ASTM C127—2012. Standard test method for density, relative density (specific gravity), and absorption of coarse aggregate

ASTM C128—2012. Standard test method for density, relative density (specific gravity), and absorption of fine aggregate

ASTM C131/131 M—2006. Test method for resistance to degradation of small size coarse aggregate by abrasion and impact in the Los Angeles machine.

ASTM C136/136 M—2014. Test method for sieve analysis of fine and coarse aggregates

ASTM C142 / C142 M—2017. Standard test method for clay lumps and friable particles in aggregates

ASTM C1777—2013. Standard test method for rapid determination of the methylene blue value for fine aggregate or mineral filler using a colorimeter

ASTM C29/C29 M - 2009. Test Method for Bulk Density ("unit weight") and Voids in Aggregate.

ASTM C295/295 M—2012. Standard guide for petrographic examination of aggregates for concrete

ASTM C535—2012. Test method for resistance to degradation of large size coarse aggregate by abrasion and impact in the Los Angeles machine

ASTM C566—2013. Standard test method for total evaporable moisture content of aggregate by drying

ASTM C666/C666 M—2008. Test method for resistance of concrete to rapid freezing and thawing

ASTM C88—2013. Test method for soundness of aggregates by use of sodium sulfate or magnesium sulfate

ASTM D1217—2015. Standard test method for density and relative density (specific gravity) of liquids by Bingham pycnometer

ASTM D2419—2009. Test method for sand equivalent value of soils and fine aggregate

ASTM D2419—2014. Standard test method for sand equivalent value of soils and fine aggregate

ASTM D3319—2011. Standard test method for accelerated polishing of aggregates using the British wheel

ASTM D3398—2006. Standard test method for index of aggregate particle shape and texture

ASTM D3744/D3744M—2011. Standard test method for aggregate durability index

ASTM D4791—2010. Standard test method for flat particles, elongated particles, or flat and elongated particles in coarse aggregate

ASTM D5821—2017. Standard test method for determining the percentage of fractured particles in coarse aggregate

ASTM D6928—2010. Standard test method for resistance of coarse aggregate to degradation by abrasion in the micro-Deval apparatus

ASTM D7428—2015. Standard test method for resistance of fine aggregate to degradation by abrasion in the micro-deval apparatus

ASTM D7698—11. Standard test method for in-place estimation of density and water content of soil and aggregate by correlation with complex impedance method

ASTM E11—2020. Standard specification for woven wire test sieve cloth and test sieves

ASTM E303—2018. Standard test method for measuring surface frictional properties using the British pendulum tester

EN 1097-1: 2011. Tests for mechanical and physical properties of aggregates—Part 1: determination of the resistance to wear (micro-Deval)

EN 1097-10: 2014. Tests for mechanical and physical properties of aggregates—Part 10: determination of water suction height

EN 1097-11: 2013. Tests for mechanical and physical properties of aggregates—Part 11: determination of compressibility and confined compressive strength of lightweight aggregates

EN 1097-2: 2020. Tests for mechanical and physical properties of aggregates—Part 2: methods for the determination of resistance to fragmentation

EN 1097-3: 1998. Tests for mechanical and physical properties of aggregates—Part 3: determination of loose bulk density and voids

EN 1097-4: 2008. Tests for mechanical and physical properties of aggregates—Part 4: determination of the voids of dry compacted filler

EN 1097-5: 2008. Tests for mechanical and physical properties of aggregates—Part 5: determination of the water content by drying in a ventilated oven

EN 1097-6: 2022. Tests for mechanical and physical properties of aggregates—Part 6: determination of particle density and water absorption

EN 1097-7: 2008. Tests for mechanical and physical properties of aggregates—Part 7: determination of the particle density of filler—pycnometer method
EN 1097-8: 2020. Tests for mechanical and physical properties of aggregates—Part 8: determination of the polished stone value
EN 1097-9: 2014. Tests for mechanical and physical properties of aggregates—Part 9: determination of the resistance to wear by abrasion from studded tyres—Nordic test
EN 12620: 2013. Aggregates for concrete
EN 1367-1: 2007. Tests for thermal and weathering properties of aggregates—Part 1: determination of resistance to freezing and thawing
EN 1367–2: 2009. Tests for thermal and weathering properties of aggregates—Part 2: magnesium sulfate test
EN 1367-3: 2001. Tests for thermal and weathering properties of aggregates—Part 3: boiling test for "Sonnenbrand basalt"
EN 1367-4: 2008. Tests for thermal and weathering properties of aggregates—Part 4: determination of drying shrinkage
EN 1367-5: 2011. Tests for thermal and weathering properties of aggregates—Part 5: determination of resistance to thermal shock
EN 1367-6: 2008. Tests for thermal and weathering properties of aggregates—Part 6: determination of resistance to freezing and thawing in the presence of salt (NaCl)
EN 1367-7: 2014. Tests for thermal and weathering properties of aggregates—Part 7: determination of resistance to freezing and thawing of lightweight aggregates
EN 1367-8: 2014. Tests for thermal and weathering properties of aggregates—Part 8: determination of resistance to disintegration of lightweight aggregates
EN 1744-1: 2009. Tests for chemical properties of aggregates—Part 1: chemical analysis
EN 932-3: 2004. Tests for general properties of aggregates—Part 3: procedure and terminology for simplified petrographic description
EN 933-1: 2012. Tests for geometrical properties of aggregates—Part 1: determination of particle size distribution-sieving method
EN 933-10: 2009. Tests for geometrical properties of aggregates—Part 10: assessment of fines—grading of filler aggregates (air jet sieving)
EN 933-2: 1995. Tests for geometrical properties of aggregates—Part 2: determination of particle size distribution—test sieves, nominal size of apertures
EN 933-3: 2012. Tests for geometrical properties of aggregates—Part 3: determination of particle shape—flakiness index
EN 933-4: 2008. Tests for geometrical properties of aggregates—Part 4: determination of particle shape—shape index
EN 933-5: 1998. Tests for geometrical properties of aggregates—Part 5: determination of percentage of crushed and broken surfaces in coarse aggregate particle
EN 933-6: 2014. Tests for geometrical properties of aggregates—Part 6: assessment of surface characteristics—flow coefficient of aggregates
EN 933-8: 2012. Tests for geometrical properties of aggregates—Part 8: assessment of fines—sand equivalent test
EN 933-9: 2009. Tests for geometrical properties of aggregates—Part 9: assessment of fines—methylene blue test
ISO 2395: 1990. Test sieves and test sieving—Vocabulary
ISO 3310-1: 2000. Test sieves—technical requirements and testing—Part 1: test sieves of metal wire cloth
ISO 3310-2: 1990. Test sieves—technical requirements and testing—Part 2: test sieves of perforated metal plate
ISO 6782: 1982. Aggregates for concrete—determination of bulk density
ISO 6783: 1982. Coarse aggregates for concrete—determination of particle density and water absorption—hydrostatic balance method
ISO 7033: 1987. Fine and coarse aggregate for concrete—determination of the particle mass-per-volume and water absorption—pycnometer method

Geological Occurrence

Contents

3.1 Introduction – 54

3.2 Sand and Gravel – 56
3.2.1 Fluvial Deposits – 62
3.2.2 Glacial Deposits – 69
3.2.3 Coastal Deposits – 72
3.2.4 Marine Deposits – 72
3.2.5 Eolian (Windblown) Deposits – 76

3.3 Crushed Stone – 76
3.3.1 Sedimentary Rocks – 77
3.3.2 Igneous Rocks – 87
3.3.3 Metamorphic Rocks – 95

3.4 Questions – 97

References – 99

M. Bustillo Revuelta, *The Basics of Aggregates*,
Springer Textbooks in Earth Sciences, Geography and Environment,
https://doi.org/10.1007/978-3-031-42961-3_3

3

Abstract

This chapter introduces the geological characteristics of aggregate deposits. This type of mineral resource, regardless of whether they are rocks for crushing or sand and gravel for direct utilization, is present in numerous geological environments, all with their own unique characteristics. Sand and gravel deposits are formed by the erosion, transportation, and deposition of fragments. They are most common in glaciated areas, alluvial basins, and fluvial deposits near rivers and streams. Sand and gravel deposits are principally Pleistocene or younger and are a reflection of the region's latest geological and climatic history. In fluvial deposits, aggregates are commonly mined directly from the river channel and/or from river terrace deposits. Glaciated regions also contain numerous deposits of this type. Marine deposits on continental shelves are large potential sources of sand and gravel, being generally sand-size and finer. Many different types of rocks are suitable for utilization as aggregates when crushed but high-quality aggregates are exclusively obtained from very hard, dense siliceous rocks. Limestone, dolostone, and sandstone are the most widely used sedimentary rocks as aggregate. Granite, basalt and andesite are typical examples of igneous rocks commonly utilized as aggregate whereas gneiss, marble, and quartzite are metamorphic rocks quarried for aggregates.

3.1 Introduction

Except for a small amount of lightweight aggregates utilized principally in special applications, almost all construction aggregates are produced by crushing strong rocks such as granite and limestone (◘ Fig. 3.1) or from particulate deposits such as sand and gravel (◘ Fig. 3.2). Aggregate materials can be grouped into two main sources: the unconsolidated deposits of rock fragments that form part of surficial deposits and the hard rocks that constitute part of the Earth's crust. Aggregates used for construction purposes can be grouped into two main categories (a) exposed or near-surface sedimentary, metamorphic, and igneous rocks to be crushed and (b) sand and gravel deposits, which can be either used directly in their natural state or further crushed, washed, and sized to meet specifications of the selected application. Both types of primary or natural aggregates are utilized in construction according to the specification standards and economic considerations. For instance, in the manufacturing of concrete, sand and gravel are selected while crushed stone is preferred in asphaltic mixes since asphalt adheres better to rough and tough surfaces. Aggregate deposits, regardless of whether they are rocks for crushing or sand and gravel for direct utilization, are present in numerous geological environments, all with their unique characteristics.

Glaciers, rivers, and streams have contributed to the formation of most sand and gravel deposits whose materials are used as aggregates (the most important transport medium is flowing water, as in rivers). On the other hand, for every region of the world, the availability of rocks for crushing relies upon the global geological history of the region, that is, the processes that formed, eroded, and exposed the rocks. The location of suitable deposits for assessing the potential of new aggregate sources strongly depends on understanding the geological processes that form them and the specific geological settings in which these new deposits occur. Consequently, the geological reconnaissance of a region is essential to establish its aggregate potential. It provides the relations between the geological properties of sediments and rocks and their expected performance as an aggregate, including information about the physical and chemical properties related to the aggregate quality. This is because these properties (i.e., grain size and texture) result from the geological origin and mineralogy of the potential source and its further alteration. In summary, if an aggregate supply is needed, geological studies may define the location, distribution, and nature of aggregates in a region. They are required to locate new deposits and extension of known deposits and to devise the most efficient extraction pattern.

Quaternary geological history for deposits of sand and gravel and an overall study of the stratigraphy, origin, and structural evolution of the area for crushed stone is the key to mapping potential sources of aggregates. Nevertheless, some regions lack sand and gravel deposits and sources of crushed rocks can be covered by overburden, which is often too thick to enable economical surface mining. This means that the availability of aggregates is not universal. "Communities lacking local aggregate sources generally face the costly alternatives of importing aggregate from outside the area or substituting another material for aggregate" (Langer et al. 2004).

Sand and gravel deposits were the main source of natural aggregates until the eighties. Since then, environmental regulations have constrained the exploitation of this type of natural aggregate in many countries, producing a decline in its utilization ("in 2017, sand and gravel accounted for 35% of the total primary aggregate supply in Great Britain"—BGS, 2019). The relative proportions of crushed stone and sand and gravel production generally do not reflect the presence of sand and gravel deposits, which need the simplest treatments and are cheaper to produce.

Fig. 3.1 Aggregate limestone quarry (image courtesy of Avoca)

Fig. 3.2 Sand and gravel sediments

3

Aggregates are obtained from materials originating from geological processes on and within the Earth's crust. Erosion processes may have created sand and gravel thousands of years ago while granite may have originated a billion years ago when molten magma cooled and solidified. For its part, basalt can originate recently as molten lava flowing from a volcano (◘ Fig. 3.3) cooled and solidified. It must be stated "that the relationship between the rock masses of which the earth is made resulted from complex sequences of interactive geological events which may have begun when the planet took on tangible form and have continued over geological time" (Smith and Collis 2001). In this sense, the geological time scale (◘ Table 3.1) is essential to place events as well as principal time divisions, names, and terminology into the established chronological context.

3.2 Sand and Gravel

Sediments such as gravels, sands, and clays are grouped as clastic deposits. Once sediment is available for transportation, different agents (i.e. gravity, water, wind, and ice) will move it from its site of formation to various sites of deposition. Gradation, sorting, and concentration processes occurs during transport, accounting for the composition and relative frequency of sediments. For this reason, transport and deposition processes control the grain-size distribution of clastic sediments.

The process of terrestrial rock decomposition due to weathering generates gravel—to mud-size particles. This type of particles is formed by individual mineral grains or sometimes aggregates of minerals, termed rock fragments. Silicates (i.e., quartz and feldspars) are the main mineral composition of these particles. The rock fragments or clasts are igneous, metamorphic, or older sedimentary rocks in origin and are mainly composed of silicate minerals (terrigenous materials can also be called siliciclastics since they are predominantly siliceous—quartz-rich—clastic sediments—Fig. 3.4). ◘ Table 3.2 shows the standard grain-size scale for clastic sediments.

Deposits of sand and gravel are generated by erosion, transport, and deposition of fragments (◘ Fig. 3.5). Ice and water, and to a lesser extent, wind are the main geological agents that affect the distribution and size of deposits of sand and gravel. Where rock erodes, it is crushed and rounded during transport in rivers and deposited as gravel. The transport distance covered by clastic components is controlled by their grain size. Particles such as silt and clay ones, suspended in water, reach the open ocean or larger inland lakes whereas vast majority of coarser grained particles of sand or gravel size are deposited along the way on riverbeds or in inland

◘ **Fig. 3.3** Solidified lava flow (image courtesy of United States Geological Survey)

Table 3.1 Geological periods

EON	ERA	PERIOD	Millions of years ago
PHANEROZOIC	CENOZOIC	Quaternary	3
		Neogene	23
		Paleogene	65
	MESOZOIC	Cretaceous	140
		Jurassic	200
		Triassic	250
	PALEOZOIC	Permian	300
		Carboniferous	360
		Devonian	420
		Silurian	440
		Ordovician	490
		Cambrian	541
PROTEROZOIC	NEO-PROTEROZOIC		1000
	MESO-PROTEROZOIC		1600
	PALEO-PROTEROZOIC		2500
ARCHEAN	NEOARCHEAN		2800
	MESOARCHEAN		3200
	PALEO-ARCHEAN		3600
	EOARCHEAN		4000
HADEAN			4500

depressions. These sediment concentrations result in relatively smooth, round particles (Fig. 3.6).

Water action is an effective mechanism for extracting weaker particles and separating different size fractions. The source area and weathering regime establish the original petrology and shape of the clasts (the weathering of basalt predominantly results in pyroxene). However, sands tend to be composed basically of quartz (Fig. 3.7), which is the most common mineral present in many types of rock and is more resistant to mechanical and chemical alteration than most other minerals. Nevertheless, "although the weathering of granite, gneiss, and sandstone results primarily in quartz, feldspars are also present when transport distances are short" (Neukirchen and Ries 2020). In some cases, other components such as limestone, dolomite, and heavy minerals (i.e., monazite, rutile, zircon, garnet, and magnetite) can be also present.

3

Fig. 3.4 Siliceous—quartz-rich—clastic sediments

Table 3.2 Standard grain-size scale for clastic sediments (modified after Blatt 1982 and Pettijohn et al. 1987)

Group	Size class	Average grain size	
Rudite = psephite = gravel	Boulder	4.096 mm	
	Cobble	256 mm	
	Pebble	64 mm	
	Granule	4 mm	
Arenite = psammite = sand	Very coarse sand	2 mm	
	Coarse sand	1 mm	
	Medium sand	0.5 mm	500 μm
	Fine sand	0.25 mm	250 μm
	Very fine sand	0.125 mm	125 μm
Argillite = pelite = mud	Coarse silt	0.062 mm	62 μm
	Medium silt	0.031 mm	31 μm
	Fine silt	0.016 mm	16 μm
	Very fine silt	0.008 mm	8 μm
	Clay	0.004 mm	4 μm

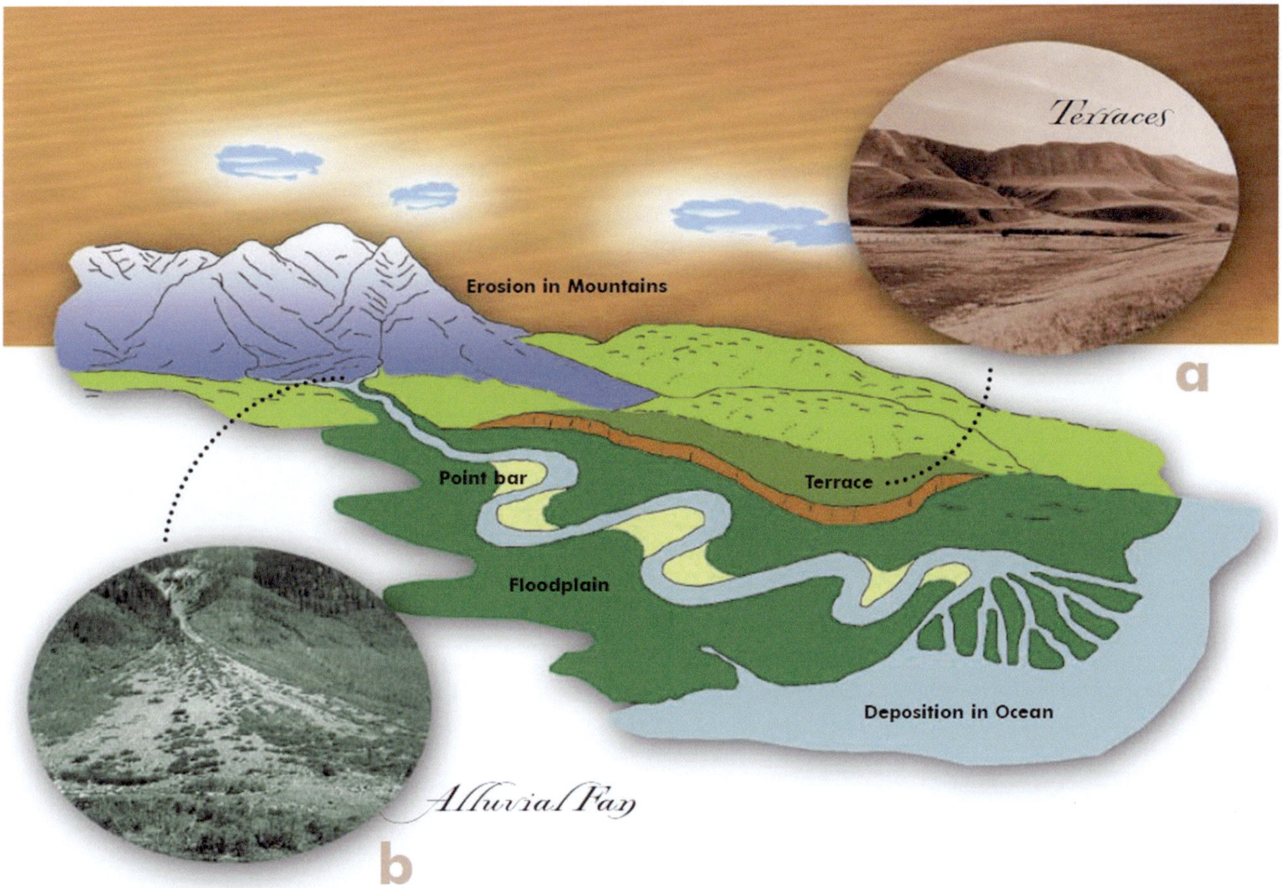

Fig. 3.5 Sand and gravel deposits can be formed by geologic processes such as "(a) rivers or streams depositing sand and gravel as stream-channel or terrace deposits, and (b) valley basins in arid and semiarid environments including thick fan-shaped deposits (alluvial fans) of unconsolidated clay, silt, sand, or gravel that were deposited during torrential floods" (Langer et al. 2004)

Fig. 3.6 Round particles

Sand and gravel deposits are principally Pleistocene or younger, and are a reflection of the region's latest geological and climatic history. The "influence of the Pleistocene glaciation has been profound and has been a major factor in ensuring that the temperate zones of the world today have had an adequate supply of aggregate to create their concrete cities" (Prentice 1990). As a result, urban development and the past mining of sand and gravel deposits are steeply decreasing their availability in areas close to larger cities (Fig. 3.8).

Older than Pleistocene deposits may not be sufficiently unconsolidated to be regarded as sand and gravel and they cannot be mined adequately; in these cases, blasting and further crushing are always needed, as with crushed stone. This is because the sediments have acquired some degree of cementation and compaction, so the sands have been converted to sandstones and the gravels to conglomerates (Fig. 3.9). Sand and gravel are found in alluvial fans, river terraces (Fig. 3.10), alluvial deltas, moraines, tills, and others. The debris from in situ weathering of some bedrock may also be a source of sand and gravel (Fig. 3.11).

3

Fig. 3.7 Quartz sand used in hydraulic fracturing; view width = 7 mm (image courtesy of Siim Sepp—sandatlas)

Fig. 3.8 Sand and gravel quarry near the large city of Madrid (Spain)

Fig. 3.9 Alternating sandstone and conglomerate (image courtesy of Jon Spencer—Arizona Geological Survey Photographic Atlas)

Fig. 3.10 Pleistocene river terraces of the Verde River (Arizona, USA) (image courtesy of Joseph Cook—Arizona Geological Survey Photographic Atlas)

3

Fig. 3.11 Debris-flow channel scar (image courtesy of Joseph Cook—Arizona Geological Survey Photographic Atlas)

In these deposits, the sand to gravel ratio is very variable, although river deposits present lower fines content than glacial deposits. Glaciofluvial deposits are thicker although the overburden thickness may also be high. Sand and gravel marine deposits occur as small patches separated or covered by extensive areas of uneconomic deposits of gravel-bearing sediments, varying in their proximity to the shore, thickness, composition, and size of the particles. "Their formation is substantially similar to those of land, but became submerged due to sea-level rise after the most recent glacial period and subsequently reworked by tidal currents" (BGS 2019). Windblow sediments (Fig. 3.12) are too fine-grained materials to be an important source of natural aggregates except possibly as blending sands, and are of little economic importance.

The existence of sand and gravel deposits is directly related to the regional geological history of each area. In the geological assessment of any sand and gravel deposit, aspects such as size and location, groundwater conditions (the level of the water table is very important because aggregates obtained from below water are more expensive than those mechanically mined in dry sediments), and deleterious constituents (i.e., iron pyrites, coal, mica, shale, flaky or elongated particles, organic impurities), among many others, are very important. The properties of sand and gravel mainly depend on the rock properties from which these sediments were derived "although during their transport prior to deposition, weathered or otherwise weaker fragments tend to be selectively worn away, and the resulting aggregate is generally stronger than the parent rock" (Smith and Collis 2001). On the other hand, if transporting distances o sediments are short, for example in glaciofluvial environments and alluvial fans, deleterious elements (i.e., mudflakes) may stay in the sediment and decrease the attractiveness of the sand and gravel deposit. In summary, the geological setting and mode of origin control not only the mineralogy of sand and gravel deposits but also their size and shape as well as the quantity and distribution of waste materials. These are the dominant factors that affect the amount and quality of the aggregate and the cost of mining it.

The depositional environment of gravel is reflected by the level of sorting, rounding, pebble size, etc. As a rule, the well-sorted gravels, which are free of clayey or silty fines, are preferred to the poorly sorted gravel and sands of some glacial deposits. This is because they contain abundant fines and are either unsuitable for some applications (i.e., aggregate for concrete) or need expensive beneficiation treatment.

3.2.1 Fluvial Deposits

Fluvial sediments (Fig. 3.13) accumulate in rivers where their capacity to carry sediment has been exceeded. Fluvial deposits are sediments transported and deposited by rivers in continental environments, representing the preserved record of one of the major nonmarine environments. During river transport, rock fragments will undergo further weathering, be slowly reduced in size, and be shaped by processes of attrition. As a consequence, the majority of river aggregates are reasonably well shaped, rounded, and smooth.

Fluvial sediments consist of channel and overbank deposits. They are present in a diverse array of continental environments and their abundance is controlled by the geological setting, which also controls the architecture of the deposits. "Channel deposits are formed mainly by sand and gravel and typically occur as lenticular bodies or thick tabular sheets with complex bedding geometries and erosive bases; Overbank deposits consist primarily of silt and clay and lesser amounts of sand, and often encase sand bodies" (Aslan 2013).

Fluvial deposits mainly embrace river sand and gravel that take the form of extensive spreads occurring along the floors of major river valleys, usually beneath alluvium, and as river terraces flanking the valley sides. Sand to gravel ratios are uncertain although this type of deposits is often relatively clean with lower fines content

Fig. 3.12 Windblow sediments located in the Coral Pink Sand Dunes State Park (Utah, USA)

than glacial deposits. The composition of fluvial sediments can range from almost all clay, through mixes of clay, silt, sand and gravel to almost all sand and gravel.

There are basically three types of fluvial deposits (a) stream deposits (channel and flood plain), (b) terrace deposits, and (c) alluvial fans. Aggregates are mined directly from the river channel and/or from river terrace deposits. Alluvial (river) sand and gravel, either in the channels or floodplains of rivers and streams (Fig. 3.14), or in terraces (lying above the level of the present flood plain) found alongside rivers or streams, are the principal sources of sand and gravel aggregates. As a rule, sand and gravel aggregates from terrace deposits are of similar quality to those existing in river channel deposits, always exhibiting considerable variation within alluvial deposits. However, terrace deposits may be more desirable than river channel deposits since they are above stream level. However, their extraction may produce numerous environmental issues. As a consequence, in many developed countries, this type of deposits is restricted by environmental constraints. Conversely, mining of these deposits is very common in many developing countries where rivers are easily accessible sources of aggregate.

- *Stream Channel and Flood Plain Deposits*

Stream channel deposits (Fig. 3.15) consist of sand and gravel deposited in stream beds along present and ancient stream courses. The deposits can occur in channels or bars of the present river or on adjacent terraces that formed where the river flowed at a higher level. Stream channel deposits are easily accessible and easily mined. They are dredged to obtain sand and gravel in many rivers all over the world. The materials only require washing and screening to produce suitable aggregates. However, in some cases harmful components can be present (i.e., chemically reactive rocks such as some silica minerals—Bustillo 2021), which makes them unsuitable for many aggregate applications (i.e., concrete production). It is important to note that the characteristics of the particles in the stream channel depend on the nature of the source rocks within its drainage region. As a consequence, the quality of the sand and gravel in the deposit is strongly dependent on the occurrence and properties of nearby bedrock sources. For example, sediments derived from metamorphic and igneous rocks are suitable whereas sediments derived from rocks rich in shale tend to be unsuitable.

The materials of these deposits are very useful as aggregates for construction purposes for many reasons. The natural abrasive action of stream transport has cleared the most soft, weak particles and retaining exclusively the harder components (i.e., quartz). Moreover, these harder particles have undergone an adequate rounding process and are sub-rounded to well-rounded, which is often desirable for some concrete uses such as pumping concrete.

Channel deposits are free of excessive amounts of silt and clay particles, with sand and gravel present in the size gradations for many aggregate applications.

Fig. 3.13 Plio-Pleistocene fluvial sediments (image courtesy of Ann Youberg—Arizona Geological Survey Photographic Atlas)

Fig. 3.14 Pleistocene Colorado River gravel floodplain deposits (image courtesy of Philip Pearthree—Arizona Geological Survey Photographic Atlas)

Fig. 3.15 Poorly sorted alluvium in the active stream channel (Scott 2016)

Nevertheless, the particle size deposited depends upon the velocity of the river and the presence of larger particle sizes is caused by larger stream velocities. Thus, in meandering streams, channel deposits are relatively fine-grained up to sand size, which results in inappropriate aggregate sizes for applications such as concrete. The channels can contain economically important deposits of sand and gravel if streams are vigorous.

During flooding, flood plain deposits are formed by sediments (sand and gravel as well as fines) deposited on plains bordering streams and caused by eventual overflow of the streams from their channels. In these cases, materials deposited are composed of silt and sand grains. However, "fine materials may mantle usable deposits of sand and gravel, particularly in areas where, in the geological past, the streams were more vigorous and transported greater volumes of coarser material" (Goldman 1990). In these deposits, the sand and gravel are similar to those present in channel deposits, and their utilization is adequate after the overlying flood plain silt layers are removed. Meandering and splitting of sediment-choked streams can also generate braided flood plain deposits with sand and gravel.

- *Terrace Deposits*

If fluvial sediments are uplifted as a part of a global elevation of a land mass, the stream speed is augmented due to its steeper gradient. The resulting higher energy water cuts through the previously deposited deposits. When a river changes gradient and downcuts its channel, the older channel and floodplain deposits can be preserved as river terraces (▶ Box 3.1: Terrace Deposits).

Box 3.1—Terrace Deposits

*River terraces (*Fig. 3.16*) are formed by river incisions into the channel and floodplain sediments, generating sediments conserved along river channel margins or valley sides. They are the dissected or eroded remnants of earlier abandoned river floodplains, being nearly ubiquitous features of river valleys. They are formed in many ways and several geologic and environmental settings. The formative processes of river terraces are highly complex, as illustrated in* Fig. 3.17.

Most of the terraces "represent a multiple change of accumulation and erosion periods, which dominantly reflect

3

Fig. 3.16 Manzanares River terrace (Madrid, Spain)

times of climatic instability, where rates of erosion, sediment transfer, and deposition dramatically change to produce terrace landforms" (Kamp and Owen 2013). *Repeated downcutting can result in a series of terraces or terrace remnants* (Fig. 3.18) *above the level of the modern stream base. Consequently, stepped terraces remain on bedrock benches that are typically present on both flanks of the valley. They can include economic sources of sand and gravel or can be formed by fine flood plain sediments. River terraces are visible within a landscape. This is because their relatively flat surfaces grade topographically toward the valley center and in a downstream direction. Older terraces may be exposed to prolonged weathering, weakening the materials and decreasing their suitability as aggregates.*

River terrace deposits are a focus of considerable scientific, archaeological, and economic interest. They may be very variable and difficult to characterize in terms of structure and lithology, specifically where deposits of multiple or dissected terraces are present. Stream terrace deposits are benchlike sand and gravel sediments that are bordering the stream, lying always above the level of the current flood plain. These deposits range in size from a few meters to tens of meters thick and should be taken into account in aggregate assessments. This type of deposit can be more desirable for extraction than stream channel deposits. This is because water in stream channel deposits water tables are shallow and abundant groundwater makes mining operations very complex.

The type and composition of the bedrock source of the clasts as well as the size and sorting of these clasts in the deposit control the potential usefulness of any specific deposit. Sand and gravel are not suitable if the grains have become coated with harmful substances (i.e., clay and iron oxides) caused by the action of ground waters. Moreover, the higher, older terraces that have been subjected to erosion processes can contain significantly reduced thicknesses of sand and gravel sediments. In fact, although flood plain and terrace sediments are particularly favored sources of aggregates they can display unpredictable discontinuities and variations in quality.

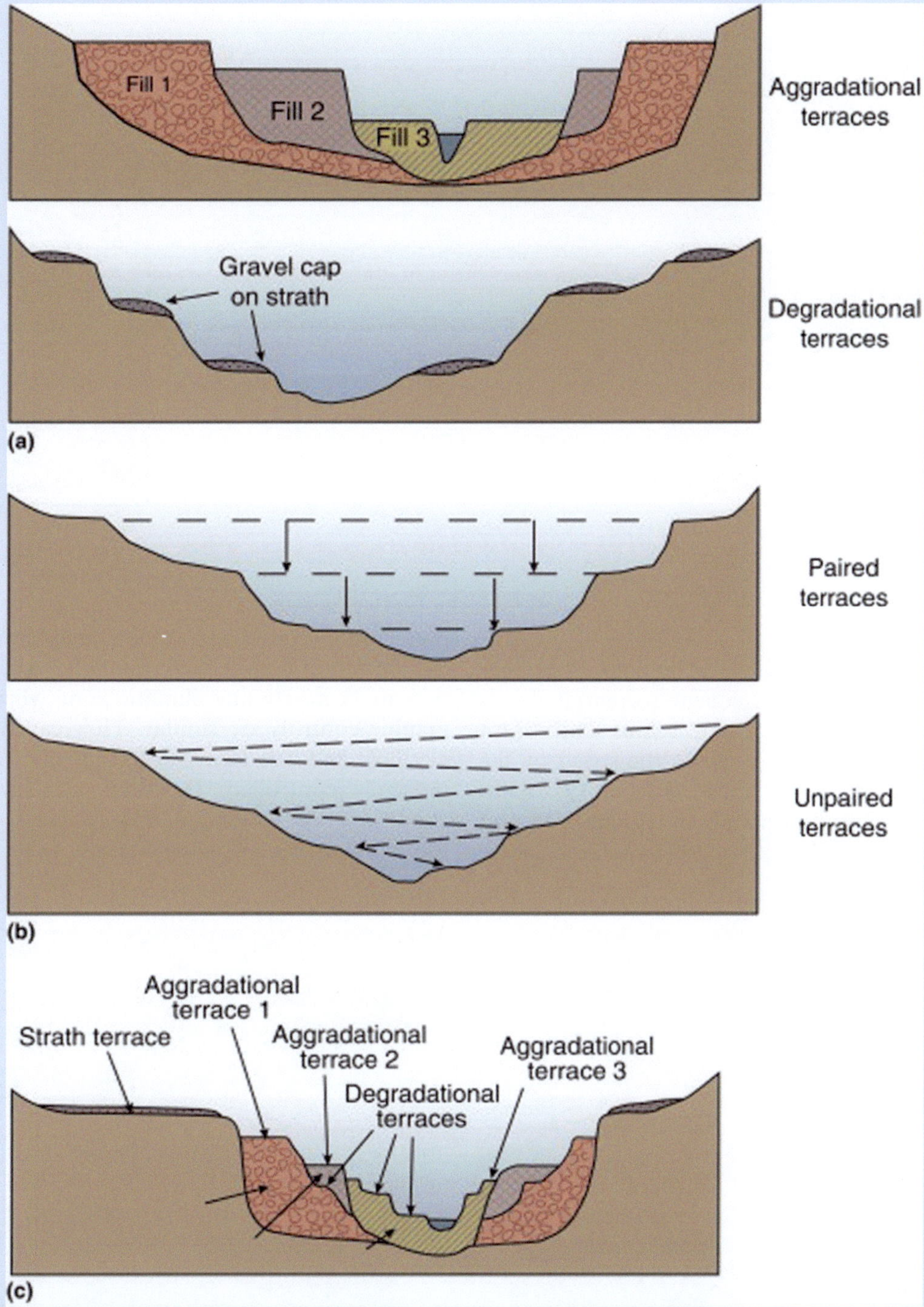

Fig. 3.17 River terrace formation: **a** aggradational and degradational fluvial terraces, **b** paired and unpaired river terraces, and **c** complex sequence of aggradational and degradational surfaces (modified from Kamp and Owen 2013)

3

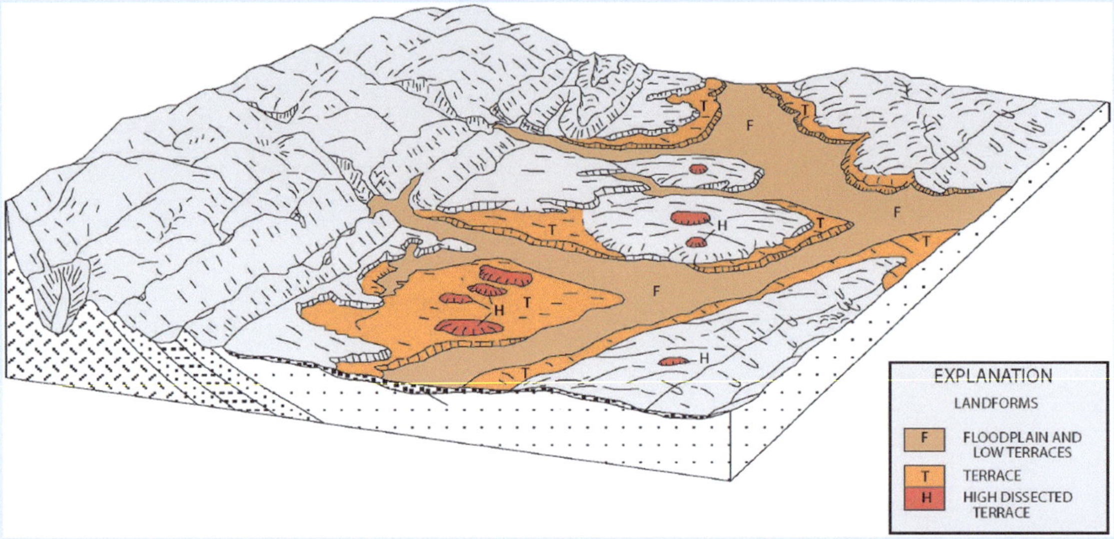

Fig. 3.18 Repeated downcutting can result in the formation of a series of terraces or terrace remnants (Langer 2003)

- *Alluvial Fans*

In mountainous arid and semiarid regions, rocks are eroded and transported by frequent torrential floods down steep-gradient streams toward basins. The abrupt change in slope causes a decrease in the speed of the stream, resulting in alluvial fans (Fig. 3.19). Alluvial fans have been studied mainly in regions with arid or semiarid climates, where they are larger and better preserved. Nevertheless, some alluvial fans can be present in more humid environments. Alluvial fans are present in nearly any climatic zone where the physiographic controls are similar. They can be classified as dry fans and wet fans, suggesting that fan type is climatically controlled. Dry fans are developed under conditions of ephemeral flow whereas wet fans are formed by streams flowing continuously.

Fig. 3.19 Alluvial fan

In alluvial fans, erosion is dominant in the upper part of the watershed while deposition occurs at its lower reaches where sediment is free to accumulate without being confined within a river valley. Currently, alluvial fans have radii from 1.5 to 10 km and some of them can reach depths up to approximately 50 m.

Alluvial fan deposits include thick unconsolidated materials ranging from large boulders to clay particles, reflecting the bedrock composition in the adjacent uplands. The particles of the sediments tend to be sub-angular to angular although sub-rounded particles can also occur in large alluvial fans. Some of this material provides useful sources of aggregates; in many cases, poor-quality aggregates may be generated since transport distances are too short to allow abrasion of deleterious components. Coarse material is deposited near the mountain front and becomes progressively finer toward the downstream edge of the deposits. However, frequent changes in stream positions cause a complex configuration of sand and gravel and fine-grained waste material.

Sand and gravel in alluvial fans may be suitable for aggregate extraction but fan deposits include lenticular beds or poorly sorted material interbedded with different amounts of silt and clay. In older fans, gravel may be highly weathered and not suitable for aggregate extraction. In addition, fan gravels can be cemented with caliche, a calcium carbonate precipitate in the soil, which makes them noneconomic to mine and process.

3.2.2 Glacial Deposits

Because the recognition in the mid-nineteenth century that glaciers had been significantly more extensive than presently, "the Quaternary (Pleistocene and Holocene series) has been synonymous with glaciation of the mid-latitudes; today evidence from both the land and ocean-floor sediment sequences demonstrates that the major continental glaciations occurred repeatedly over what are now temperate regions of the Earth's surface" (Ehlers et al. 2018). The glaciation period generated advances and recessions of the continental ice sheets. Glaciers play an essential role for denudation and transport of large rock fragments in high mountains and polar regions.

In the Northern Hemisphere, evidence of glaciation is widespread throughout the Quaternary and Neogene. According to thickness and areal extent, glacial processes and sediments dominate glaciated continents and ocean basins. The most recent episodes of glaciation occurred over the last 2.5 million years during which much of the temperate zones of the world were successively covered by glaciers and uncovered during warm interglacial periods. Glaciated regions contain numerous sand and gravel deposits. For instance, sand and gravel as aggregate products are produced commercially from glacial deposits in the midwestern and northeastern United States; the bulk of these materials were deposited from continental ice sheets generated in Canada and spread south into the United States. Yeend (1973) states that "the most important commercial sources of sand and gravel in the United States are river channels and glaciated terrain".

The characteristics of glacial deposits are much less predictable than those of other deposit types such as fluvial deposits, in almost every respect. This is because glacial erosion and deposition are complex dynamic processes. Deposition of sand and gravel from a glacier can take place in several ways. In glacial environments, sediments occur as a function of "ice dynamics, the type of ice mass, basement and subjacent geology and sedimentology, the temporal and spatial variability of sediment discharge (flux), associated hydrological regimens and topography" (Menzies et al. 2016). Sorting action is minimal and chemical weathering does not take place during glacial transport, with crushing being the dominant method of mechanical breakdown. For these reasons, the sediment load carried by glaciers ranges in size from clay to boulders and consists of angular clasts. It is clear that the sediments transported by ice masses reflect the underlying and surrounding lateral and up-ice bedrock geology. Although ice masses can transport sediments to long distances, a significant proportion of the sediment load in glacial environment comes from a few kilometers up-ice (i.e., 10–15 km).

In the areas over which the glaciers pass, rock fragments are picked up and incorporated into the glacier. Later, different types of sediments (i.e., clay, silt, sand, pebbles, and boulders) are deposited as glacial till when the glacial ice melted. Till is the end product of glacial erosion, transport, comminution and deposition (◘ Fig. 3.20). In general, it is formed by two particle sizes (a) rock fragments greater than 1 mm and (b) matrix (< 2 mm) constituted by mineral fragments. In summary, till is a poorly sorted mixture of clay-size to boulder-size particles. As a result, till deposits ten to be not economical.

The debris deposited along the margin of the glacier formed irregular ridges called moraines. They contain material of changing size, shape and quality, and are only available after expensive treatments. On the other hand, kames (hills) and eskers (ridges) constituted by sand and gravel sediments can be deposited by melt water within the ice sheet. Kames are irregular sand and gravel accumulations deposited by water running alongside, within, or on ice whereas eskers are long, sinuous landforms that were deposited as poorly stratified sand and gravel in stream channels under or on ice sheets. Kames are found as shoulders often at

Fig. 3.20 The surface of the ground is covered with a coarse glacial till (Chugach National Forest, Alaska, USA) (image courtesy of United States Geological Survey)

a constant elevation at the edges of valleys. Eskers are generally less than 10 m high and typically formed by poorly sorted, coarse to very coarse debris emplaced by streams under or within glacier ice (Fig. 3.21). Sediments in eskers can be quite coarse with boulders weighing several tons. They are "the premier gravel deposits in glaciofluvial valleys because they contain the greatest percentage of boulders and coarse gravels with lesser portions of sand" (Sutphin et al. 2002).

The materials in kames and eskers, although showing a broad variation in size and physical characteristics, are common sources of sand and gravel and can conform to major deposits of aggregates. At times, kames occur with eskers and each type of deposit can grade into deltas that form in ponded glacial lakes. They can be economical sources of sand and gravel although the sediments in deltas are principally composed of sand, silt, and clay (where rivers are vigorous, gravel can also be deposited in deltas).

When glacial streams flow into bodies of water (i.e., a glacial lake), their transporting power and velocity are quickly stemmed, dropping thus their coarsest sediments. A delta will form if conditions are favorable. This happens "in an orderly pattern with sand being deposited in foreset beds, which are the steeply dipping strata forming the main body of the delta; foreset beds are made chiefly of sand with minor amounts of gravel. Finer material (silt and clay) is swept further along and settles on the lake floor as lake-bottom sediments (bottomset beds)" (Sutphin et al. 2002).

Glacial deposits subjected to stream action are termed glaciofluvial (also known as glacifluvial or fluvio-glacial) deposits (Fig. 3.22). As glacier ice melts, crushed, abraded, and carried by the ice rock fragments are further transported, abraded, and rounded by melt water. Finer elements are deposited in lakes and ponds whereas coarser sand and gravel materials are located along stream channels, forming so-called glaciofluvial deposits. These deposits occur in different configurations, giving rise to conspicuous and readily recognized land-forms. They are associated with till and can show complex relationships (Fig. 3.23). Much of the sand and gravel deposited by glacial melt water streams may be utilized as aggregates. Most of the economic sand and gravel deposits found in glaciated areas are constituted by sediments deposited from water associated with ice.

Glaciofluvial deposits show better degree of sorting and contain fewer clay-size fragments than glacial deposits. Many times, this type of deposit is of high enough quality to produce satisfactory aggregates, usually comparable to stream sand and gravel. The particle size in glacial melt water deposits ranges from boulder to sand, silt, and clay, and can change abruptly, both with depth and laterally. This is because melt water streams are often torrential and deposition is very rapid. Lateral and vertical changes in grading produce difficulties in quarrying. As a rule, glaciofluvial deposits are thinner than stream deposits in areas of significant topographic relief. Thicknesses of approximately 30 m have been reported, although overburden thicknesses can also be high.

Fig. 3.21 Coarse-grained esker core, Gracefield, Outaouais Region, Canada (image courtesy of Géologie Québec, Mines Service Centre, Québec)

Fig. 3.22 Quaternary glaciofluvial deposit in Scotland (image courtesy of British Geological Survey)

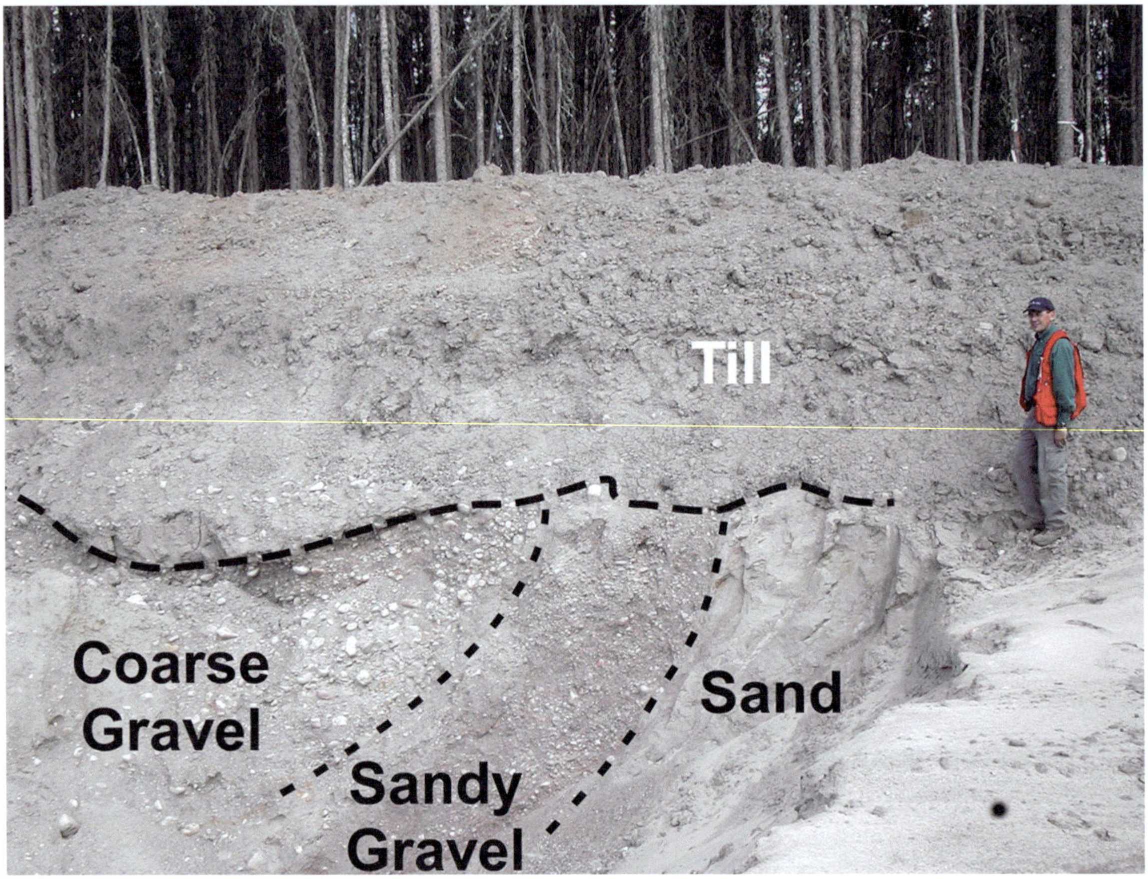

Fig. 3.23 Glaciofluvial sediments overridden by till sediments (Smith et al. 2005)

3.2.3 Coastal Deposits

Coastal deposits (Fig. 3.24) are found near the margins of permanent water bodies, resulting from wave actions. Beach materials fall within the influence of waves and tidal action. Sand and gravel originated by currents and waves on a beach may generate good aggregates since the gravel and sand particles are well-rounded and well-sorted both in terms of particle size distribution and shape. Gravel tends to be swept high onto the coast (backshore) and the sand and finer particles out to sea. Beach gravels are strongly localized vertically but laterally extensive along the shore. Movements of material also occur along the shore by the process of long-shore drifting and sediments accumulate in ridges and bars. The particles of beach deposits are much more uniform in size than those of glacial deposits.

The sands are composed mainly of resistant quartz and tend to be relatively free of deleterious materials due to the abrading action of waves; varying amounts of shell fragments may also be present and, in some tropical areas, these deposits are principally formed by calcium carbonate derived from the accumulation of marine shells.

Beach deposits can be utilized as aggregate if they meet the required specifications; however, many beach deposits are thin although many of them extend along the coast for great distances. They lack proper size gradation since a stable beach will rarely contain material finer than medium sand (approximately 0.2 mm). The poor grading of coastal sediments, resulting from the sorting action of waves, is the major deficiency of beach sands for aggregate production. In any case, exploitation of beach deposits as aggregates generates intense and severe environmental impacts.

3.2.4 Marine Deposits

Marine sediments cover approximately 70% of the Earth's surface. About 85% of all coastal marine sediments are clastic sediments, which are used as aggregates (▶ Box 3.2: Marine Aggregates). Most clastic sediments in marine sediments are terrigenous materials eroded from land, and the size of the sediments ranges from cobble to fine clays. Marine deposits on continental shelves are large potential sources of sand and gravel, being generally sand-size and finer.

Fig. 3.24 Coastal deposit

Box 3.2—Marine Aggregates

Marine aggregates (Fig. 3.25) *are naturally occurring sand and gravel materials found on the inner continental shelf. In the 1960s, marine aggregates as a resource began to attract the focus of the construction industry all over the world, and this industry started to mine materials in great quantities. Although most of the demand for aggregates is satisfied by those extracted from land-based sources, marine production is growing. Thus, "they could become more significant as other sources of aggregates are depleted, and if economic, regulatory, and environmental concerns can be adequately addressed"* (Bustillo 2021). *However, it is crucial to remember that marine aggregates are a nonrenewable resource, that is, their amounts are finite. Therefore, they should be utilized sustainably and adequate policies must be defined to address future demand. Aggregates are currently the most extracted mineral resource in the marine environment and their demand is growing. "Due to the low value of the product, most marine aggregate extractions are carried out at short distances from landing ports close to the consumer base and at water depths of less than 50 m"* (UNEP 2014).

Offshore mining of sand and gravel resources is well established in European countries such as Germany, France, Denmark, the Netherlands, and the United Kingdom. They are also extracted in other regions of the world such as "the tidal channels of the Yellow River China, the west coast of the Republic of Korea, tidal channels between the islands south of Singapore and in a range of settings in the waters surrounding Hong Kong, China" (United Nations 2016).

According to the British Marine Aggregate Producers Association (BMAPA), "the marine aggregate industry is one of the UK's key suppliers of sand and gravel. Marine supplies provide approximately 25% of sand and gravel sales in England and 48% of sales in Wales". In United Kingdom, the supply of marine aggregates is essential in major urban centers such as London, where 50% of the aggregates needed in construction are obtained from offshore. Russell (com. pers.) points out that "The Netherlands and Belgium are heavily reliant upon imports of construction aggregate from adjacent nations, especially Germany and France, along with crushed rock by sea from Scandinavia; while both countries have significant volumes of fine-medium marine sand which are widely exploited for beach nourishment, land reclamation and construction, neither has any significant resource of the coarse sand or gravel required for concrete production on their continental shelves".

3

Fig. 3.25 Marine aggregates (image courtesy of BMAPA)

Marine sand and gravel resources are unevenly distributed on the continental shelf but, from a geological point of view, they are very similar to the sands and gravels found on land, in terms of origin and mineral composition. Marine sand and gravel resources occur as small portions divided or covered by extensive areas of uneconomic deposits and exhibit the same mechanical, geometrical, and chemical properties as their land-based equivalents. This is because the continental shelf is a natural extension of the land into the marine environment. Thus, the shape, strength, and many other characteristics of marine aggregates are similar to those of high-quality land-based aggregates.

A common characteristic of marine sand and gravel deposits is their low silt content (mainly in dredged aggregates since dredging washes out much of the fine material) while "the presence of salt is a disadvantageous feature since chloride ions attack the steel reinforcement in concrete" (Bustillo 2021). *The existence of chloride ions from sea salt and biogenic material (shell/shell debris) are the principal differences between most land-based aggregates and marine aggregates. Siliciclastic marine aggregates are formed by harder material than land-won aggregates. This is because these materials have been subjected to strong abrasion in coastal and inner shelf marine environments. Consequently, the final uses of marine aggregates are no different from those of land-won aggregates: production of concrete and asphalt, manufacturing of masonry and paving blocks, etc.*

In general, marine aggregate deposits can be subdivided into relict and modern deposits. "The former consist of sedimentary material deposited in the past and under different

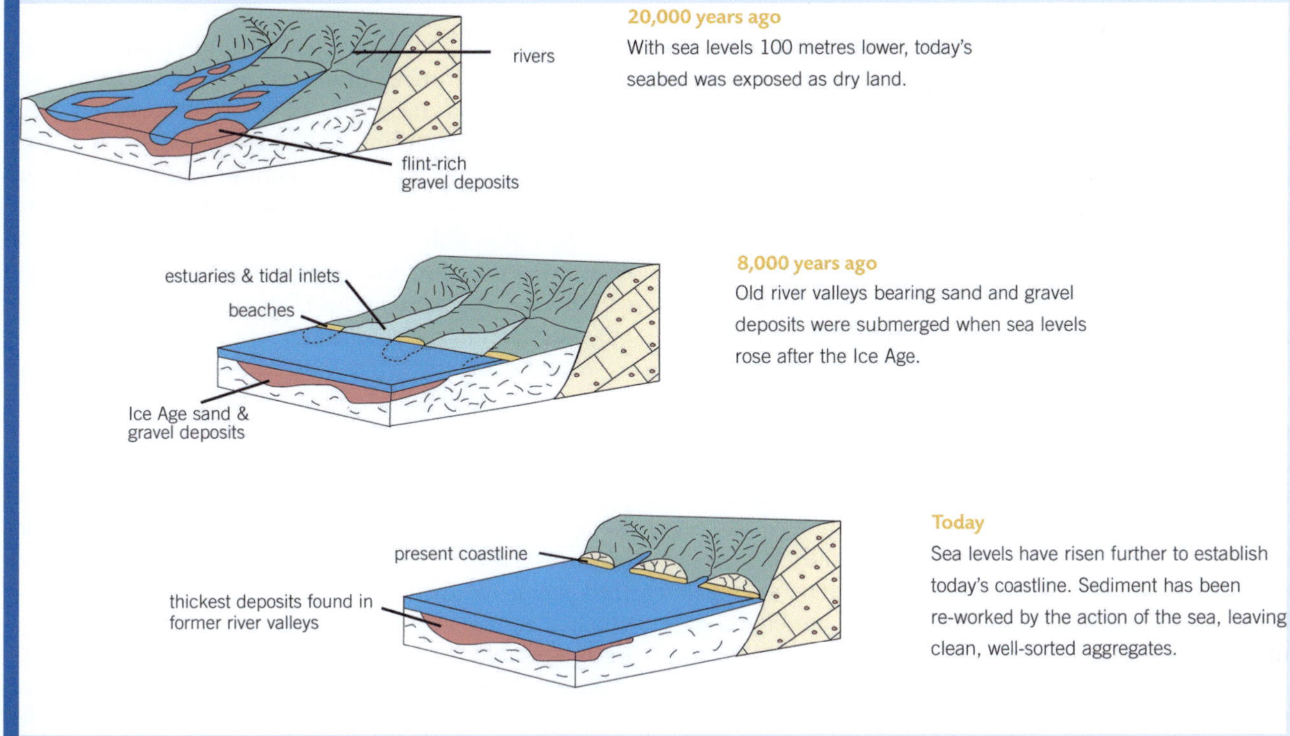

Fig. 3.26 Genesis of marine aggregates in UK (illustration courtesy of BMAPA)

hydrodynamic and sedimentological regimes than those existing actually (i.e., the gravel/sand deposits of the Pleistocene buried river valleys of the northwestern European shelves). The latter are deposits, which have been formed and controlled by the modern hydro-and sediment dynamic conditions (i.e., the linear sand banks of the southern North Sea)" (Velegrakis et al. 2010).

Relict deposits are of particular interest for the marine aggregate industry since they can include large amounts of gravels. "They have been mainly formed during the Pleistocene climatic changes and sea level oscillations" (Velegrakis et al. 2010). *In the sea level falls (lowstands), deposition of unconsolidated coarse-grained material occurred at the riverine thalwegs, terraces, point bars and fans. Other good relict deposits for aggregate mining are associated with drowned beaches and barrier islands. These are "coastal sedimentary bodies left behind during the last marine transgression, as a result of their protected position and/or the fast rate of the transgression"* (Velegrakis et al. 2010).

The distribution of marine sand and gravel that can be utilized by the aggregate industry is limited to discrete geological deposits that reflect the physical processes that have formed them (Fig. 3.26). *In the Quaternary period, several glacial episodes took place from about 2.6 million years to approximately 11,000 years before present. Although these sand and gravels are now submerged, they were originally deposited by large river systems that flowed across the sea floor, when it was exposed as dry land during glacial stages, which in turn corresponded to periods of sea level lowering. The sands and gravels were deposited by fast-flowing rivers fed by snow and ice melt, which poured out across the dry shelf. In turn, these deposits were subsequently reworked by wave and tidal processes as sea levels rose during warmer interglacial stages. This warm-cold cycle was repeated many times producing sand and gravel deposits that are essentially immobile and locked in ancient river terraces and channels. The availability of suitable resources for commercial dredging is defined not only by the presence of ancient paleovallies but also by the extent to which more recent marine processes have reworked these deposits.*

3

3.2.5 Eolian (Windblown) Deposits

Dune deposits (◘ Fig. 3.12) and other forms of accumulation formed by wind action are examples of very well-selected granular sediments, with particles of extremely fine size. This is due to the attrition processes in wind transport and normal wind speeds, which are only adequate to transport the finer particles. Dune sands are located either adjacent to the coast or in dry desert-like areas where vegetation is sparse and sands are transported and deposited by the action of wind. Large accumulations of sand dunes are present in desert areas and along sea shores (◘ Fig. 3.27).

The particles of dune sediments are very similar in terms of particle shape, relative density, and even mineral composition. The sediments in the deposit are typically sands with predominate particle sizes less than 1 mm and range from 0.02 to 0.1 mm in the case of loess deposits (loess is dust deposited by the wind). Eolian deposits are generally of limited value to mine aggregates because they are deficient in both silt and gravel-size particles. Therefore, their use can be restricted as correctors of certain grain size distributions. On the other hand, there are strong objections to the exploitation of coastal dunes due to their role as coast-protection elements.

◘ **Fig. 3.27** Beach sand dunes in Santander, Spain (image courtesy of María Ángeles Morán)

3.3 Crushed Stone

Crushed stone is "the product resulting from the artificial crushing of rock, boulders, or large cobblestones, substantially all faces of which have resulted from the crushing operation" (ASTM standards). It is produced by drilling, blasting, loading, and crushing bedrock, but its production may also involve crushing large boulders or cobbles (◘ Fig. 3.28) obtained in sand and gravel deposits. For this reason, angular edges are typical in this type of aggregate because all of its surfaces are generated in the crushing operation.

Many different types of rocks are suitable for utilization as aggregates when crushed. Physical and chemical characteristics can provide essential information to define the technical suitability of each type of rock for the different aggregate applications. Rock weathering, grain size, foliation, or mineral orientation affect rock durability and must be considered when identifying potential sources of rocks for crushed stone. Although most hard rocks are potentially suitable for coarse aggregates, high-quality aggregates are exclusively obtained from very hard, dense siliceous rocks. High-quality aggregate is needed for applications such as railway ballast whereas lower quality aggregates can be utilized in applications such as concrete or mortar.

Crushed stone can also be classified based on the rock type from which it was produced. Three broad categories of rocks can be distinguished: sedimentary, igneous, and metamorphic; this classification is based on the mode of formation of rocks. Sedimentary rocks are formed by mechanical processes (clastic or detrital rocks) or by direct chemical or biochemical precipitation (i.e., limestone). Limestone and dolostone are the most widely used sedimentary rocks as aggregate. In clastic rocks, hard, dense sandstone is the only type that is a source for crushed stone. Igneous rocks are formed from molten material generated below the Earth's surface, which is further solidified at or near the surface. Typical examples of igneous rocks commonly used as aggregate are granite, basalt (◘ Fig. 3.29), and andesite.

Metamorphic rocks are formed from preexisting rocks (both igneous and sedimentary in origin) subjected to heat and pressure within the Earth. Some examples of typical metamorphic rocks are slate, schist, gneiss, marble, and quartzite although the only metamorphic rocks used as aggregates are gneiss, marble, and quartzite. This is because the slates tend to be platey and are unsatisfactory for their utilization as aggregates (the crushing process tend to cause flat particles). Schists are sounder than slates but can form flat particles after the crushing process, and often are of marginal quality.

Fig. 3.28 Cobbles obtained in a sand and gravel deposit

3.3.1 Sedimentary Rocks

Sedimentary rocks and sediments ranging in age from Precambrian to modern cover around three-fourths of the surface of the Earth. They are formed at low temperatures and pressures at the Earth's surface. Sedimentary rocks are formed "under specific processes derived from other preexisting rocks of igneous, sedimentary or metamorphic origin and/or as products of life activities of organisms or chemical secretion" (Haldar and Tisljar 2014). As a rule, sedimentary rock formation includes (a) physical and chemical weathering of previous rocks, (b) transport of materials in solid or dissolved states, (c) deposition or sedimentation and (d) diagenesis and lithification. The term diagenesis comprizes "all surface-near processes by which a loose, subaquatically deposited sediment is transformed, at low pressure and temperature, into a solidified sedimentary rock" (Okrusch and Frimmel 2020). Sedimentary rocks are characterized especially by the presence of layers (Fig. 3.30) and by certain textures and structures.

Sedimentary rocks are subdivided into two main groups concerning the type of physical, chemical and/or biochemical, and geological processes (a) those that have been chemically or biochemically (organic) formed (calcareous and evaporate rocks—i.e., limestone), and (b) those that have been mechanically deposited (clastic or detrital or terrigenous rocks—i.e., sandstone and shale). The rocks that form Earth's sedimentary shell are principally shales, sandstones, and carbonate rocks. Sedimentary rocks show a varied range of properties, from heavy to light, strong to weak, hard to soft, and dense to porous. Therefore, their suitability for use as crushed stone for aggregates changes accordingly.

- *Limestone and Dolostone*

Chemical and biochemical processes operating within depositional basins "may lead to extraction from basin water of soluble constituents to form minerals such as calcite and gypsum as well as formation of calcareous and siliceous tests or shells of organisms; these

Fig. 3.29 Vesiculated basalt cobble (image courtesy of Mike Conway—Arizona Geological Survey Photographic Atlas)

chemical/biochemical constituents are the materials that make up intrabasinal sedimentary rocks such as limestones and evaporates" (Boggs 2012).

Carbonate rocks, mainly limestones and dolostones, form around one-fifth to one-quarter of all sedimentary rocks in the stratigraphic record. The range of physical and chemical properties of these rocks is large and their color ranges from almost pure white through changing shades of gray to black. Red tones are produced by the presence of iron oxides, being only necessary very minor amounts of iron oxides (i.e., 0.5%) to generate an intense red color in the rock.

Carbonate rocks are so-called because they are composed primarily of carbonate minerals. Limestone is a sedimentary rock formed mainly by calcium carbonate ($CaCO_3$) and, when augmenting the magnesium carbonate ($MgCO_3$) content, limestone grades into dolomite ($CaMg(CO_3)_2$). Dolostone is the dominant carbonate rock in Precambrian and Paleozoic sequences whereas carbonate units of Mesozoic and Cenozoic age are dominated by limestones. The vast majority of limestone and dolomite rocks are hard, durable, and useful for producing aggregates, being common rock types and consequently broadly mined for this application. However, in some cases limestone may be friable, soft, and absorptive when it contains a significant amount of clay.

Although limestones are mineralogically quite simple, they are highly varied in composition. Limestones range in color from dark gray to white and can be tinted by small quantities or ferrigenous oxides. They are typically bedded and layered rocks including numerous fossils (Fig. 3.31).

Regarding limestone classification, the most broadly used classifications are those of Folk and Dunham. Folk (1959, 1962) defined three principal components in limestones: allochem (grains), matrix (micrite) and sparite (cement). This classification includes a system of naming each rock with sparites and micrites and the abbreviated allochem name (Figs. 3.32 and 3.33). However, the simplest limestone classification is that of Dunham (1962). This author classifies limestones based on the rock or sediment fabric and the presence of any biological binding. Main divisions in this classification are the following: matrix-supported

Fig. 3.30 Layered sedimentary rocks (image courtesy of Mike Conway—Arizona Geological Survey Photographic Atlas)

Fig. 3.31 Limestone with fossils (orbitolinid)

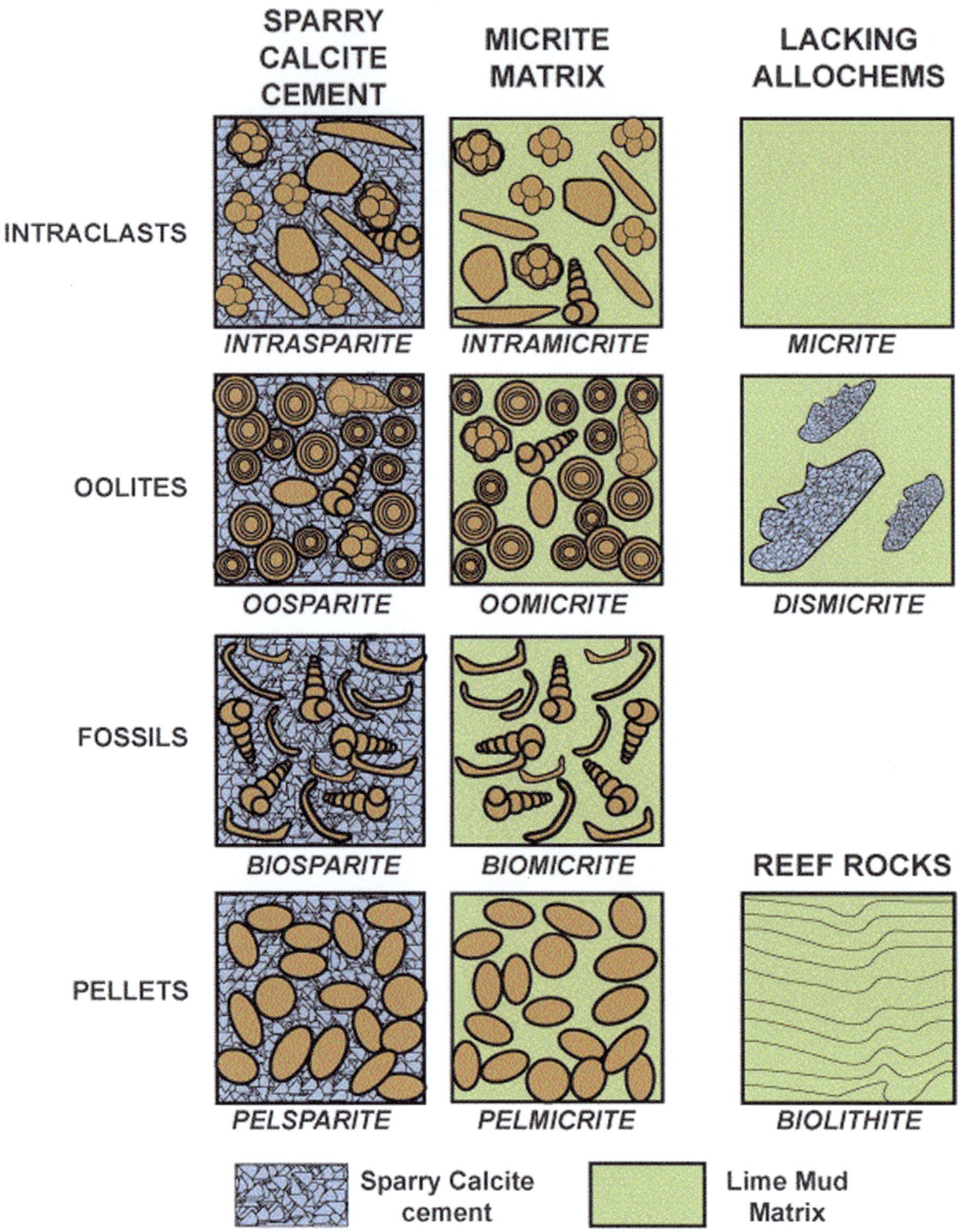

Fig. 3.32 Some examples of limestone types according to Folk's (1959) classification

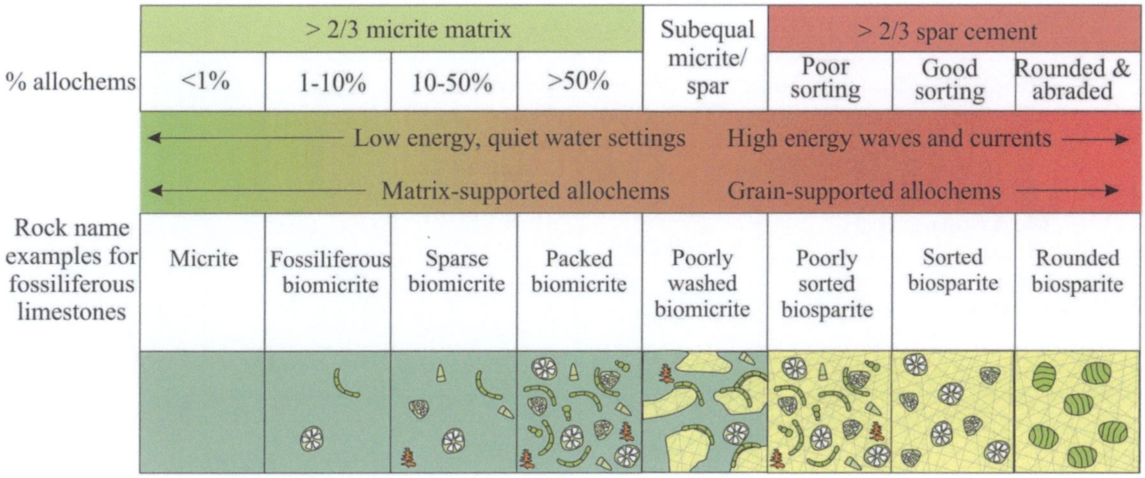

Fig. 3.33 Basic limestone types of Folk's (1959) classification

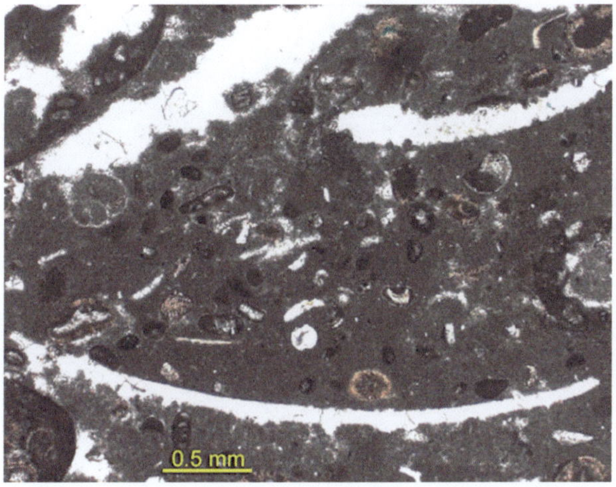

Fig. 3.34 Wackestone according to Dunham classification

(mudstones and wackestone) (Fig. 3.34), grain-supported (packestones and grainstones) and biologically bound (boundstones). This classification was extended by Embry and Klovan (1971) (Fig. 3.35) to include the types of rocks encountered in reef deposits. With respect to dolostones, they occur in a wide range of crystal forms, fabrics and mosaics). The classification of Sibley and Gregg (1987) is the most common system to classify dolomite rock textures (Fig. 3.36).

Limestone and dolostone cover roughly half to three quarters of the rocks utilized for crushed stone, with granite and other igneous rocks making up the bulk of the remainder. Limestone is also extensively utilized for cement and lime production. Limestone deposits are thickly bedded and consistent, which allows them to be mined thoroughly and economically. This type of sedimentary deposit produces durable aggregates suitable for road stone (subbase and base layers) and concrete. The quality of the limestone deposits can change according to the different geological factors (i.e., waste content, dolomitization processes, and degree of faulting and folding).

Actually, dolomite is precipitated "within shallow sediments on the seafloor, in marginal marine environments, in meteoric groundwaters, during deep burial and from hydrothermal fluids; it also replaces carbonate sediments and rocks across the geological spectrum" (James and Jones 2016); it is common that ancient limestones have been partially or even completely dolomitized. Dolomitization process may occur as long as sufficient pore space is available. With respect to the geological environments for limestone formation, carbonate sediments are formed in a variety of settings, and the two ends of the environmental spectrum are marine and terrestrial settings (▶ Box 3.3: Geological Environments for Limestone Formation).

3

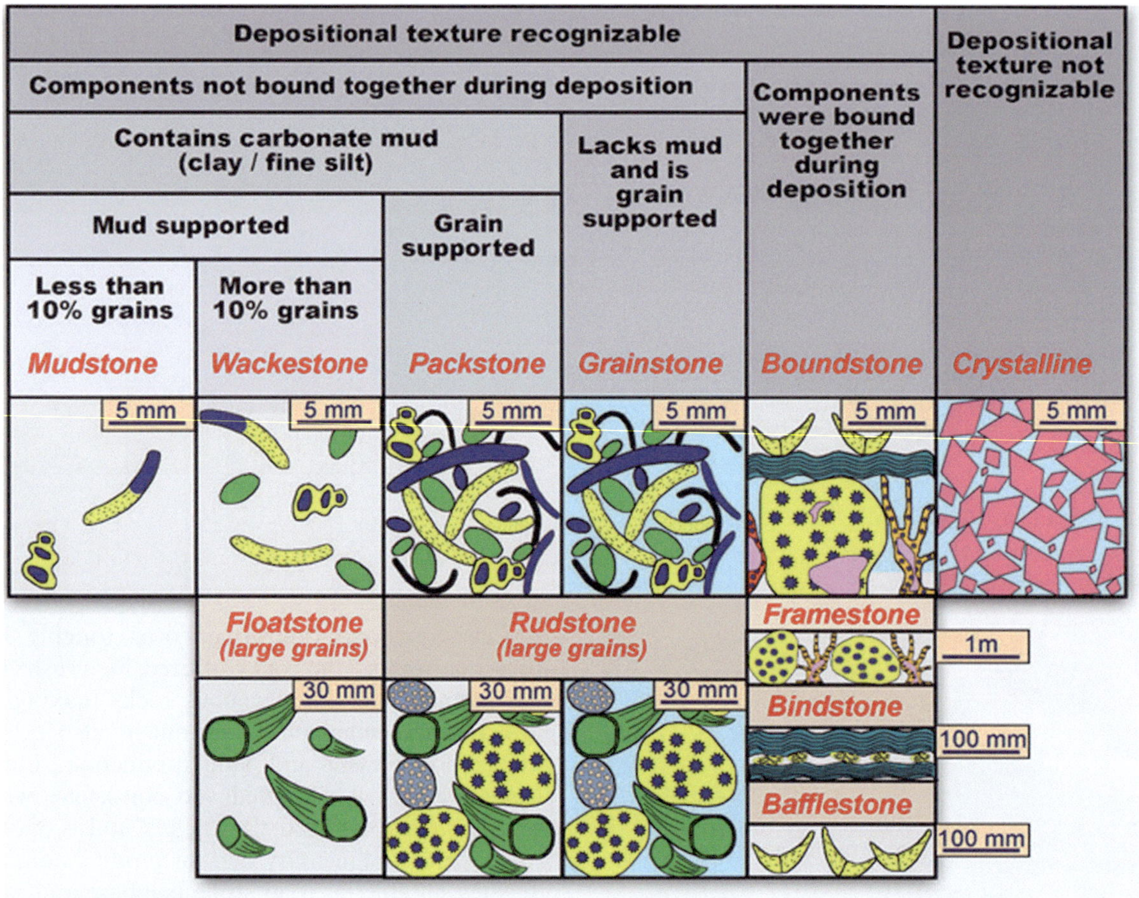

Fig. 3.35 Dunham (1962) carbonate rock classification with modifications by Embry and Klovan (1971) (illustration courtesy of Bureau of Economic Geology)

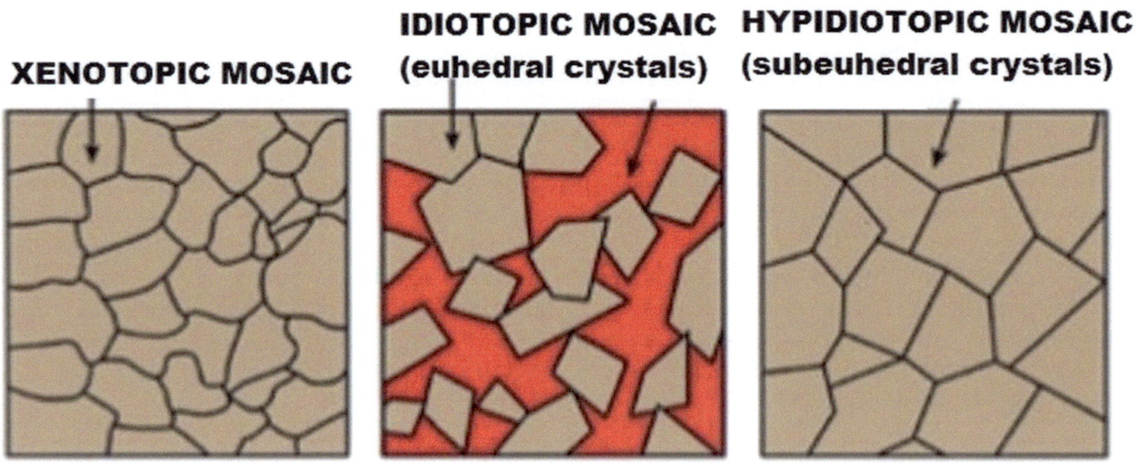

Fig. 3.36 Dolomite textures according to the Sibley and Gregg classification

Box 3.3—Geological Environments for Limestone Formation

The geological definition of limestone is a sedimentary rock with more than 50% $CaCO_3$, calcite (trigonal) and/or aragonite (orthorhombic). At ordinary temperatures, calcite is the stable form of calcium carbonate and can be considered the main mineral of limestones. Aragonite and high-magnesian calcite are abundant primary minerals in recent carbonate sediments but both are metastable and tend to transform to stable calcite or low-magnesian calcite with burial and/or time. Limestones can also contain several other carbonate minerals such as those listed in Table 3.3.

The carbonate mineral system is relatively easy to understand because it is controlled by only four minerals: calcite, aragonite, Mg-calcite, and dolomite. It is essential to remember that carbonate formation is a biogeochemical system in operation since the Archean. Limestones can be classified based on their origin, chemical composition, structure, and geological formation. In general, all limestones are crystalline and the crystal size increases with the degree of crystallization that occurred during the formation of the carbonate rock.

The environment of deposition is significant since it determines the size, shape, purity, and other essential characteristics of the carbonate rocks used as aggregates. Most components of carbonate sediments are either biogenic or nonbiogenic or a combination of both, and they precipitate from seawater or freshwater saturated with respect to $CaCO_3$. The formation of carbonate sediments is controlled by biological and biochemical processes, although inorganic precipitation of calcium carbonate from seawater is also possible. Limestone formed in high-energy zones contains few noncarbonate components and hence can be the source of high-purity carbonate material. In contrast, limestones accumulated in low-energy areas are likely contaminated with clay and other noncarbonate materials (i.e., quartz grains).

Table 3.3 Summary of the minerals commonly associated with limestones

Mineral	Formula	Crystal system
Aragonite	$CaCO_3$	Orthorhombic
Calcite	$CaCO_3$	Hexagonal
Magnesite	$MgCO_3$	Hexagonal
Dolomite	$CaMg(CO_3)_2$	Hexagonal
Ankerite	$Ca(MgFe)(CO_3)_2$	Hexagonal
Siderite	$FeCO_3$	Hexagonal

Marine carbonates "have the potential to accumulate anywhere in the ocean even in deep areas but most carbonate sediments are produced in bright, relatively clear, near-surface subtropical ocean waters" (James and Jones 2016). *The two major environments for carbonate production are the neritic zone, from the shoreline to the continental shelf edge, which is now located at approximately 200 m water depth, and the abyssal zone. In these environments, the most important organisms are benthic and planktic, respectively. Zones of neritic carbonate sedimentation are either in intracontinental basins or in the open oceans as shallow banks. In fact, the marine neritic system has probably originated most of the carbonates in geologic history, depending on water temperature, salinity, siliciclastic input, etc. In tropical zones, carbonate production and accumulation are impressive because the water is warm (> 20 °C), supersaturated with carbonate, and clear and brightly illuminated. Classical examples of these environments at present time are located in the Caribbean, the Persian Gulf, and Australia.*

Carbonate rocks are also formed "in a variety of terrestrial settings such as lakes, springs, and caves; limestones and dolostones generated in lakes and marshes are by far the most extensive and economically important terrestrial accumulations" (James and Jones 2016). *Carbonate deposits form in freshwater or saltwater lakes under different climate regimes, from humid regions to evaporitic in arid regions, and are present in both perennial and ephemeral lakes. Lacustrine carbonate sediments are diverse in terms of their mineralogy, facies, and geochemistry. Carbonate spring deposits are also widespread and spectacular deposits formed of calcite and/or aragonite characterize many spring systems* (Fig. 3.37). *They develop where subterranean water comes to the Earth's surface and is discharged subaerially or subaqueously. The size of these carbonate deposits is controlled by the volume, the physical and chemical characteristics of the discharged water, the flow paths, and the biota. Some examples can cover areas of 700 km² with up to 250 m thick carbonate spring deposits.*

Fig. 3.37 Carbonate spring deposits at Pamukkale (Turkey) (image courtesy of José Pedro Calvo)

- *Sandstones*

Sandstones, an abundant and important class of sedimentary rocks, are lithified accumulations of sand grains. Although they constitute 5 to 15% of all sediments, they may make up 25% of the continental stratigraphic record. They are formed by particles with an average size between 2.00 and 0.0625 mm in diameter. This term is reserved for siliciclastic and related sedimentary rocks, rather than clastic carbonate rocks. This is because a group of special terms is utilized to designate the latter. Sandstones have four constituents: grains, matrix, cement and, sometimes, porosity. Sand-sized particles constitute the framework of the rock while the matrix (finer grained material infilling the space between the framework grains) is deposited at the same time as the framework grains; it is composed of clay, silt and/or iron oxide. Cement is the term used to describe minerals precipitated in pores after deposition of the sediment, commonly carbonate and silica cements. Clay minerals, especially kaolinite, are present in sandstone, and originate during diagenesis. Table 3.4 shows the most common minerals and rock fragments present in siliciclastic sedimentary rocks.

Grains of sandstones are composed of different quantities of the mineral quartz (SiO_2), including single grains or polycrystalline aggregates (easily observable under the microscope); the main characteristics of the clasts of quartz are related to the type of source rock they are derived from. As an example, typical polycrystalline quartz grains are derived from metamorphic rocks. In many sandstones, the clastic quartz grains are overgrown by clear quartz crystals. It is very common that carbonate minerals (i.e., calcite and dolomite) may fill the former pores between detrital quartz grains.

In order of decreasing abundance, sandstones also contain feldspar (alkali feldspars and plagioclases), micas, ferromagnesian minerals, and heavy minerals (i.e., hornblende, garnet, tourmaline, and zircon—Table 3.5). The percentage of plagioclase over alkali feldspar may be a source indicator; for instance, sandstones in which alkali feldspar is dominant have probably had a granitic source. Sandstones can also contain grains formed by more than one mineral or crystal, which are termed rock fragments or lithic grains (i.e., shale fragments, volcanic rock clasts, plutonic rock clasts, and metamorphic rock clasts). Any rock-forming mineral can be present in a sandstone, given the appropriate conditions of erosion and deposition.

Table 3.4 Common minerals and rock fragments in siliciclastic sedimentary rocks (Boggs 2012)

Major minerals (abundance > ~1.2%)
Stable minerals (greatest resistance to chemical decomposition)
Quartz makes up approximately 65% of average sandstone, 30% of average shale: 5% of average carbonate rock
Less stable minerals
Feldspars include K-feldspars (orthoclase, microcline, sanidine, anorthoclase) and plagioclase feldspars (albite, oligoclase, andesine, labradorite, bytownite, anorthite); make up about 10–15%, of average sandstone, 5% of average shale, < 1% of average carbonate rock Clay minerals and fine micas clay minerals include the kaolinite group, illite group, smectite group (montmorillonite a principal variety), and chlorite group: fine micas are principally muscovite (sericite) and biotite: make up approximately 25–35% of total siliciclastic minerals, but may comprise > 60% of the minerals in shales
Accessory minerals (abundances < ~1–2%)
Coarse micas principally muscovite and biotite Heavy minerals (specific gravity > ~2.9) Stable nonopaque minerals—zircon, tourmaline, rutile, anatase Metastable nonopaque minerals—amphiboles, pyroxenes, chlorite, garnet, apatite, staurolite, epidote, olivine, sphene, zoisite, clinozoisite, topaz, monazite, plus about 100 others of minor importance volumetrically Stable opaque minerals hematite, limonite Metastable opaque minerals—magnetite, ilmenite, leucoxene
Rock fragments (make up about 10–15% of the siliciclastic grains in average sandstone and most of the gravel-size particles in conglomerates; shales contain few rock fragments)
Igneous rock fragments may include clasts of any igneous rock, but fragments of fine-crystalline volcanic rock and volcanic glass arc most common in sandstones Metamorphic rock fragments—include metaquartzite, schist, phyllite, slate, argillite, and less commonly gneiss clasts Sedimentary rock fragments any type of sedimentary rock possible in conglomerates; clasts of fine sandstone, siltstone, shale, and chert are most common in sandstones; limestone clasts are comparatively rare in sandstones
Chemical cements (abundance variable)
Silicate minerals predominantly quartz; others may include chalcedony, opal, feldspars, and zeolites Carbonate minerals principally calcite; less commonly aragonite, dolomite, siderite Iron oxide minerals—hematite, limonite, goethite Sulfate minerals anhydrite, gypsum, barite
Note Stability refers to chemical stability

Table 3.5 Detrital heavy minerals in sandstones (Boggs 2012)

Nonopaque heavy minerals	
Ultrastable	Rutile, tourmaline, zircon, anatase (uncommon)
Stable	Apatite, garnet (iron-poor), staurolite, monazite, biotite
Moderately stable	Epidote, kyanite, garnet (iron-rich), sillimanite, sphene, zoisite
Unstable	Hornblende, actinolite, augite, diopside, hypersthene, andalusite
Very unstable	Olivine
Relative stability not well established	Ankerite, barite, brookite, cassiterite, chloritoid, chrondrite, clinozoisite, corundum, chromite, dumortierite, fluorite, glaucophane, lawsonite, magnesite, monazite, phlogopite, pitotite, pleonaste, pumpellyite,siderite, spinel, spodumene, topaz, vesuvianite, wolframite, xenotime, zoisite (many of the minerals in this group are uncommon as detrital grains in sandstones)
Opaque heavy minerals	
Stable to moderately stable	Magnetite, ilmenite, hematite, limonite, pyrite, leucoxene

Two parameters are utilized to name and classify sandstones: chemical mineralogy and physical texture. Since quartz (Q), feldspar (F) and rock (lithic) fragments (L) are the most common components of sandstones, they are often used to classify sandstones through the so-called QFL plot (i.e., Dott 1964) (◘ Fig. 3.38). The proportions of these components can be used to define petrofacies that reflect the provenance of the sediment. Additional factors influencing the quantities of different framework grains include physical and chemical processes in the environment of deposition and diagenetic processes. As a consequence, it can be stated that the mineral composition of the sandstones is reflected in their chemistries, which in turn reflect the provenance.

From a textural viewpoint, sandstones are divided into two major groups, that is, whether the sandstones are composed of grains only, termed arenites, or contain more than 15% matrix, forming the so-called wackes. Both types of sandstones occur in a wide spectrum of continental to deep marine environments, and locally can occur together. Another popular classification of sandstones emerged using the matrix content, such as that of Folk (1974) (◘ Fig. 3.39). This classification makes a fundamental division between rocks with >15% matrix, which are termed greywackes, and those with <15% matrix. Folk's classification subdivides the second group of rocks into arkoses and arenites (arkoses are feldspar rich while arenites are poor in feldspar content). Sedimentary structures of sandstones include different types of depositional, erosional, chemical, and deformational structures, with bedding and laminations almost ubiquitous (◘ Fig. 3.40). All textures and structures in the sandstone reveal something of its history.

Sandstones, when hard and dense, are the only type of clastic rocks used for aggregate production (◘ Fig. 3.41), being a major source of aggregate in

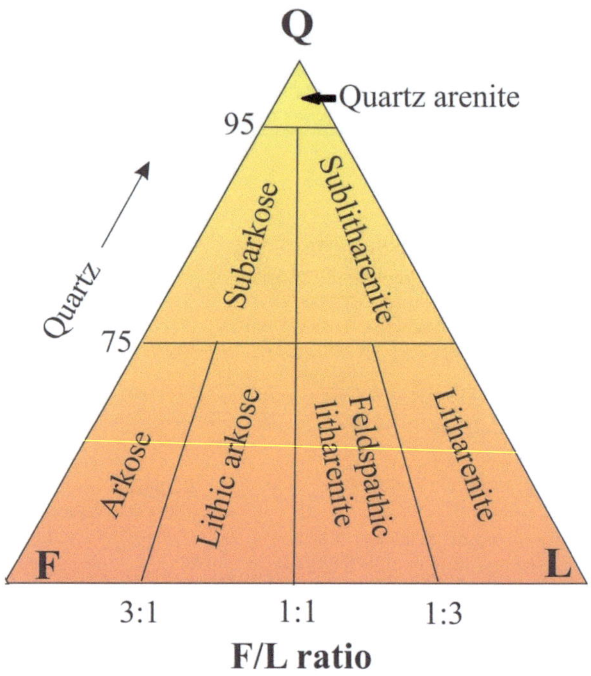

◘ **Fig. 3.39** Sandstone Folk (1974) classification

some regions. Sandstones subjected to tectonic compression show higher strengths and are good materials for some aggregate applications. However, sandstones are commonly friable or excessively porous materials due to imperfect cementation of the constituent grains. On the other hand, sandstones can also contain clay that renders the rock friable, soft, and absorptive. Sandstone consisting of sand grains bound together by clay is unsatisfactory since it is weak and porous.

In summary, sandstone suitability for aggregate utilization is principally based on its strength, porosity, and durability. These properties depend on the mode of formation and geological history. Therefore, "the

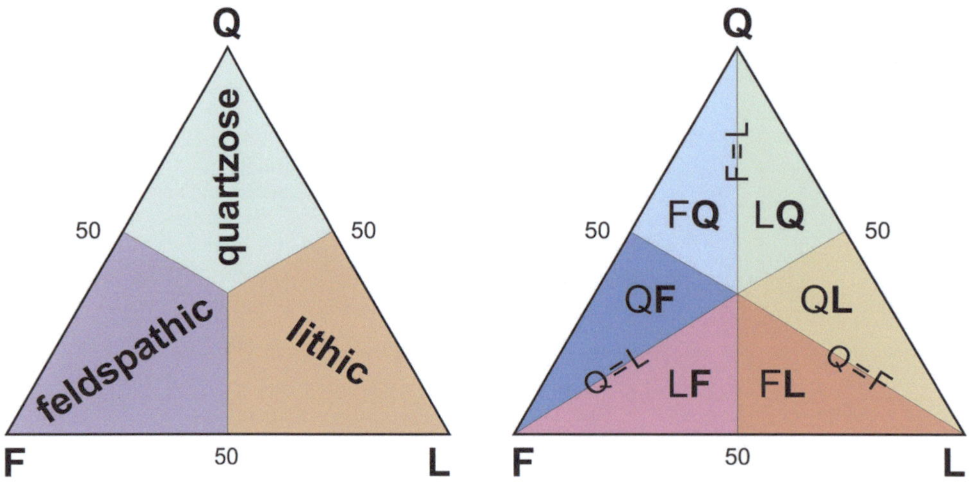

◘ **Fig. 3.38** Sandstone classification using the QFL (quartz, feldspar, rock fragments) system (Dott 1964)

■ **Fig. 3.40** Sandstone showing bedding/lamination (Navajo Sandstone, Arizona, USA) (image courtesy of Carolina Bustillo)

mineralogical composition, grain size, degree of grain sorting, nature and degree of cementation, degree of compaction and weathering state are fundamental rock properties that directly affect the end-use performance of the sandstone and its economic potential" (BGS 2019). To a much lesser extent, siltstone and conglomerate are also utilized as crushed stone for aggregates.

3.3.2 Igneous Rocks

All phenomena involving rock in the liquid (molten) state are igneous processes (ignis means fire in Latin). They are an essential part of the rock cycle (■ Fig. 3.42), which leads to the broad variety of observed igneous rocks (■ Fig. 3.43). The igneous processes form the fundamental process by which Earth and the planets differentiate into crust and mantle (■ Fig. 3.44) (three different layers in spherical cells form the Earth: core, mantle and crust—Table 3.6). They involve "making magma (molten or partially molten mass in the Earth's interior that penetrates the lithosphere) from previously solid source rocks, moving magma from one place to another and freezing magma to produce igneous rocks" (Asimow 2005).

Igneous processes are closely linked to the igneous rocks that are their final products. Thus, "origin by crystallization from a melt seems a simple enough criterion for considering a rock to be igneous" (Winter 2010); igneous rocks can be recognized either by field setting or by textural criteria. From a textural viewpoint, there are two main types of igneous rocks (a) phaneritic: most of the crystals are readily visible with the naked eye (> 0.1 mm), which means that the rock crystallized slowly beneath the surface of the Earth (plutonic or intrusive), and (b) aphanitic: the majority of crystals are too small to be seen readily with the naked eye (< 0.1 mm), which means that the rock crystallized rapidly at the Earth's surface (volcanic or extrusive).

Igneous rocks comprize the bulk of the Earth and are the precursors of sedimentary and metamorphic rocks. On land, crystalline rocks, which make up 34% of the exposures, are more or less equally divided into

Fig. 3.41 Sandstone quarry in the UK (image courtesy of Clive Mitchell)

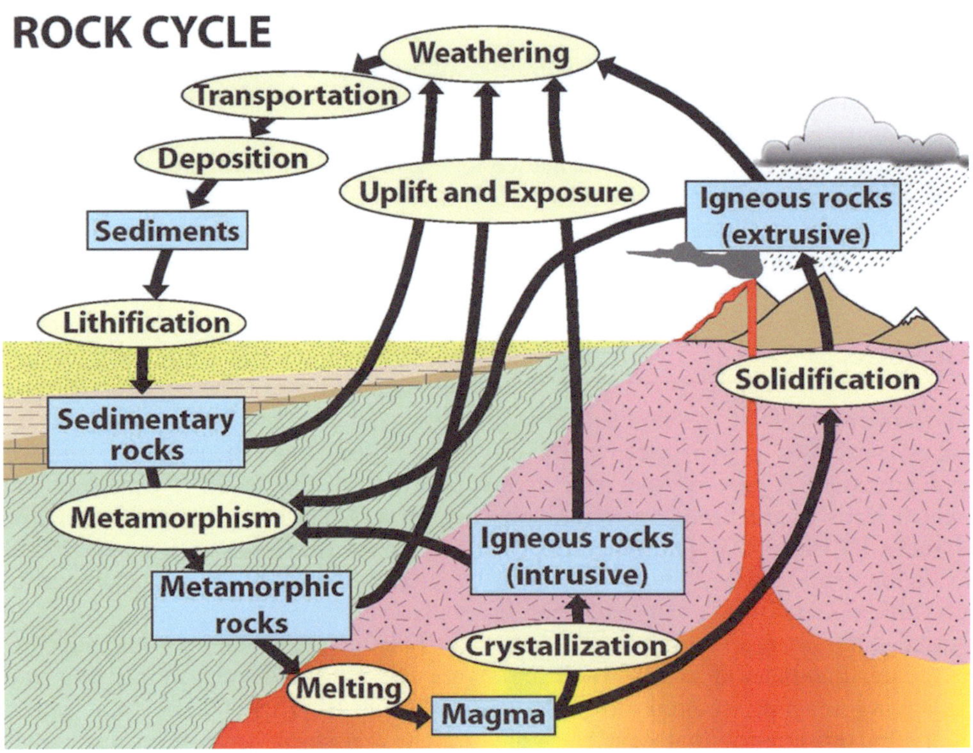

Fig. 3.42 Rock cycle

Fig. 3.43 The famous Devil's Tower is an igneous rock (Devil's Tower National Park, Wyoming, USA) (image courtesy of Manuel Jorge Bustillo)

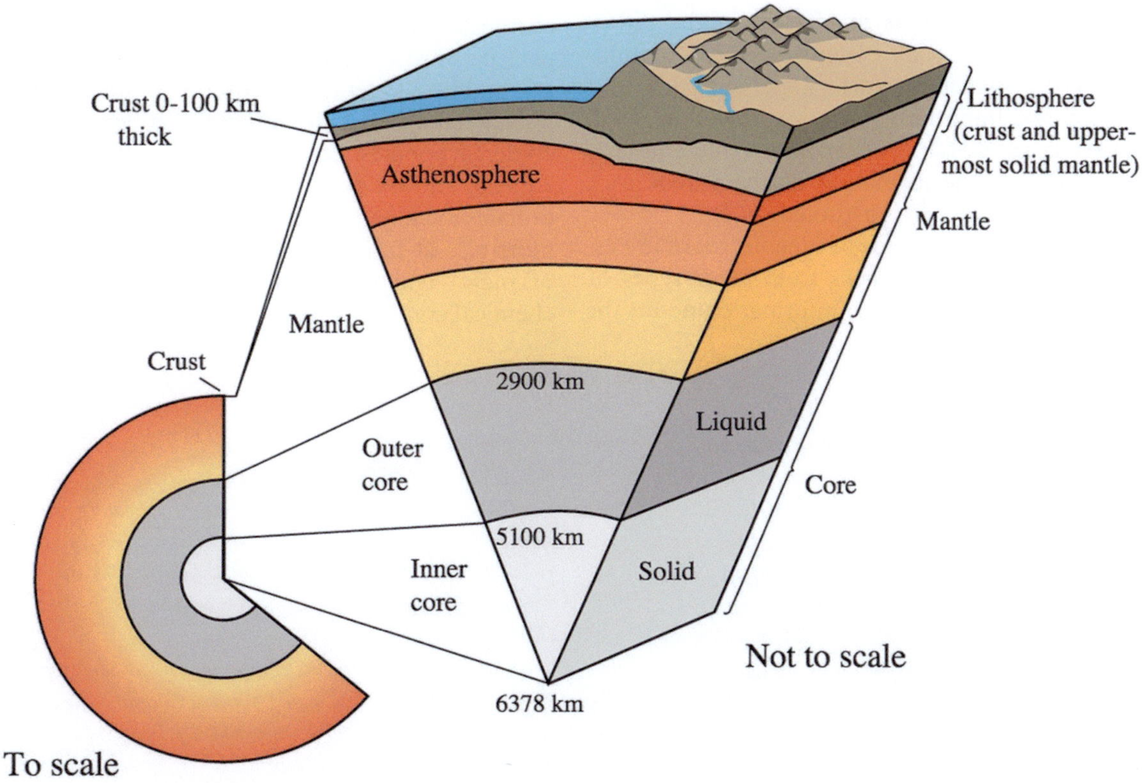

Fig. 3.44 Internal structure of the Earth (illustration courtesy of United States Geological Survey)

3

Table 3.6 Volume, mass and density of the Earth's crust, mantle and core (Okrusch and Frimmel 2020)

	Volume (%)	Mass (%)	Average density (g/ cm^3)
Crust	0.8	0.4	2.8
Mantle	83.0	67.2	4.5
Core	16.2	32.4	11.0
Bulk earth			5.53

igneous and metamorphic rocks; beneath the sedimentary cover, igneous and metamorphic rocks dominate the oceanic crust. Igneous rocks are present "on the surface of the Earth due to tectonic movements to rise near or at the very surface of the Earth or with strong erosion of the existing rocks that cover the surface" (Haldar and Tisljar 2014).

Igneous rocks are formed at temperatures well above those common at the Earth's surface and are classified according to their structure, texture, and composition; the latter includes both mineral composition and chemical composition. In some cases, the chemistry alone is the criterion for classification. Nevertheless, mineral composition plays a key role in the overall distribution of igneous rocks. Minerals in igneous rocks can be defined as major, minor (or accessory), and secondary. These definitions are based on the different proportions in the rock composition. Major minerals are those by which the rocks are classified, minor minerals are not important and are present in small amounts (typically $< 1\%$), and secondary minerals are non-present during the formation of the parent rock and are later introduced or substituted.

The textures and structures of igneous rocks are essential to distinguish these rocks from other types of rocks and they provide valuable information about the history of formation of igneous rocks. In fact, certain minerals, textures, structures, and chemical characteristics are unique to igneous rocks. Moreover, each of the main genetic groups of igneous rocks is characterized by a specific shape, appearance, and characteristic structures and textures. This complex mixture is reflected in the complexity of the nomenclature and classification of igneous rocks.

The magma, formed in the mantle and lower crust, can or cannot be entirely molten. This is because important quantities of solids, both crystals and rock fragments, as well as gases, are present in the magma. The magma tends to rise toward the surface because it is less dense than the surrounding rocks. The products of crystallization of these magmas are igneous rocks. If the rock is formed by slow cooling of the magma at relatively high depth, it shows coarse crystals and includes the general rock types of granite (▶ Box 3.4: Granite), tonalite, diorite—Fig. 3.45—and gabbro (plutonic igneous rocks). Volcanic igneous rocks were extruded onto the Earth's surface or on the sea floor, where they cooled and solidified relatively quickly from the lava of volcanoes, promoting the formation of small or microscopic crystals. Examples of these include rhyolite, andesite, and basalt, among others. In an igneous rock, chemical composition of the magma and physical–chemical conditions during crystallization process control the formation of minerals, with silicates being the most important rock-forming minerals. The chemical composition of magma or lava is very complex because the magma or lava from different places are highly different. In addition to silicates, other important rock-forming minerals in igneous rocks include carbonates, certain sulfides, sulfates, phosphates, halides, and oxides. Oxides such as hematite and carbonates such as calcite are also common as secondary minerals. Table 3.7 shows the mineral composition of major igneous rocks and Table 3.8 shows their chemical composition.

Box 3.4—Granite

Granite has a Latin origin, granum, which means grain, in reference to the coarse-grained texture of the rock. Although the term granite means different things to different petrologists, "granite is a crystalline plutonic igneous rock consisting essentially of quartz and feldspar, in which the relative proportion of the former is between 20 and 60%; yet, this simple definition belies the remarkable variety of color, grain size, texture, mineralogy and composition that is intrinsic to the granitic family" (Kemp 2005). *Most parts of the continents on Earth are covered by sediments and sedimentary rocks but the rocks below (the basement) are almost always granites, metamorphosed granites, or closely related rocks; together with granodiorite and tonalite, granite constitutes the most common plutonic rock. Charles Leclerc, Count of Buffon, in his first volume of Histoire Naturalle des Minéraux (1783), states that "at present granites cover most parts of the globe". Granite is also well known for its many world-famous exposures such as those present at Yosemite National Park (USA)* (Fig. 3.46) *or the Mount Rushmore sculptures in South Dakota (USA)* (Fig. 3.47).

Plutonic rocks are classified and named according to mineral content. The following typical nomenclature is taken into account: Q = quartz, A = alkali feldspar (incl. albite), P = plagioclase, F = feldspathoids, M = mafic and related minerals. In this classification, granites, in their broadest sense, are composed of 20–60% quartz as well as variable proportions of alkali feldspar and plagioclase. The simplest

Fig. 3.45 Diorite intruding limestone (image courtesy of Mike Conway—Arizona Geological Survey Photographic Atlas)

Table 3.7 Mineral composition of major igneous rocks (Haldar and Tilsljar 2014)

	Intrusive rocks	Extrusive rocks	Main minerals
Felsic	Granite	Rhyolite	Quartz 20–40% K—feldspar > NA—plagioclase and mica
	Adamellite	Dellenite	Quartz, K—feldspar = NA—plagioclase
	Grandiorite	Dacite	Quartz 10–30%, NA—plagioclase + NA—Ca plagioclase > K—feldspar, biotite and hornflende
	Tonalite and quartz diorite	Dacite	Quartz, NA—plagioclase + NA—Ca plagioclase > K—feldspar, biotite and hornflende
Intermediate	Monzite	Latite	Quartz, K—feldspar = NA—plagioclase biotite, hornblende and pyroxene
	Diorite	Andesite	Na—plagiocose and Na—Ca—plagiocose 60–80%, amphibole and pyroxene
	Syenite	Trachyte	K—feldspar 60–80%, Na—Ca plagiocose, hornblende, biotite, pyroxene and arfvedsonite
	Nephelene syenite	Phonolite	Nepheline, leucite, aegirine, k—feldspar, riebeckite, biotite, pyroxene and arfvedsonite
Mafic	Gabbro	Basalt diabase spilite	Ca—plagiocose (40–70%) pyroxene (augite, hypersthene), small quantities of hornblende and biotite, with or without olivine
	Norite	Basalt	Ca—plagioclase, pyroxene (hypersthene) with or without olivine

(continued)

3

Table 3.7 (Continued)

	Intrusive rocks	Extrusive rocks	Main minerals
	Anorthosite		Ca—plagioclase (90–100%) with pyroxene, ilmenite, magnetite (0–10%) ± olivine
Ultramafic	Peridotite		Olivine, one or more pyroxene
	Dunite		Mostly Mg—olivine with little pyroxene
	Lherzolite		Olivine, bronchite, and dialage
	Serpentine		Serpentine derived from olivine
	Pyroxenite		Monoclinic pyroxene (augite, diopside, and dialage)

Table 3.8 Average chemical composition of major igneous rocks: **a** plutonic and **b** volcanic (Okrusch and Frimmel 2020)

(a) Average chemical composition of major plutonic rocks (in element oxides, wt%) H_2O stands for water that is chemically bound in minerals as (OH) group, i.e. not adsorbed

Oxide	Peridotite	Gabbro	Diorite	Monzonite	Granodiorite	Granite
SiO_2	43.54	48.36	51.86	55.36	66.88	72.08
TiO_2	0.81	1.32	1.50	1.12	0.57	0.37
Al_2O_3	3.99	16.84	16.40	16.58	15.66	13.86
Fe_2O_3	2.51	2.55	2.73	2.57	1.33	0.86
FeO	9.84	7.92	6.97	4.58	2.59	1.67
MnO	0.21	0.18	0.18	0.13	0.07	0.06
MgO	34.02	8.06	6.12	3.67	1.57	0.52
CaO	3.46	11.07	8.40	6.76	3.56	1.33
Na_2O	0.56	2.26	3.36	3.51	3.84	3.08
K_2O	0.25	0.56	1.33	4.68	3.07	5.46
P_2O_5	0.05	0.24	0.35	0.44	0.21	0.18
H_2O	0.76	0.64	0.80	0.60	0.65	0.53
Total	100.00	100.00	100.00	100.00	100.00	100.00

(b) Average chemical composition of major volcanic rocks (in element oxides, wt%)

Oxide	Basalt	Andesite	Dacite	Rhyolite	Phonolite
SiO_2	50.83	54.20	63.58	73.66	56.90
TiO_2	2.03	1.31	0.64	0.22	0.59
Al_2O_3	14.07	17.17	16.67	13.45	20.17
Fe_2O_3	2.88	3.48	2.24	1.25	2.26
FeO	9.05	5.49	3.00	0.75	1.85
MnO	0.18	0.15	0.11	0.03	0.19
MgO	6.34	4.36	2.12	0.32	0.58
CaO	10.42	7.92	5.53	1.13	1.88
Na_2O	2.23	3.67	3.98	2.99	8.72
K_2O	0.82	1.11	1.40	5.35	5.42
P_2O_5	0.23	0.28	0.17	0.07	0.17
H_2O	0.91	0.86	0.56	0.78	0.96
Total	99.99	100.00	100.00	100.00	100.05[a]

[a]Including 0.23% CI and 0.13 SO_3

Fig. 3.46 Granite relief in Yosemite National Park (Califormia, USA) (image courtesy of Carolina Bustillo)

subdivision of granite rocks is based on the ratio of alkali feldspar to plagioclase. Thus, this ratio generates four main categories: alkali feldspar granite, granite (stricto sensu), granodiorite and tonalite. Granite (stricto sensu) can also be subdivided into monzogranites (35–65% alkali feldspar) and syenogranites (65–90% alkali feldspar). On the other hand, granite and granitic can be used only in reference to granite sensu stricto while granitoid "is generally used for granular, plutonic, quartz-feldspar rocks of unspecified or a wide range of compositions (lato sensu)" (Raymond 2002).

The volume of granitic bodies can range from less than 1 km^3 *for single intrusions to more than* 10^6 km^3 *for batholitic structures; great masses of plutonic igneous rocks hundreds of kilometers long and tens of kilometers wide form the cores of many Phanerozoic mountain ranges. Granitic bodies are present in all tectonic settings but their compositions, volumes, and characteristics of the associated igneous rocks change correspondingly.*

Granite is nearly always massive, that is, lacking any internal structures. Although it contains quartz, feldspars, and biotite, the presence of additional aluminous mineral species must be remarked because they participate in defining the chemical signature of the rock. For instance, two-mica granite contains biotite and muscovite. Iron oxides such as ilmenite and magnetite are accessories showing the oxygen fugacity that characterizes the magma.

Regarding the grain-size in granites, it is defined as "fine-grained (< *1 mm diameter), medium-grained (1–5 mm diameter), coarse-grained (5–50 mm diameter) and very coarse-grained (*> *50 mm)"* (Winter 2010) *(pegmatitic is also used for very coarse-grained granites but this term has genetic implications). The texture in granite can change sharply within the limits of an individual body. Nevertheless, the color and texture in granites are usually uniform for large volumes of rock.*

Although granites are probably the best-known and most common plutonic igneous rocks found at Earth's surface, "the origin of granites has been a matter of controversy for more than a century" (Nédélec and Bouchez 2015). *This is because how large plutonic masses are emplaced in the rigid upper crust has long been contentious. At present, there are two basic theories to explain the genesis of granites: anatexis of the crust and fractionation of magma derived from the mantle. The vast majority of ascending granitic magmas do not reach the surface. Nevertheless, they can finally intrude into various levels of the continental crust and originate plutons or stocks.*

Fig. 3.47 Granite sculptures in Monte Rushmore (South Dakota, USA) (image courtesy of María Ángeles Morán)

In terms of chemical composition, all magmas have a silicate composition and silica (SiO_2) is the dominant component; it comprizes around 40–80% by weight. The silica percentage forms the basis of a threefold classification into acid (> 66% wt. SiO_2), intermediate (52–66% wt. SiO_2) and basic (45–52% wt. SiO_2) categories (rocks with lower SiO_2 contents do exist—ultrabasic, <45% wt. SiO_2, but they are relatively rare). Alumina (Al_2O_3) is a distant second component (13–18% by weight). SiO_2, TiO_2, Al_2O_3, CaO, MgO, MnO, FeO, Fe_2O_3, Na_2O, K_2O and P_2O_5 are termed major oxides or major elements since together they comprize approximately 99% of any igneous rock. Many other elements (i.e., V, Sr, Ni, Ce, Nd, and many more), whose abundances are measured in parts per thousand or parts per million, are present in igneous rocks; they are referred to as trace elements. These trace elements are essential because they provide very important information about the origin of igneous rocks. For this reason, modern chemical analyses of igneous rocks are separated into two parts: a major element analysis and a minor or trace element analysis.

Based on the depth of emplacement of the magma, the igneous rocks can be subdivided into intrusive and extrusive types. The former are those that solidify below the surface whereas the latter includes all the hardened eruptive products. Silica-rich igneous rocks, which usually include free silica as clear, colorless quartz crystals, and opaque but pale-colored feldspars are lighter in weight than intermediate and basic rocks. As a rule, they have increasing quantities of dark-colored ferromagnesian minerals of higher density.

Many igneous rocks such as those mentioned above are hard, tough, and dense, and are excellent crushed stone aggregates. For instance, granite is hard and resistant to abrasion due to its mineral composition and interlocking of crystals. In the aggregate industry, light-colored intrusive igneous rocks are referred to as granite whereas extrusive igneous rocks, commonly dark-colored, are frequently referred to as traproc. Although igneous rocks exhibit a broad spectrum of chemical compositions and their adequacy for aggregate depends on their mineral composition and crystalline fabric and texture, they tend to produce hard aggregates with a high degree of skid resistance. Igneous rocks are thus very appropriate in road surfacing applications and for utilization in the lower parts of road pavement. Nevertheless, there are some exceptions: certain extrusive rocks, especially those with high silica content, are too porous to make good aggregate, and other igneous rocks with a very high content of silica tend to react chemically with alkali when used as aggregate in cement concrete.

Some coarse grained rocks can be less attractive in some applications because of their low crushing strength. For example, "coarse-grained granites are less resistant to temperature changes and weathering processes that the finer textured varieties" (NSSGA 2013). Substantial quantities of coarse aggregate are used as railway track ballast. Because ballast requires hard, clean, and angular aggregates with a very high resistance to abrasion, most of this type of aggregate application is produced from igneous rocks (i.e., andesite—Fig. 3.48—or basalt).

3.3.3 Metamorphic Rocks

The term metamorphism is derived from the Greek meta (meaning change) and morphe (meaning form). Metamorphic rocks result from the alteration of preexisting rocks (sedimentary, igneous, and metamorphic) by heat, pressure, and chemical activity, changing their mineralogy, texture and/or chemical composition (the two dominant changes are in mineralogy and texture). The preexisting original rocks, affected by metamorphic changes, are termed protoliths. The metamorphic changes in protoliths can be partial, gradational, or complete and the sequence of transformation or metamorphism of minerals and rocks takes place at a constant and gradual increase in temperature and pressure conditions. The main phenomenon of metamorphism

Fig. 3.48 Andesite quarry for railway ballast

3

is the recrystallization of the preexisting original mineral components. For example, limestones with different textures (i.e., different grain sizes and matrices) are transformed to coarse crystalline calcite mosaics during progressive metamorphism, resulting in a metamorphic rock called marble. Simultaneously, metamorphism involves no chemical changes in the bulk rock composition, except for an increase or loss of volatiles.

Depending on "the chemical composition and pressure–temperature conditions of metamorphism (a) granite, granodiorite or rhyolite are converted to orthogneisses or felsic granulites, gabbro and basalt to greenschist, blueschist, amphibolite, mafic granulite or eclogite (b) mudstone and greywacke are transformed into phyllite, micaschist, paragneiss or felsic granulite, (c) limestone and dolomite are converted to either calcitic or dolomitic marble, and (d) quartz–rich sandstone to quartzite" (Okrusch and Frimmel 2020). ◘ Table 3.9 summarizes some common metamorphic rocks derived from igneous and sedimentary rock equivalents.

Metamorphic processes occur predominantly in a solid-state and a great variety of geological environments under conditions between diagenesis and large-scale melting; "the boundary between sedimentary diagenesis and the lower limit of metamorphism is placed at around 150 °C ± 50 °C" (Sen 2014) (a gradual transition exists between diagenesis and metamorphism). As a consequence, they comprize a complex range of rock types with different physical characteristics. The broad diversity of metamorphic rocks is generated by the diversity of metamorphic processes and preexisting rocks that are available for metamorphism. Metamorphic rocks usually retain some of their parental heritage (i.e., chemical composition) as well as up to a point primary structural features (i.e., bedding). Thus, the rocks of the metamorphic family are crystalline and resemble sedimentary rocks. Like other rocks, the metamorphic rocks can be classified based on texture and composition (either mineralogical or chemical).

Two basic types of metamorphism, local versus regional, are recognized (◘ Table 3.10). In the first group, contact metamorphism takes place adjacent to igneous intrusions, principally as a result of the thermal effects of hot magma intruding cooler shallow rocks (the rocks surrounding a pluton are termed country rocks or host rocks) at relatively low pressure. This process is limited to the area of contact of rocks with the igneous body and, for this reason, it is called contact metamorphism. The effect of heating propagates in several concentric rims. Regional metamorphism (lato sensu)

◘ **Table 3.9** Some common metamorphic rocks derived from igneous and sedimentary rock equivalents (NSSGA 2013)

Sediments	Sedimentary rock equivalents	Metamorphic rock equivalents
Gravel	Conglomerate	Gneiss and various Schists
Sand	Sandstone	Quartzite and Quartz Schist if from Pure Quartz sand; Mica Schist if certain impurities are present
Clay	Shale	Slate and Schists, especially Mica Schist
Lime deposits such as chalk or shells	Limestone	Marble

Igneous rock	Metaphoric rock equivalents
Granite, Syenite and other rock with much Felspar	Gnesiss
Fine Grained Felspar rock such as felsite and tuffs	Slate and Schists
Diorite, Basalt and other basic rock	Hornblende Schist and Schist

◘ **Table 3.10** Types of metamorphism (Bucher and Grapes 2011)

Regional extent	Local extent
Orogenic metamorphism (regional metamorphism)	Contact (igneous) metamorphism
Subduction metamorphism	Cataclastic metamorphism
Collision metamorphism	Hydrothermal metamorphism
Ocean-floor metamorphism	*Exotic local metamorphism*
Burial metamorphism	Impact metamorphism
	Lightning metamorphism
	Combustion metamorphism

Fig. 3.49 Mylonite quarry (image courtesy of Benito Arnó e Hijos, S.A.U)

is any metamorphism that affects a large body of rock. For this reason, it often covers a great lateral extent (usually tens of kilometers or more).

The consequence of regional metamorphism changes not only the mineral composition of the original rock (protolith) but also the texture and structure of the newly formed metamorphic rocks. Typical products of regional metamorphism are foliated rocks such as phyllite, micaschist, and gneisses. Contact metamorphism generates textures lacking any preferred orientation whereas regional metamorphism generates the preferred orientation of the minerals.

Metamorphic rocks that are hard, tough, and dense may be utilized as aggregate (i.e., mylonite—Fig. 3.49). Gneiss (banded crystalline rock) (Fig. 3.50), marble (metamorphosed limestone) and quartzite (metamorphosed sandstone—Fig. 3.51) are typically the most characteristic metamorphic rocks. However, few of them are or more than locally important as aggregate. Some metamorphic rocks (i.e., highly foliated schist—Fig. 3.52) "have planar foliation that produces undesirable planes of weakness, not being suitable for crushed stone" (Langer 2006). On the other hand, some metamorphic rocks such as quartzite can cause excessive wear on crushing equipment.

3.4 Questions

Short Questions

- Define the terms crushed stone and sand and gravel
- Enumerate the three main types of fluvial derived deposits
- How are river terraces formed?
- Explain the concept of marine aggregates
- List the main divisions in the carbonate Dunham classification
- Enumerate the main grain types of sandstone
- Explain the origin of igneous rocks
- On what basis are igneous rocks classified?
- What textures can be generated as magma cools and crystallizes to form igneous rocks?
- What are the major subdivisions of the Earth?
  Long Questions

3

Fig. 3.50 Folds in pegmatitic gneiss (image courtesy of Jon Spencer—Arizona Geological Survey Photographic Atlas)

Fig. 3.51 Quartzite quarry (image courtesy of Cuminer S.A.)

Fig. 3.52 Schist (image courtesy of Mike Conway—Spencer—Arizona Geological Survey Photographic Atlas)

- Explain the geological environments of limestone formation
- Explain the plate tectonics theory

References

Aslan A (2013) Fluvial environments | Sediments. In: Elias SA, Mock CJ (eds) Encyclopedia of quaternary science, 2nd Ed. Elsevier, Amsterdam, Netherlands, pp 663–675

Asimow PD (2005) Igneous processes. In: Selley RC, Cocks LRM, Plimer IR (eds) Encyclopedia of geology, vol 3. Elsevier Academic Press, Oxford, UK, pp 209–217

BGS (2019) Construction aggregates. British Geological Survey, Mineral Planning Factsheet, London, UK, 32 p

Blatt H (1982) Sedimentary petrology. WH Freeman and Co., San Francisco, USA, p 514

Boggs S (2012) Petrology of sedimentary rocks, 2nd edn. Cambridge University Press, Cambridge, UK, p 612

Bucher K, Grapes R (2011) Petrogenesis of metamorphic rocks, 8th ed. Springer—Verlag, Berlin Heidelberg, 428 p

Bustillo M (2021) Construction materials—geology, production and applications. Springer, Cham, Switzerland, 603 p

Dott RH Jr (1964) Wacke, greywacke and matrix–what approach to immature sandstone classification? J Sediment Petrol 34:625–632

Dunham RJ (1962) Classification of carbonate rocks according to depositional texture. In: Ham WE (ed) Classification of carbonate rocks. The American Association of Petroleum Geologists, Mem., 1, pp 108–121

Ehlers J, Gibbard PL, Hughes PD (2018) Quaternary glaciations and chronology. In: Menzies J, Van der Meer JM (eds) Past glacial environments, 2nd ed. Elsevier, Amsterdam, Netherlands, 859 p

Embry AF, Klovan JE (1971) A late Devonian reef tract on northeastern Banks Island, N.W.T. Bull Can Pet Geol 19:730–781

Folk RL (1959) Practical petrographic classification of limestones. Bull Am Assoc Pet Geol 43:1–38

Folk RL (1962) Spectral subdivision of limestone types. In: Ham WE (ed) Classification of carbonate rocks. The American Association of Petroleum Geologists, Mem., 1, pp 62–84

Folk RL (1974) Petrology of sedimentary rocks. Hemphill, Austin, TX, 182 p

Goldman HB (1990) Sand and gravel. In: Kennedy BA (ed) Surface mining, 2nd ed. Society for Mining, Metallurgy and Exploration Inc., Littleton, Colorado, USA, pp 237–241

Haldar SK, Tisljar J (2014) Introduction to mineralogy and petrology. Elsevier, Amsterdam, p 338

James PN, Jones B (2016) Origin of carbonate sedimentary rocks. John Wiley & Sons, West Sussex, p 320

Kamp U, Owen LA (2013) Polygenetic landscapes. In: Shroder J, Owen LA (eds) Treatise on geomorphology, vol 5. Academic Press, San Diego, CA, pp 370–393

Kemp AIS PD (2005) Granite. In: Selley RC, Cocks LRM, Plimer IR (eds) Encyclopedia of geology, vol 3. Elsevier Academic Press, Oxford, UK, pp 233–247

Langer WH (2003) A general overview of the technology of in-stream mining of sand and gravel resources, Associated potential environmental impacts, and methods to control potential impacts. U.S. Geological Survey Open-File Report 02-153, 44 p

Langer WH (2006) Crushed stone. In: Kogel JE, Trivedi NC, Barker JM, Krukowski ST (eds) Industrial minerals and rocks—commodities, markets and uses. Society for Mining, Metallurgy and Exploration, Inc., Colorado, USA, pp 171–180

Langer WH, Drew LJ, Sachs JS (2004) Aggregate and the environment. AGI, Environmental Awareness Series, American Geological Institute, Alexandria, VA, USA, 68 p

Menzies J, Hess DP, Rice JM, Wagner KG, Ravier E (2016) A case study in the New York Drumlin Field, an investigation using microsedimentology, resulting in the refinement of a theory of drumlin formation. Sed Geol 338:84–96

Nédélec A, Bouchez JL (2015) Granites—petrology, structure, geological setting and metallogeny. Oxford University Press, Oxford, UK, 348 p

Neukirchen F, Ries G (2020) The world of mineral deposits. Springer, Cham, Switzerland, p 371

NSSGA (2013) The aggregates handbook, 2nd ed. The National Stone, Sand and Gravel Association, Alexandria, Virginia, USA, 906 p

Okrusch M, Frimmel HE (2020) Mineralogy—an introduction to minerals, rocks, and mineral deposits. Springer-Verlag GmbH Germany, Berlin, p 719

Pettijohn FJ, Potter PE, Siever R (1987) Sand and sandstone, 2nd edn. Springer, New York, USA, p 580

Prentice JE (1990) Geology of construction materials. Topics in the earth sciences. Springer Netherlands, Amsterdam, Netherlands, 216 p

Raymond LA (2002) Petrology—the study of igneous, sedimentary and metamorphic rocks, 2nd edn. McGraw Hill, New York, USA, p 720

Scott PW (2016) The geological setting of aggregate resources in some Caribbean islands. In: Hunger E, Brown TJ, Smith G, Anderson P (eds) Proceedings of the 19th extractive industry geology conference 2016 and technical meeting 2017. EIG Conferences Ltd, London, UK, 9 p

Sen G (2014) Petrology—principles and practice. Springer-Verlag, Berlin Heidelberg, p 368

3

Sibley DF, Gregg JM (1987) Classification of dolomite rock texture. J Sediment Petrol 57:1987–1100

Smith MR, Collis L (eds) (2001) Aggregates: sand, gravel and crushed rock. Geological Society, London, Engineering Geology Special Publication vol 17, London, UK, 360 p

Smith IR, Paulen R, Plouffe A, Kowalchuk C, Peterson R (2005) Surficial mapping and granular aggregate resource assessment in northwest Alberta. British Columbia Ministry of Energy and Mines, Summary of Activities 2005, pp 80–95

Sutphin DM, Drew LJ, Fowler BK, Goldsmith R (2002) Techniques for assessing sand and gravel resources in glaciofluvial deposits—an example using the surficial geologic map of the Loudon Quadrangle, Merrimack and Belknap Counties, New Hampshire. US Geological Survey Professional Paper 1627, 26 p

UNEP (2014) Sand, rarer than one thinks. UNEP Global Environmental Alert Service (GEAS), 15 p

United Nations (2016) The First Global Integrated Marine Assessment. Cambridge University Press, 972 p

Velegrakis AF, Ballay A, Poulos S, Radzevicius R, Bellec V, Manso F (2010) European marine aggregates resources: origins, usage, prospecting and dredging techniques. J Coastal Res 51:1–14

Winter JD (2010) Principles of igneous and metamorphic petrology, 2nd ed. Prentice Hall, Pearson Education, Inc., New Jersey, USA, 702 p

Yeend W (1973) Sand and gravel. In: Brobst DA, Pratt WP (eds) United States mineral resources. US Geological Survey Professional Paper 820:561–565

Exploration and Evaluation of Deposits

Contents

4.1 Exploration Process – 102
4.1.1 Introduction – 102
4.1.2 Phases of Exploration – 102
4.1.3 Methods of Exploration – 107
4.1.4 Sampling – 128

4.2 Evaluation Process – 133
4.2.1 Introduction – 133
4.2.2 Classical Methods – 134
4.2.3 Geostatistical Methods – 137
4.2.4 Economic Evaluation – 138

4.3 Questions – 141

References – 141

M. Bustillo Revuelta, *The Basics of Aggregates*,
Springer Textbooks in Earth Sciences, Geography and Environment,
https://doi.org/10.1007/978-3-031-42961-3_4

4

Abstract

This chapter is concerned with the process of analyzing an area to find aggregate deposits as well as their resource/reserve evaluation. The information collected during exploration is utilized for assessing the size and quality of the aggregate deposit and for determining if there is the potential for it to be extracted. Two principal stages are outlined in aggregate exploration: reconnaissance exploration and detailed exploration. The main methods applied at different phases are discussed, especially geophysical methods because they are essential in the exploration of a sand and gravel deposit. The methods discussed are organized in order of scale and stage, from photogeology to drilling, through remote sensing and geophysical surveys. Technical and economic/financial methods used for evaluating an aggregate mining project are considered. The technical aspects are related to the geological setting of the deposit and the main characteristics of the material (i.e., tonnage). For this, a general introduction to geostatistics from a conceptual point of view and the principal classical methods utilized in the evaluation process are provided. The economic/financial methods cover the estimation of the principal economic inputs and outputs in the project and the amount, type, and cost of capital forthcoming for a project. Net present value, internal rate of return, payback period values, and risk analysis permits are briefly explained.

4.1 Exploration Process

4.1.1 Introduction

Mineral exploration can be defined as "the process of analyzing an area of land to find mineral deposits" (Bustillo 2018). It covers the group of processes that gather information about the presence or absence of a mineral deposit. The information collected during exploration is used to assess the size and quality of a mineral deposit and for determining if there is the potential for it to be mined. The objective of mineral exploration is the discovery of a mineral deposit susceptible of economic exploitation. It is important to note that the mining process can occur now or in the future, depending on economic factors.

Very often, the terms prospection and exploration are used in a misleading way. For some authors, exploration sounds similar to prospecting but "other authors consider prospecting simply as the search for ores or other valuable minerals (first stage) while exploration (second stage) determines as accurately as possible the size and value of a mineral deposit, by using techniques similar to but more refined than those used in prospecting (prospecting covers large areas in an attempt to see if mineral deposits are present while exploration involve more detailed data gathering over smaller and specific areas)" (Bustillo 2018). Thus, "the line differentiating prospecting and exploration is not sharp and often is not possible" (Hartman and Mutmansky 2002). However, as early as 1932, Thoenen in his work entitled "Prospecting and exploration for sand and gravel", states that "Prospecting is herein defined as the act of searching for, with the object of discovering, deposits of mineral. Exploration is defined as the act of examining a deposit when found, with the object of determining roughly its size, shape, and value". In this chapter, both terms are utilized indistinctly to avoid problems of interpretation.

As with all mineral resources (◘ Fig. 4.1), the mining cycle of aggregate production starts with exploration, continues through mining and processing and finishes with closure and postmining land use. Exploration for aggregate deposits involves locating a suitable resource near where it is to be used. In urban regions, the maximum economically feasible shipping distance from the market (i.e., 25 km) typically defines a crude target area for aggregate exploration. As a rule, exploration technicians (i.e., geologists) try to find a viable aggregate deposit for supplying certain market at a competitive price. This price must include "the cost of development of the aggregates mining facility, hauling, processing, regulatory expenses, sales and administration costs along with a reasonable profit" (NSSGA 2013). These costs must be estimated and taken into account to evaluate the economics of an aggregate deposit. Otherwise, the failure of the project is almost unavoidable.

4.1.2 Phases of Exploration

It is difficult to define exactly the number of stages in an aggregate deposit exploration. This is because the exploration process depends on different factors such as the type of aggregate, the area for exploring, the overall costs of the different stages and many others. It is possible to find in the literature up to five phases in mineral exploration: "programme design, reconnaissance exploration, detailed exploration, prospect

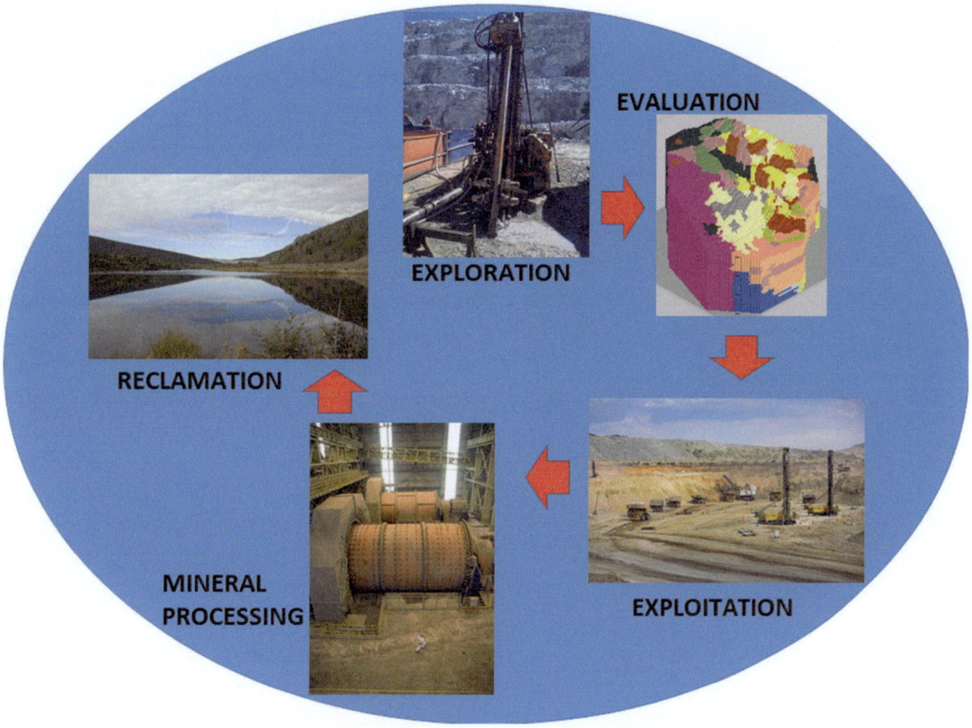

Fig. 4.1 The mining cycle

evaluation, and preproduction" (Bustillo 2018). Nevertheless, two principal stages are defined: reconnaissance exploration and detailed exploration. An aggregate project can be subdivided into two main phases, which involve the utilization of successively more discriminating (and more expensive) techniques, to successively smaller land areas. The final objective is to identify, develop, and produce an economic aggregate deposit.

During the phases of exploration, the so-called weakest point can exist and it must be taken into account prior to other aspects of exploration. The following are some examples of the weakest point concept: "(1) to explore an area extensively when a zoning restriction makes it impossible to open a new operation is a waste of money. Therefore, exploration should proceed only after the zoning situation, which is the weakest point, is better defined and understood, (2) conversely, to go to the expense of obtaining total assurance that all zoning or political obstacles are removed before the quality of the aggregates materials has been determined at least preliminarily also may be money badly spent and (3) to explore a piece of property extensively without obtaining at least an option to purchase the land is another case of not addressing a weak point in proper sequence" (NSSGA 2013). Each weak point should be addressed and solved to an acceptable probability. Many times, at the end of the exploration process, previous weak points must be readdressed and resolved to produce a lower level of risk. In other words, exploration needs judgment and experience and should proceed in a logical sequence.

An adequate exploration process depends on efficiently managing an exploration campaign. It must be based on the correct implementation and interpretation of geology and geophysics for generating the exploration targets. With the aim of discriminating and selecting the best terrains at the regional scale, having good baseline exploration data (geological and geophysical) and drilling data is essential.

- Reconnaissance Exploration

The first stage of aggregate exploration is reconnaissance exploration (also termed prospection, target identification, early reconnaissance, and others). As a rule, reconnaissance exploration classically includes two steps: regional appraisal and reconnaissance of the region. The objective of reconnaissance exploration is to identify an aggregate deposit that will be the target for further exploration; the number of tons in the deposit has a low level of confidence and is based solely

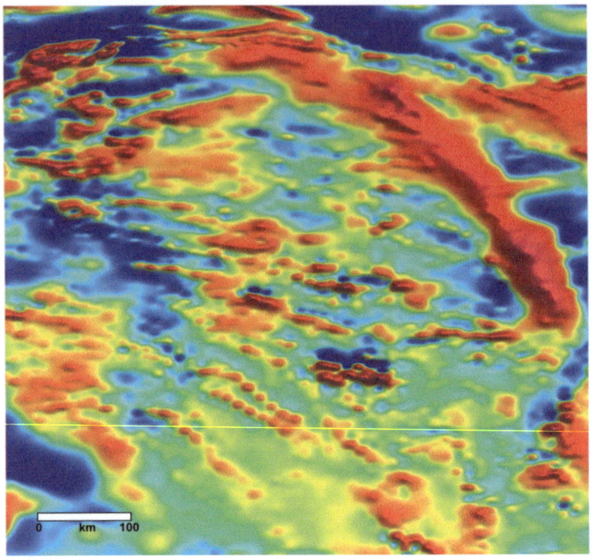

◘ **Fig. 4.2** Regional magnetic map in mainland Europe (image courtesy of Getech)

on the interpretation of geological and geophysical results. In this phase, the areas are studied through field inspections as well as geophysical surveys. The use of regional geophysics (◘ Fig. 4.2) is also an invaluable addition to surface regional geology. In areas of reasonable outcrops, the first phase can include checking the geology by driving along roads or from the air if ground access is very difficult.

In the first stage of reconnaissance exploration (regional appraisal), the following procedure is generally performed (1) review of all available information on the area such as geological mapping and geophysical surveys as well as other works (i.e., national geological surveys, reports of academic/research institutions—thesis, mineral deposit models and accessing company reports containing prior proprietary knowledge), (2) photogeological study of available air photographs, (3) analysis of available remote sensing data, (4) field inspection from the air or on the ground and (5) petrographic rock studies for determining the main rock types and mineral assemblages.

To understand and forecast where a mineral deposit can occur, scientists have developed mineral deposit models (► Box 4.1: Sand and Gravel Mineral Deposit Model). The use of mineral deposit models is an integral part of any quantitative mineral-resource assessment. For this reason, the sand and gravel models are necessary for resource-supply analyses of aggregates. A working definition of a model in the context of mineral deposits, "the overriding purpose being to communicate information that helps mankind find and evaluate mineral deposits, is the systematically arranged information describing the essential attributes (properties) of a class of mineral deposits" (Cox and Singer 1986).

Box 4.1—Sand and Gravel Mineral Deposit Model

Mineral deposit models are basically based on existing knowledge of the features of previously known mineral deposits. They are developed from the combined information of numerous very similar deposits. The grouping of deposits based on common features constitutes the basis for classification. Successful modeling of a sand and gravel deposit must address, among others, the following items: (a) deposit volume and geometry, (b) sediment particle size, (c) physical characteristics of the material, and (d) chemical composition and reactivity of the sediments.

Bliss (1993) *revealed an initial strategy to model a sand and gravel deposit. A broad range of sand and gravel deposits data were firstly used for modeling. Sand and gravel deposits mainly included those found in the following geological locations: stream beds, stream terraces, glacial terraces, eskers, kame terraces, outwash plains and beaches, both modern and raised. Data were either from California or the United Kingdom.*

The models produced were an initial attempt to characterize sand and gravel deposits. Bliss (1993) *generated models for three sand and gravel deposit variables: volume, area, and thickness.* ◘ Figure 4.3 *shows the general volume model of sand and gravel deposits,* ◘ Fig. 4.4 *shows the general area model of sand and gravel deposits, and* ◘ Fig. 4.5 *shows a general model of the thickness of sand and gravel deposits (in the three figures, the letter n means the number of deposits studied). These models "can be used to predict qualities expected in sand and gravel deposits either undiscovered or incompletely defined"* (Bliss 1993).

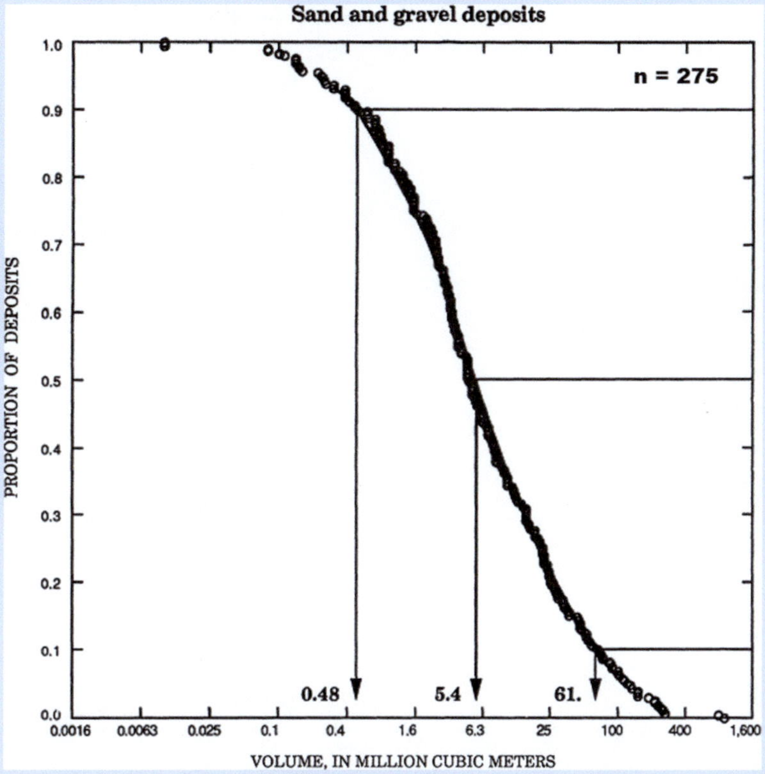

Fig. 4.3 General volume model of sand and gravel deposits

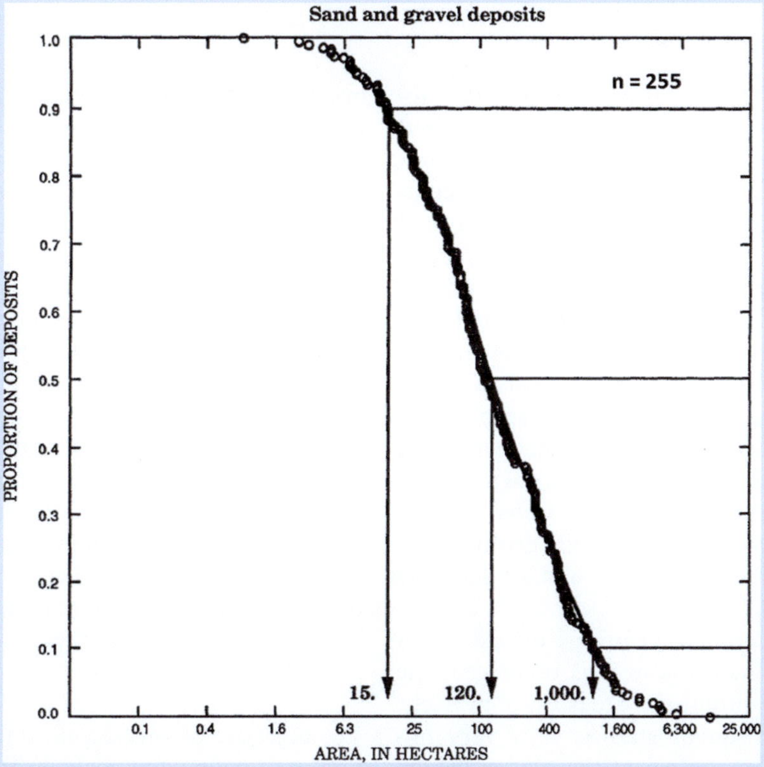

Fig. 4.4 General area model of sand and gravel deposits

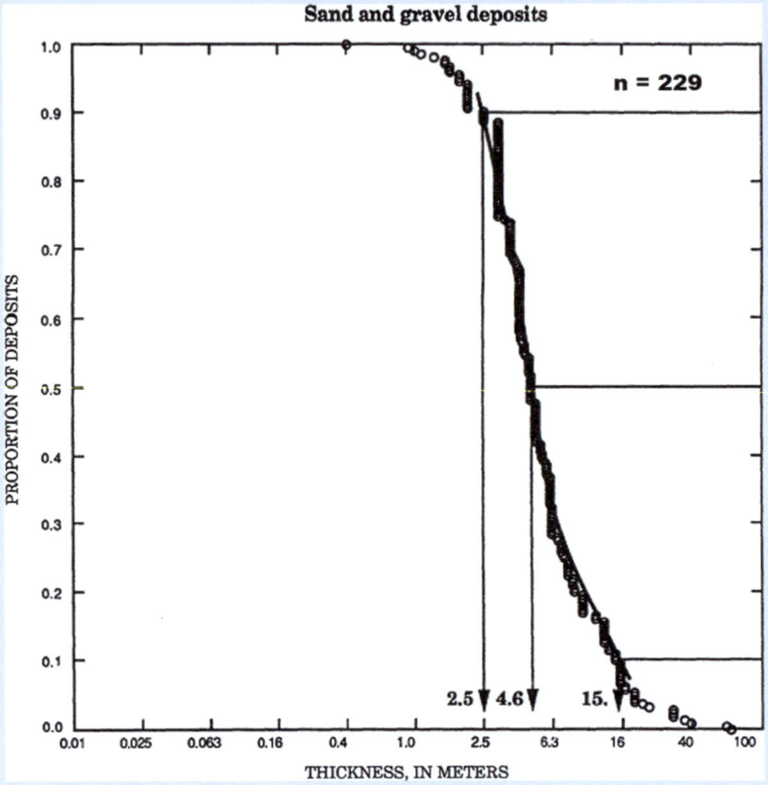

■ **Fig. 4.5** General model of the thickness of sand and gravel deposits

In the second phase of reconnaissance exploration (reconnaissance of the region), the main techniques used are the following: (1) geological mapping, (2) geophysical surveys (airborne or ground), (3) drilling, including geophysical borehole logging if possible, (4) field inspection of outcrops and sampling, and (5) petrographic and mineralogical studies. The main advantages of airborne techniques over ground-based surveys are the following: (1) cover large areas quickly and efficiently (cost-effective), (2) reduce costs and improve efficiency, helping development to take place on shorter time scales, (3) can cover regions inaccessible to ground-based exploration, (4) mitigate the environmental impacts, (5) reduce the area to be covered by further detailed exploration, and (6) provide geology at a regional scale for better interpretation and for selecting the target areas.

In the preliminary evaluation of the deposit, geologic and topographic maps as well as other geologic reports aid in locating promising areas or, conversely, rejecting zones in subsequent studies. It is very important to note that the geological knowledge gained through education and experience and the geological information of the regions to explore are very important, that is, aggregate exploration is based on a knowledge of the foundations of geological processes and the Earth's history.

These investigations can be followed by detailed studies using techniques such as satellite imagery, aerial photography, geophysical studies, and field reconnaissance studies of target areas. The main objective is to define the limits of the potential sources of aggregates more accurately. For this purpose, the field studies focus on natural exposures (i.e., stream cuts, cliffs, and other natural outcrops) and on artificial exposures (i.e., highway and railroad cuts and abandoned or active quarries). These are complemented with taking some samples collected utilizing hand-sampling methods and drilling for obtaining subsurface samples. If a reconnaissance study indicates a target area deserving further investigation, more detailed field studies should be developed at specific sites.

- Detailed Exploration

The objective of this second stage of aggregate exploration is to define and evaluate the deposit in detail. For this purpose, it is necessary to determine the geological setting, geometry, tonnage, and extent of the aggregate deposit identified. Detailed exploration of an identified source of aggregates can change according to the nature of the potential resource and the intended uses. In general, similar techniques to those applied in reconnaissance exploration are utilized but in a more comprehensive manner over a much smaller area. Detailed exploration is restricted to relatively small areas and is intensive and expensive, especially where drilling is carried out. In this sense, in increasing order of cost per km^2, the principal techniques utilized for aggregate exploration are remote sensing, geological mapping, geophysical surveys, and drilling.

Typical techniques for this phase are comprehensive geological mapping and sampling, detailed geophysical surveys, usually on the ground, drilling, logging, trenching, down-hole geophysical surveys, and recovery of samples in a systematic manner. Drilling starts with moderately wide spacing of holes. In areas of poor to moderate outcrop, a trench is invaluable. Backhoes can also be used to collect bulk samples, and truck-mounted power augers or drill rigs can be employed to collect deeper subsurface samples. In addition, geophysical studies may be used to determine the thickness of overburden (overlying material), to determine important changes within the deposit (i.e., changes from gravel to sand or shale to sandstone) and to provide continuity between drillholes. Once the samples have been collected, bagged and labeled, they must be sent to the laboratory for analysis. In this phase, economic and social factors (i.e., current land use, regulations of the areas, and property ownership) must be investigated in detail since it will be necessary to obtain permission from landowners to access sites for further investigations.

4.1.3 Methods of Exploration

- Photogeology

World War I was the onset of the development of aerial photography and photointerpretation. Photointerpretation is the study of the character of the ground surface utilizing aerial photographs (◘ Fig. 4.6). The two main pieces of photointerpretation equipment are field stereoscopes and mirror stereoscopes. The former are capable of being used into the field for quick checking whereas the latter are principally utilized in the office because they may observe complete 23 cm × 23 cm photographs without overlap. The advantages of the study of aerial photographs are (a) save time and provide the opportunity to observe a larger area, (b) have a more detailed ground surface than maps, (c) can be studied anytime and anywhere, and (d) the studies carried out on photographs are cheaper and easier than studies in the field.

In turn, photogeology is the name given to the utilization of aerial photographs for geological studies. The utilization of aerial photographs in geology includes the following: "(a) outline the structure and structural relationship in an area, (b) outline the stratigraphic succession, (c) preparation of a geological map, (d) measurements of stratigraphic sections, (e) measurements of dip and strike and thickness of formations, and (f) inferences about rock types present in the area" (Dirik 2006).

Aerial photograph interpretation is based on systematic observation and evaluation of key elements such as relief, tonal and textural variations, drainage patterns and texture, erosion forms, and vegetation. The in-

◘ **Fig. 4.6** Aerial photograph (image courtesy of Surdex Corporation)

Fig. 4.7 Drone for topographic study (image courtesy of Eduardo Revuelta)

itial interpretation made from the images will provide "(a) definition of areas of outcrop and areas of superficial cover, (b) preliminary geological interpretation based on topographic features, drainage patterns, colors and textures of rocks, soils and vegetation, trend lines of linear features, etc., (c) geological hypotheses for field checking, (d) selection of the best areas to test these hypotheses, and (d) familiarity with the topography and access routes to assist in logistic planning of the field programme: access roads and tracks, fording points for streams, potential helicopter landing sites, etc." (Marjoribanks 2010). In this sense, topographic studies using drones are common in mineral exploration (Fig. 4.7).

Aerial photograph interpretation constitutes a source of great value to identify characteristics associated with sand and gravel and rocks for crushed aggregates. Such studies can indicate the development of geological processes associated with erosion, transport, and sedimentation of sand and gravel, allowing the realization of terrace maps and analyzing gradual changes in the longitudinal profile of the valley as well as possible changes in the composition of alluvium materials.

Photogeology can reveal the presence of faults, folds, karstic processes in carbonate rocks, modifications of the land, etc., which can be very useful for the location of materials used to produce crushed rocks. For instance, in selecting broad target areas of limestone and dolomite, photogeological mapping is essential. Thus, "the presence of sinkholes and karsts in a generally flat and subdued topography indicates the presence of limestone; limestone terrains also encourage subsurface drainage due to the presence of solution cavities; these features are very easily recognized in aerial photographs" (Roonwal 2018). The ideal photographic scale for this type of work is 1:50,000.

In addition, aerial photographs are the basis for the reconnaissance of geology in areas where there are no geological maps or their precision is very low. In such areas, formational contacts and structural relationships are easily established, and much routine mapping can be done directly on the aerial photographs. As much as possible, it is necessary to start with small-scale aerial photographs and later to photographs at a scale of 1:20,000 or even larger.

Fig. 4.8 Satellite image (Landsat) used in mineral exploration (pixel = 14.5 m × 14.5 m)

While satellite imaging (see next heading) covers very large areas of the Earth's surface, aerial photography provides the topographic and geological basis for exploration work of smaller areas of tens of square kilometers or less. Most exploration studies involve multimage interpretation. Satellite images are examined first at scales up to 1:250,000 for the regional view. If high-altitude photography is available (1:120,000 to 1:60,000), then this can also be examined. Hereafter, target areas are selected from the above studies and, together with literature reviews, will allow the selection of areas for detailed photogeological studies.

- Remote Sensing

Remote sensing is "a process to acquire, prepare, and decipher information of spectral and spatio-temporal nature on objects, phenomenon or areas under investigation without being in direct physical contact" (Gandhi and Sarkar 2016), with satellites being the main observation platforms. The components of the remote sensing system are linked through electromagnetic energy, that is, the transfer of information is carried out utilizing electromagnetic radiation. In addition to the capability for covering large and inaccessible areas quickly, the data acquired enable their utilization as a powerful tool for mineral exploration. Remote sensing works best in arid areas whereas the options are reduced in areas with dense vegetation.

Remote sensing has gained importance due to its applications in various mineral exploration scenarios (Fig. 4.8). It can help exploration geologists to distinguish different rock types, investigate morphological features, and analyze structural features. Data from other sources combined with remote sensing provide complementary measurements. Remote sensing can be used as a device for finding potential aggregate (mainly sand and gravel) exploration targets. The quality of remotely detected information has expanded as the level of technology has increased. Satellite imagery supplies a very cost-effective means of exploration, which offers the unique chance for investigating most important geological characteristics of remote areas of the Earth's surface without the requirement of accessing the area on the ground.

In the context of geological mapping, two types of electromagnetic methods are distinguished "(1) passive optical methods: use the sunlight as the source and measure the reflectance of the Earth's surface in the visible and infrared spectral bands, and (2) active microwave radar methods: use a microwave source onboard the satellite and measure the back-scatter from the Earth" (Bustillo 2018). The most famous satellite used in geological studies is Landsat (▶ Box 4.2: Landsat Program).

4

Box 4.2—Landsat Program

The Landsat Program is a series of Earth-observing satellite missions jointly managed by NASA and the U.S. Geological Survey. In the mid-1960s, stimulated by U.S. successes in planetary exploration using unmanned remote-sensing satellites, the Department of the Interior, NASA, and the Department of Agriculture embarked on an ambitious effort to develop and launch the first civilian Earth observation satellite. The goal was achieved on July 23, 1972, with the launch of Landsat 1, originally named "ERTS" for Earth Resources Technology Satellite. The Landsat Program, a joint effort of the U.S. Geological Survey (USGS) and the National Aeronautics and Space Administration (NASA), was established to routinely gather land imagery from space. NASA developed remote-sensing instruments and spacecraft, and then launched and validated the performance of the instruments and satellites. The USGS assumes ownership and operation of the satellites, in addition to managing all ground reception, data archiving, product generation, and distribution.

*Since 1972, Landsat satellites have continuously acquired space-based images of the Earth's land surface (*Fig. 4.9*). Thus, the use of satellite imagery is now a standard technique in mineral exploration, including aggregate deposits. Landsat 7, 8 and 9 orbit the Earth at 705 km altitude. They complete an orbit every 99 min, approximately 14 full orbits each day, crossing each point on Earth once every 16 days.*

For most geologists and other Earth scientists, multispectral imagery is synonymous with NASA's Landsat series. The primary sensor onboard Landsats 1, 2, and 3 was the Multispectral Scanner (MSS), with an image resolution of approximately 80 m in four spectral bands ranging from the visible green to the near-infrared (IR) wavelengths. In July 1982, the launch of Landsat 4 saw the inclusion of the Thematic Mapper (TM) sensor with a 30-m resolution and 7 spectral bands. Although the Landsat series was designed initially to provide multispectral imagery for the study of renewable and nonrenewable resources, geologists immediately recognized the geological potential of the Landsat images, and bands 5 and 7 in the TM were chosen specifically for their geological applicability. The approximate scene size of TM images is 170 km north–south by 183 km east–west and the radiance measured by the Landsat sensor is a measure of the integration of soil, rock, and vegetation characteristics. Landsat 7 carries the Enhanced Thematic Mapper Plus (ETM+), with 30-m visible, near-IR, and SWIR bands, a 60-m thermal band, and a 15-m panchromatic band.

Landsat 9 (a rebuild of its predecessor Landsat 8) is the latest satellite (2021) in this series with a design life of 5 years. It orbits in a near-polar, sun-synchronous manner. Landsat 9 incorporates Operational Land Imager 2 (OLI-2) and Thermal Infrared Sensor 2 (TIRS-2), which are two instruments that largely replicate the instruments on Landsat 8.

Fig. 4.9 Artist concept of Landsat 8 (image courtesy of NASA's Goddard space flight center)

The information obtained from remote sensors can be classified as (1) information based on the physical properties of surface materials (i.e., reflectance and thermal emissivity) and (2) spatial information of surface configuration (i.e., landforms, geological structures, and materials distribution on the surface). Both types of information are very important to study potential sites for sand and gravel deposits as well as materials that can be used for crushed rocks. Furthermore, the ability to detect certain mineral species can serve as a basis for quality studies of potential crushing aggregate deposits (i.e., carbonate rocks) or to carry out lithological discriminations (i.e., separation of mafic rocks from felsic rocks based on silica content).

- Geological Mapping

A well-prepared geological map of the area to be investigated is an absolute necessity in a prospecting expedition. If a geological map is not available, a geologist can firstly have to map the region. A geological map can be defined as a visual record of geological characteristics such as rock units and associated structures on a plane surface (◘ Fig. 4.10). Once an area of interest is selected, geological field mapping is performed. This process includes the identification of all geological aspects of the area with the purpose of preparing a detailed geological study, which further produces an accurate geological map. The geological map aims to show the geological setup of an area in terms of the occurrence and distribution of rock types, their mutual relationships, attitudes, structure, association with topography, and drainage. In general, geological mapping should be a continuous process to develop geological modeling and draw inferences to favorable areas for aggregate search. Moreover, mapping the geology is a way of systematizing field observations, which is significant to plan the drilling and other sampling operations.

Geological mapping has been used extensively for mineral exploration for more than 100 years, and the tools for recording, compiling, and synthesizing data are evolving rapidly. Beyond the use of traditional paper-based mapping methods, recent technological advances include global positioning systems, pen tablet computers, palm computers, and laser ranging devices that all support direct (paperless) field-based digital geological mapping. Portable digital-electronic instrumentation allows field geologists rapid access to digital databases that include geological maps and photographic and remote sensing imagery with automatic registration and scale independence. Geographic infor-

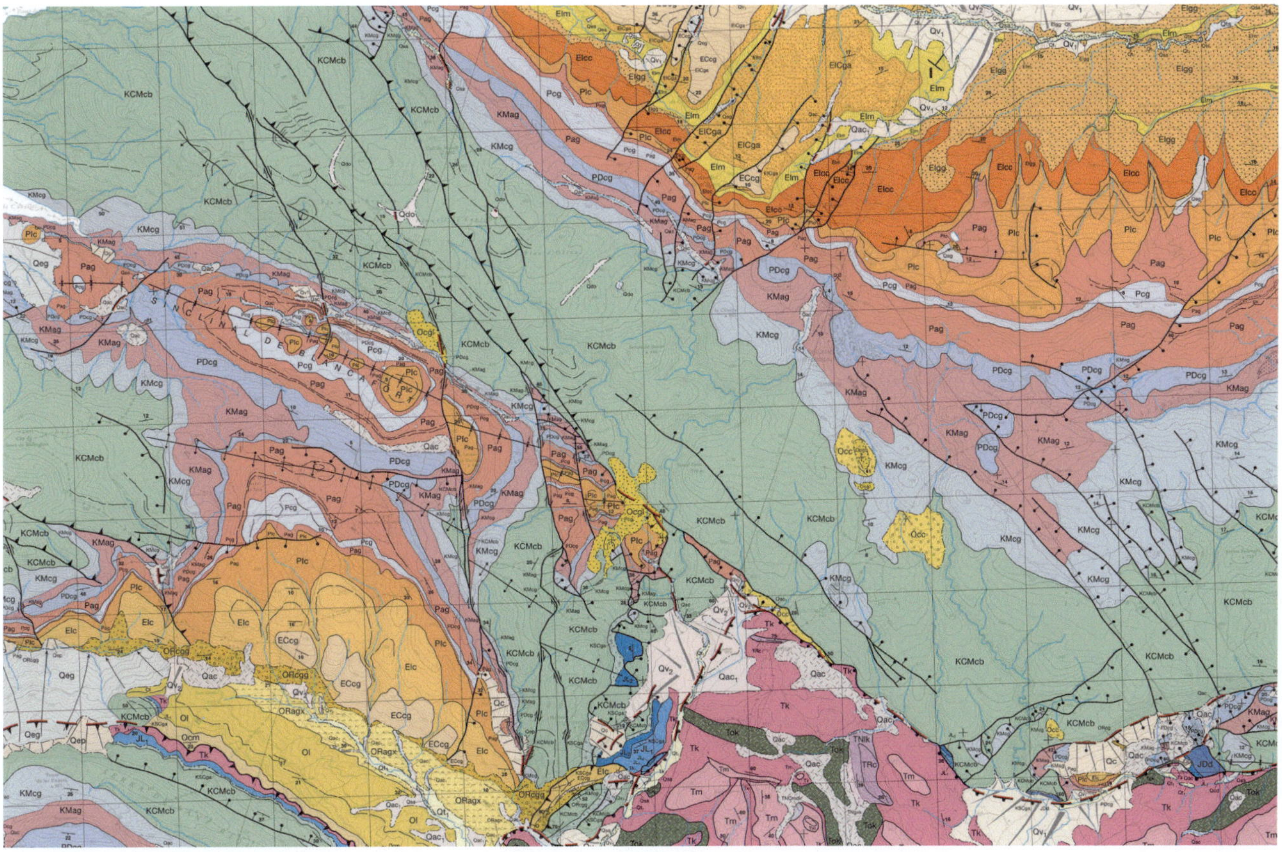

◘ **Fig. 4.10** Part of a 1:25,000 geological map (IGME, Spain)

mation systems (GIS) revolutionized exploration practices. GIS is essential in customizing and integrating a broad range of mineral exploration data consisting of information on drillholes with summary stratigraphic logs, rock sample and drill hole sample geochemistry, mineral occurrences, magnetic and gravity images, digital geology, current and historic exploration details, and more (▫ Fig. 4.11).

- Geophysical Methods

Geophysical methods are very useful in the exploration of aggregate deposits, especially when drilling can be prohibitively expensive, landowners can prohibit invasive exploration activities, or it can be difficult to get equipment into zones of the property of interest. In any case, geophysical surveys are of great value when performed early in the exploration program and they must be conducted before to intensive drilling to locate auger or drill holes and decrease the number of drill holes needed. They may provide continuity between sampling sites and three-dimensional information that can be very important for accurate aggregate resource/reserve estimations. Geophysical methods can be faster and less expensive than additional drilling and trenching. Nevertheless, geophysical methods must be refined by other techniques (i.e., drilling). This is because a specific group of field measurements cannot be linked to a unique model of subsurface conditions. Additional information such as driller log, geologic mapping, and/or the use of multiple geophysical methods is required to translate geophysical properties into meaningful geologic properties (i.e., the thickness of sand or gravel). Geophysical surveys can be taken into account as a complement to, but not a replacement for, classical methods of exploration.

Each technique is applicable depending on the specific site conditions. These methods are used where measurable physical contrasts exist between a rock (for crushed rock aggregate) or sand and gravel (for sand and gravel aggregate) and their surrounding materials (i.e., soil types). For instance, seismic methods are utilized if there are measurable differences in the sonic velocities of different materials and resistivity techniques are utilized where measurable differences exist in electrical conductivity. ▫ Table 4.1 shows selected physical properties and their related field or phenomenon as well as use of the generalized geophysical method associated with each one. ▫ Table 4.2 shows different values for some properties of some common geologic materials are shown in ▫ Table 4.2.

The primary goal of these techniques in aggregate exploration is "to identify the areal extent and thick-

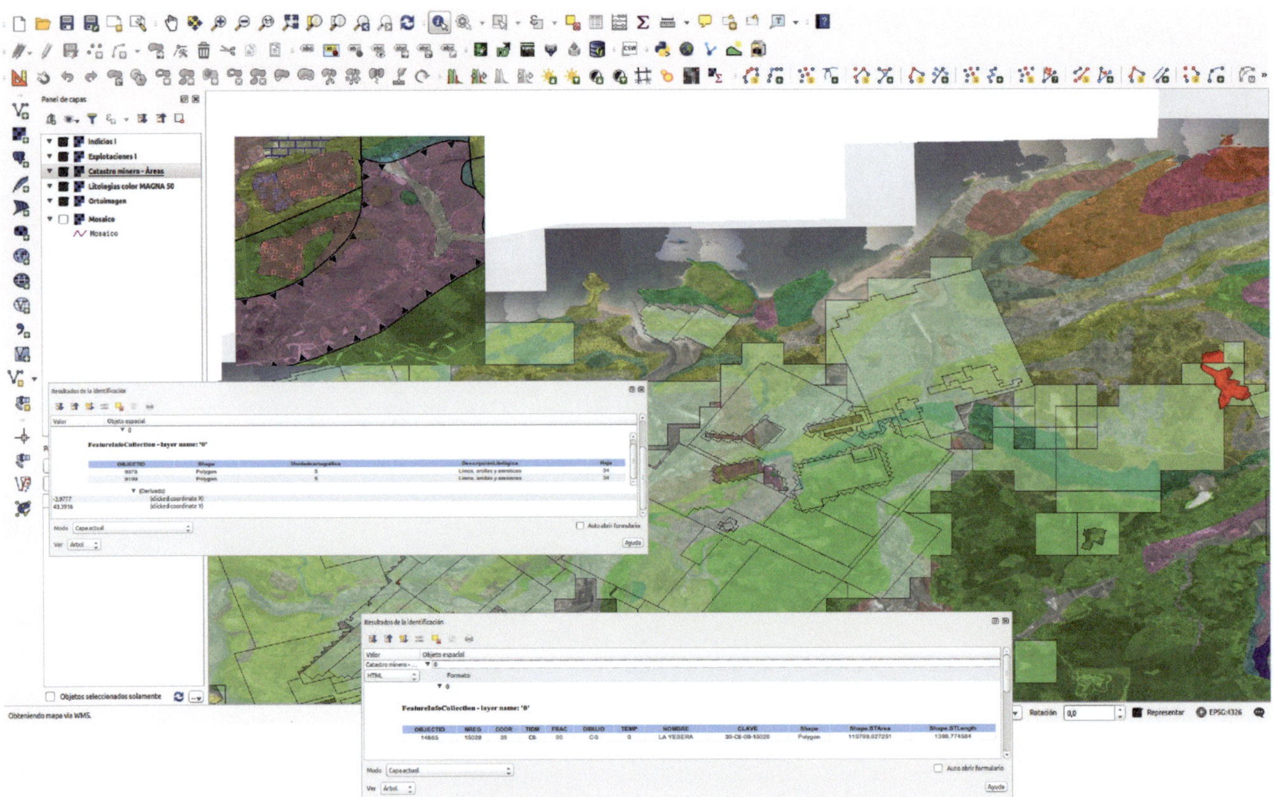

▫ **Fig. 4.11** GIS image released with QGIS (a free and open source Geographic Information System) including different types of information such as geology or mining data (image courtesy of Miguel Ángel Sanz)

Table 4.1 Physical properties and associated geophysical methods (Lucius et al. 2006)

Physical property	Field or phenomenon	Geophysical method
Density	Gravity	Gravity
Magnetic susceptibility	Magnetism	Geomagnetic
Electrical conductivity	Electricity, low-frequency electromagnetism	Electrical, electromagnetic
Dielectric permittivity	High-frequency electromagnetism	Electromagnetic
Elasticity	Elastic waves	Seismic
Radioactivity	Alpha, beta particles and gamma rays	Radioactivity

Table 4.2 Typical approximate limits and ranges of selected properties of some common geologic material (Lucius et al. 2006)

Material	Conductivity (mS/m)	Resistivity (ohm-m)	Relative dielectric permittivity[1]	P-wave velocity (m/s)
Soil	< 10 to > 50	< 20 to > 100	Low to high	100–500
Dry sand and gravel	< 5	> 200	Low	450–950
Saturated sand and gravel	> 20	< 50	Medium	1250–1850
Clay	10–200	5–100	Medium to high	900–2700
Sandstone	1–20	50–1000	Low	1800–4000
Shale	20–200	5–50	Low	2500–4300
Limestone, dolomite	< 1	> 1000	Low	2100–6100
Igneous rock	< 1	> 1000	Low	4500–6000
Metamorphic rock	< 1	> 1000	Low	3000–7000

mS/m, miilisiemens/meter; ohm-m, ohm-meter; m/s meter per second; <, less than; >, greater than
[1]For RDP, low is generally less than 10, medium is 10–36, and high is greater than 36

ness of the deposit, thickness of overburden, depth to the water table, location of zones of intense faulting or fracturing, critical geologic contacts and location and correlation of geologic features" (Lucius et al. 2006). In some cases, geophysical studies may establish deposit quality concerning the quantity of fine material and porosity. The results of geophysical surveys are presented in the form of maps or cross sections showing the spatial variations in geophysical properties.

Bedrock sources of crushed stone can be discovered using seismic, radar, gravity, and electrical geophysical surveys. None of these techniques are particularly useful but they can be applied to characterize specific properties of rocks. For instance, seismic studies can identify weathering zones and the water table and radar are useful in cavernous limestone areas. For its part, electrical methods (ERT) are essential in sand and gravel exploration although other techniques (i.e., electromagnetism) are also very useful. According to Lucius et al. (2006), "five well-established and well-tested geophysical methods can be considered for characterizing sand and gravel deposits: seismic refraction and reflection, resistivity, ground penetrating radar, time-domain electromagnetism, and frequency-domain electromagnetism".

- Seismic

Seismic methods were developed early in the twentieth century but it was not until the 1980s and the advent of affordable digital signal processors and personal computers that seismic investigations could be conducted for environmental and engineering purposes. Seismic methods are based on measurements of the time interval between the initiation of a seismic (elastic) wave and its arrival at detectors to obtain an image of the subsurface. Elastic waves originate from an external force being applied to a medium (i.e., sledgehammer). The seismic wave is detected by a geophone (Fig. 4.12) on land. Seismic wave detectors are termed geophones and convert ground vibrations into electrical signals. As a general rule, geophones are distributed in straight lines across the study area.

Fig. 4.12 Geophones for receiving a seismic signal (image courtesy of International Geophysical Technology)

Since seismic waves, which propagate with different velocities in different rock types, are reflected and refracted at bedding or fault contacts, reflection and refraction are the most used seismic techniques. Refraction methods use simpler equipment (Fig. 4.13) and require less processing than reflection methods. For both methods, interpretation is in terms of subsurface boundaries between regions or layers with different seismic velocities that result from changes in elastic properties and/or density. Seismic techniques are employed for determining site geology, stratigraphy and rock quality, providing detailed description about subsurface layers and rock geomechanical characteristics. One obvious application of these techniques is to search for offshore placers. "Subsea resources of bulk minerals such as sands and gravels can be evaluated with seismic techniques" (Milson 2006). The results of seismic surveys can be very useful to characterize sand and gravel deposits, especially for determining the depth to bedrock. This type of geophysical survey can be an alternative method for exploration in areas that are too conductive for adequate ground-penetrating radar utilization and/or where electromagnetic methods do not provide effective resolution.

- Ground Penetrating Radar

Ground penetrating radar (GPR) is a relatively new technique that was commercially developed in the 1970s for use in shallow earth exploration. It uses radar pulses to image the subsurface. This nondestructive technique utilizes the electromagnetic signals in the microwave band (UHF/VHF frequencies) of the radio spectrum for detecting the reflected signals from subsurface structures. Because of this, the ground must have reasonably low conductivity. The electromagnetic wave is radiated into the ground by a transmitting antenna. When the electromagnetic wave encounters a change in electromagnetic properties, some of the waves are reflected toward the receiving antenna, where the amplitude of the electric field and arrival time are detected. The antennas are mounted on a sled or in a cart and moved by hand or with an all-terrain or other small vehicle along a straight path on the ground. The size of the antennas is related to their frequency, with larger antennas having lower frequencies (Fig. 4.14).

The most common cause of reflections is a change in water content, whose variations are related to features such as sedimentary structures, lithology, grain size, density and saturation. Ground penetrating radar is utilized in different media, including rock, soil, ice, or freshwater and may detect objects, changes in material, and voids and cracks. Figure 4.15 is an example of a GPR survey.

Ground penetrating radar cannot be generally utilized for characterizing sand and gravel deposits due to its limitations. However, there may be occasions where its higher resolution compared to other geophysical methods may be useful at certain locations. Thus, "in some cases this geophysical method can be utilized to map the vertical and horizontal limits of sand and

Fig. 4.13 Seismic refraction survey (image courtesy of International Geophysical Technology)

gravel for determining the volume of the deposit; however, GPR cannot be used to differentiate sand from gravel unless there is additional geologic information" (Lucius et al. 2006).

- Gravity

In gravity surveying, subsurface geology is investigated based on variations in the Earth's gravitational field arising from differences in the density distribution of local masses as well as deep geologic structures. The mean value of gravity at the Earth's surface is around 9.8 m/s^2 and variations in gravity caused by density variations in the subsurface are of the order of 100 μm/s^2. The measurement of relative values of gravity, which is the difference in gravity between locations, is the standard procedure in gravity surveying. The instrument used in gravity surveying is called a gravimeter (Fig. 4.16), an extremely sensitive weighing machine.

Gravity surveying is rarely used in aggregate exploration because it is relatively slow to execute and expensive. However, the gravity method can find application as a follow-up technique utilized on a target defined by another, more cost-effective method or where electrical, electromagnetic, and seismic methods fail owing to electrical or acoustic noise or cultural features such as buried or overhead utilities. The gravity technique can demonstrate lateral changes in the density and therefore of the thickness of overburden and aggregate material.

- Electromagnetism

Electromagnetic induction (EM) utilizes the principle of induction to measure the electrical conductivity of the subsurface (Fig. 4.17). EM surveys are based on variations in electric conductivity in the rock mass either "by making use of naturally occurring electromagnetic fields in the crust or more commonly by applying an external electromagnetic field, the primary field, and inducing a current to flow in conductive rocks below" (Marjoribanks 2010). Electromagnetism induction surveys are conducted from the air (AEM), on the ground surface, and in drillholes (downhole electromagnetics—DHEM) (Fig. 4.18).

There are two fundamental categories of electromagnetic measurements: time domain and frequency domain measurements. In time-domain electromagnetics (TDEM), also called transient electromagnetics or TEM, the change in the primary magnetic field is produced by either abruptly turning off or turning on a steady current. It is a powerful transmitter and

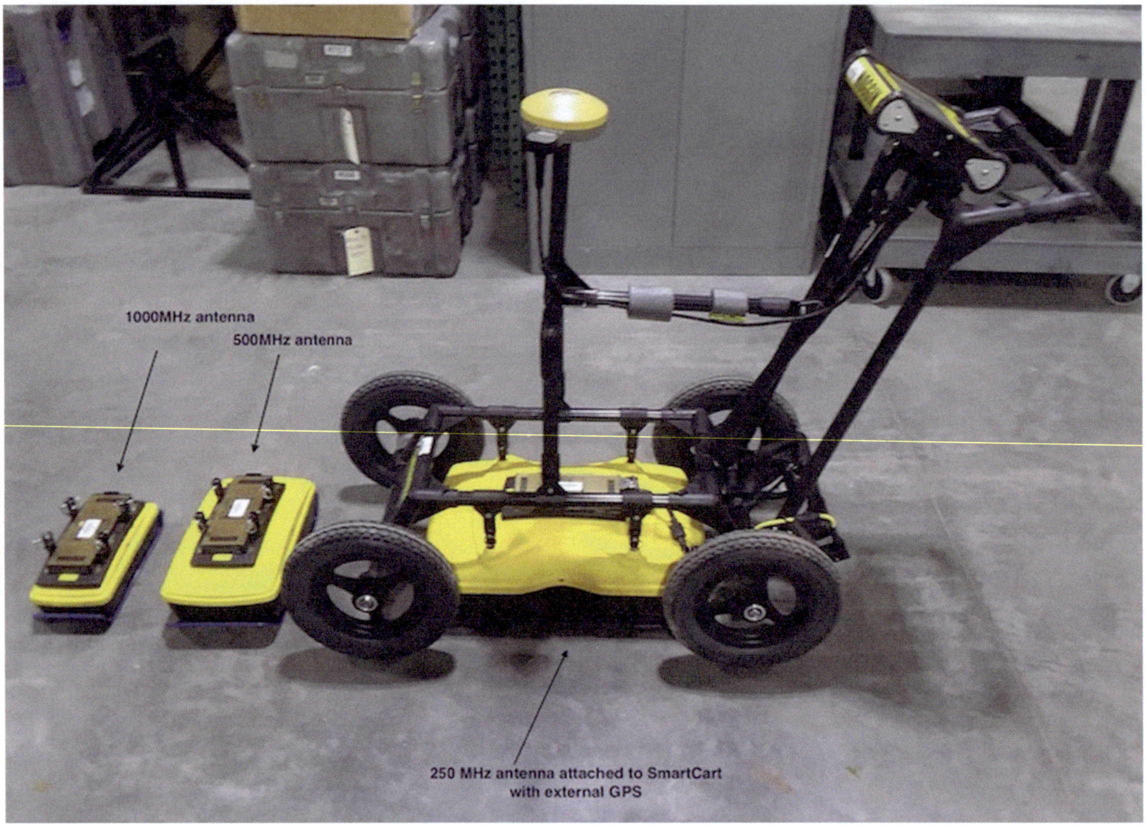

Fig. 4.14 Different sizes of antennas (image courtesy of IRIS/PASSCAL)

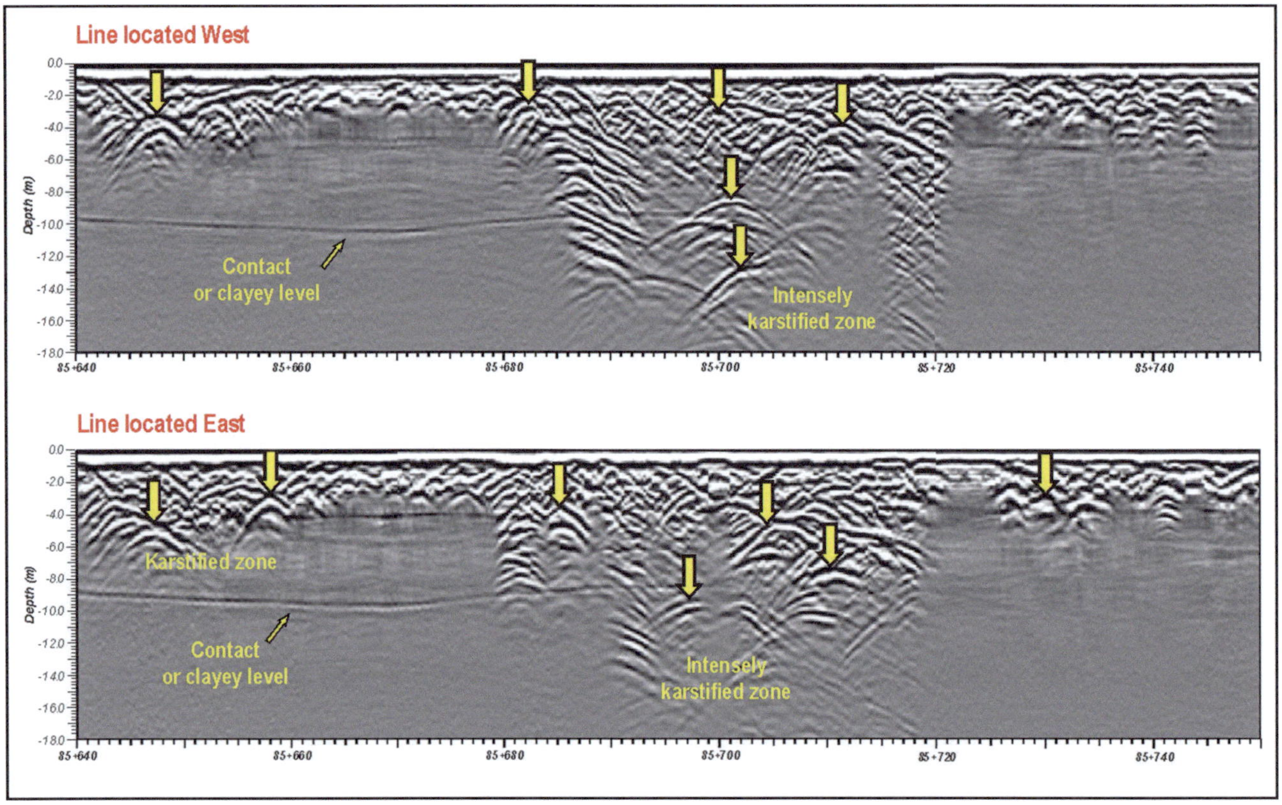

Fig. 4.15 Results of a GPR survey (illustration courtesy of International Geophysical Technology)

Fig. 4.16 Gravimeter (image courtesy of International Geophysical Technology)

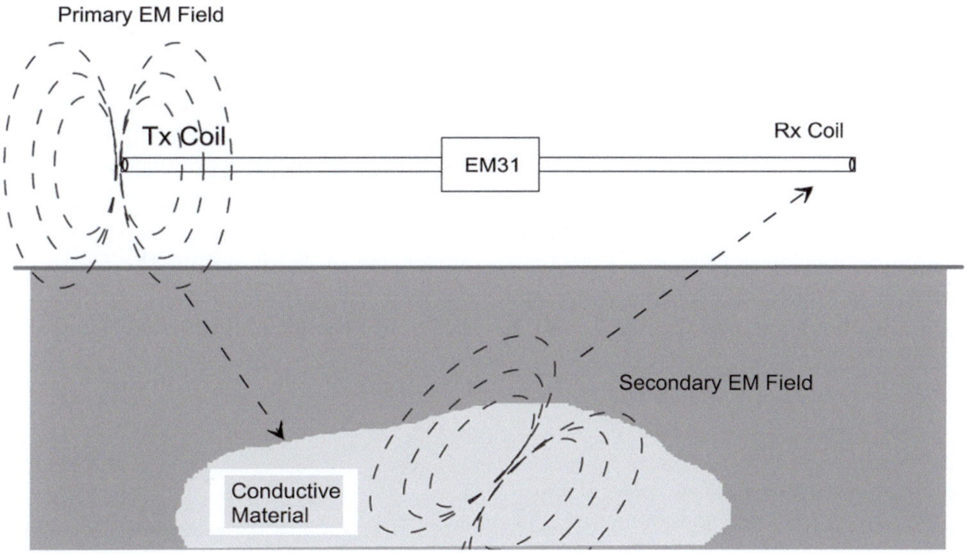

Fig. 4.17 Electromagnetic induction

receiver, and the system can sound the depth, thickness, and conductivity of layers down to 300 m below the surface. It was developed in the 1920s and 1930s in Europe. TDEM is used in ground-water and environmental studies, mineral exploration (including sand and gravel deposits), and geologic mapping. TDEM equipment is quite portable and easy to operate.

In the frequency domain electromagnetic (FDEM), a continuous sinusoidal current is used. It is an extremely sensitive and accurate tool, capable of detecting variations in conductivity of as little as 3%. Similar to TDEM, frequency domain electromagnetic methods were developed in the 1920s and 1930s in Europe. Ground conductivity meters can be useful in the exploration stage for

Fig. 4.18 DHEM survey (illustration courtesy of International Geophysical Technology)

sand and gravel deposits. The frequency domain electromagnetic equipment is portable, relatively inexpensive, and easy to operate. Data are easy to interpret.

- Electrical

Electrical methods use direct currents or low-frequency alternating currents for investigating subsurface electrical properties. The most measured property is electrical conductivity (Siemens per meter, S/m) or its reciprocal, resistivity (Ohm). Electrical surveys are used "(a) to locate mineral deposits at shallow depth, (b) to map geological structures to determine the depth of overburden and bedrock, and (c) to locate groundwater table in hydrogeological surveying" (Bustillo 2018). There are different methods of electrical surveying: resistivity (▶ Box 4.3: Electrical Resistivity Tomography), induced polarization (IP), and self-potential (SP). Some make use of fields within the Earth (SP) while others require the introduction of artificially generated currents into the ground (resistivity and IP).

Box 4.3—Electrical Resistivity Tomography (ERT)

Electrical resistivity tomography (ERT) is a quickly developing geophysical imaging technique broadly utilized for visualizing subsurface geological structure, groundwater and lithological variations. ERT introduces a direct electrical current into the ground surface, which measures the remaining voltage at several distances from the injection point. A tomogram (subsurface diagram) of resistivities can be produced because different materials have different resistances to electrical current. The electrical properties of silt, gravel and bedrock (i.e., granite) are distinct and easily definable because fine-grained/clay-rich regions are relatively conductive whereas coarse-grained materials are relatively resistive. Concerning sand and gravel deposits, "with supportive geologic or borehole information, the resistivity method, in a favorable environment, can detect the contact between aggregate and bedrock, the presence of clay layers, and the water table" (Lucius et al. 2006).

*This geophysical technique was developed early in the 1900s and was widely utilized due to its simplicity and good results. Electric current is introduced using electrodes and the electric potential is measured among other electrodes. With the utilization of computer modeling software, it is possible to determine a representative resistivity structure of the earth. Varying the position of the electrode array and the separation of the electrodes (*Fig. 4.19*), lateral and vertical variations in electrical properties can be mapped* *and used to produce data pseudosections (◘ Fig. 4.20), volumes (◘ Fig. 4.21), and maps. The depth of investigation is largely a function of electrode separation; the depth of investigation is increased by increasing the separation of the electrodes, with larger electrode separations providing information to a greater depth.*

◘ **Fig. 4.19** Electrode arrangement (image courtesy of International Geophysical Technology)

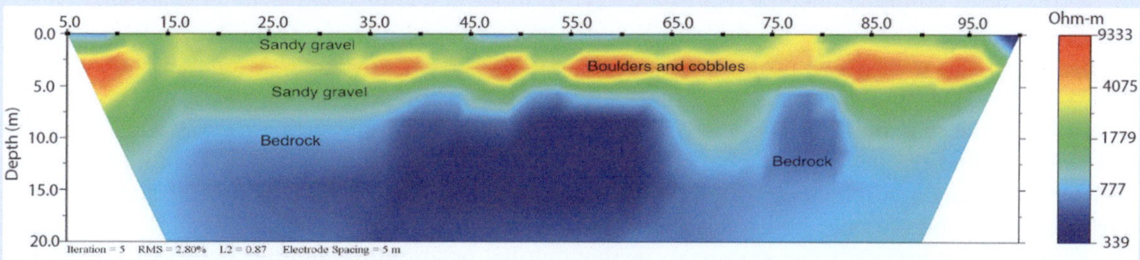

◘ **Fig. 4.20** Pseudosection obtained using ERT (Laxton and Coates 2015)

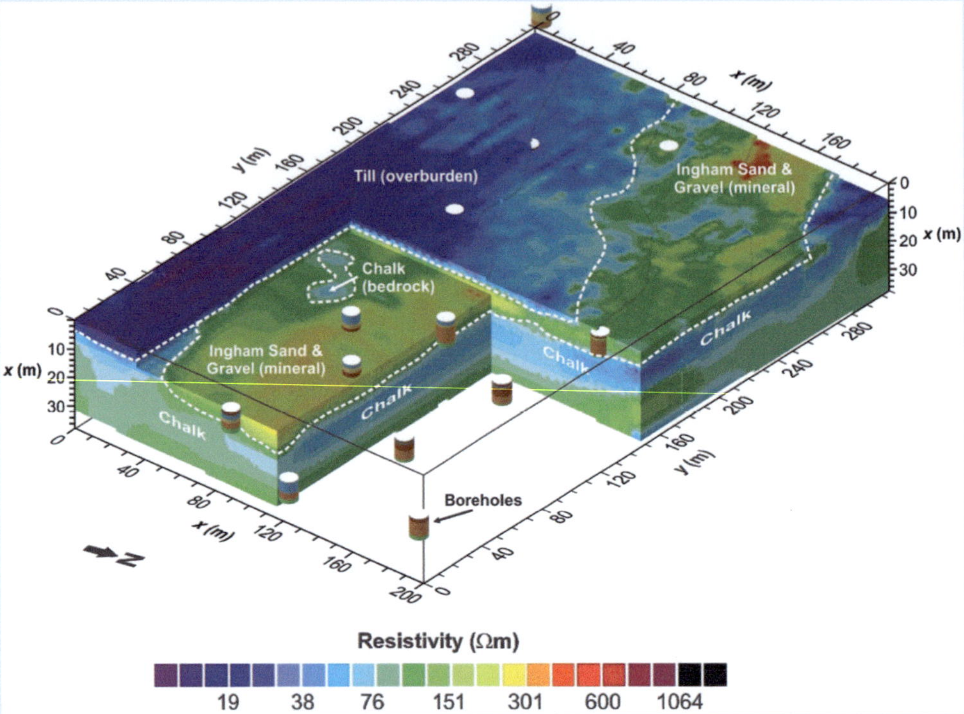

Fig. 4.21 3D ERT model (Chambers et al. 2011)

Many times, induced polarization is conducted simultaneously with ERT. As the electrical current is introduced into the ground, a charge is retained. This charge, as a function of time, later decays. The rate of decay changes according to the electrical properties of the ground materials and is very useful in differentiating subsurface material types and boundaries.

- Borehole Geophysical Logging

Because exploration drillholes are cored completely, the industry has been slow to recognize the importance of geophysical borehole logging (Fig. 4.22) (borehole geophysics is the science of recording and analyzing measurements of physical properties made in boreholes). However, the high cost of drilling makes it essential to obtain the maximum possible information from each drillhole, and the geophysical characteristics of the rocks surrounding a borehole can often be better guides for discovering the presence of aggregate source materials. Downhole geophysical surveys increase the radius and depth of investigation and provide greater resolution of buried aggregate deposits. Probes that measure different properties are lowered into the borehole to collect continuous or point data that are graphically displayed as a geophysical log (Fig. 4.23). Borehole geophysical measurements are made by sensors (receivers/detectors) that are housed inside the probe; a series of continuous measurements are made with the data transmitted to the surface. “Multiple logs typically are collected to take advantage of their synergistic nature because much more can be learned by the analysis of a suite of logs as a group than by the analysis of the same logs individually” (Bustillo 2018).

- Airborne Geophysical Surveys

Airborne geophysical surveys are undertaken employing low flying helicopters (Fig. 4.24) or light aircraft that fly in a grid pattern. The main advantages of airborne surveying relative to ground surveying are the greater speed of data acquisition and the completeness of the survey coverage. After their introduction in the 1950s, airborne geophysical surveys became used as a first step in geophysical exploration. They provide the quickest, and often the most cost-effective, ways of obtaining geological information about large areas. Airborne work has been completely transformed using global positioning satellites.

- Drilling

After the selection of the suitable location and obtaining the necessary permit, exploration drilling facilitates the definition of the aggregate deposit geometry through the study (i.e., microscopy study of rock types and chemical analysis) of the samples obtained.

Fig. 4.22 Acoustic and optical televiewer probes (image courtesy of Enviroscan)

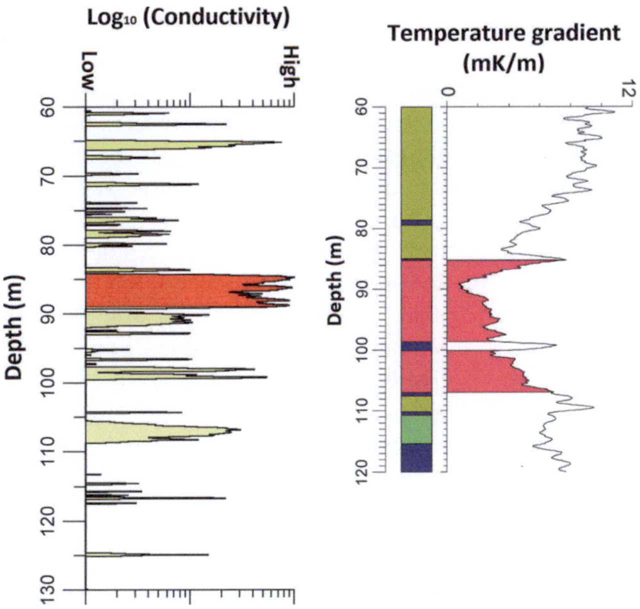

Fig. 4.23 Borehole geophysical logs

The rock types are defined using the study of the samples and portions of them are chemically analyzed with the aim of further characterization of rock types. Drilling is the precise method for geological assessment of the aggregate deposit since it is the most accurate method for predicting the reserves of the deposit correctly. However, drilling is carried out only when preliminary studies such as geological mapping and geophysical surveys indicate a high probability of finding an aggregate deposit in a particular location. This is because drilling is the most expensive method of mining exploration. The density of drilling will be set up based on the wanted level of geological confidence and project economics.

There are many drilling techniques although the four main types used in aggregate exploration are reverse circulation (RC) drilling, rotary drilling using tricone roller bits, auger drilling, and diamond core (DC) drilling. Each drilling method has its own characteristics that affect the quality of the collected samples. From a sampling viewpoint, there are two types of drilling methods in mineral exploration: drilling methods that originate rock chips or cuttings (particles of soil or rock resulting from the cutting action of drilling) and those that generate rock cores (Fig. 4.25). Some drilling methods use special fluid (termed drilling mud) for many purposes: removing the rock chips, cooling the drill bit, reducing the friction, and stabilizing the borehole. The fluid is generally a bentonite-water mixture.

Reverse circulation drilling, rotary drilling using tricone, and auger drilling produce rock chips or cuttings while diamond core drilling produces rock cores. Obviously, diamond core drilling is slower and more

Fig. 4.24 Electromagnetic airborne survey (image courtesy of Geotech)

Fig. 4.25 Rock chips and core samples (image courtesy of Atlas Copco)

expensive than the other three methods. However, the characteristics of the samples obtained are radically different because rock core samples are the only type of samples that allows a complete study of the rock types present in the deposit. There are many aspects to select an adequate drilling method: target, host-rock, water presence, sample needed, and access. A three-key-factor selection process can be established: "the time needed, the cost of getting the job done, and confidence in the quality of the samples brought to the surface" (Gustaffson 2010). Finally, it is important to note that another method of drilling, termed sonic drilling (▶ Box 4.4: Sonic Drilling), is particularly useful in areas with thick glacial sediments that can be utilized as sand and gravel aggregates.

Box 4.4—Sonic Drilling

Sonic drilling technology was first applied over 40 years ago in Canada. In mineral exploration, it is typically used to provide continuous core samples of softer or even harder rock formations of aggregate deposits. Sonic drilling has become a utilized drilling and sampling method in sand and gravel exploration. This is because some alluvial formations are often very difficult to advance utilizing conventional drilling methods. Sonic drilling performs very well in difficult conditions such as those containing cobbles and boulders.

Sonic drilling (rotasonic, rotosonic, sonicore, vibratory, and resonance drilling are other terms used) is a unique technology that generates vibrational frequencies, usually between 50 and 150 Hz (cycles per second), transferring the vibrations down the drill pipe to its tungsten carbide bit while rotating the pipe at the same time. This frequency range falls within the lower range of sound vibrations that the human ear is capable of hearing. The term sonic drill has been applied to this class of rotary-vibratory drilling machines. Instead of using a diamond bit rotating at the end of a drill rod, the sonic drill head sends high-frequency vibrations throughout the length of the entire drill pipe and onto the bit (◘ Fig. 4.26).

In sonic drilling, the head contains the mechanism necessary for rotary motion, as well as an oscillator, which causes a high-frequency force to be superimposed on the drill string. The drill bit physically vibrates up and down in addition to being pushed down and rotated. These three combined forces allow drilling to proceed quickly through most geological formations, including most types of rocks. The operator can vary the frequency and drill bit weight to match the material that is going through, ensuring that the best penetration rate and most accurate sampling are obtained.

The sonic drilling method can produce almost completely undisturbed core samples (and without the contamination caused by drilling fluids) from both solid and unconsolidated materials with a high percentage of core recovery rates. It can obtain 100% core recovery (the total linear amount of physical core sample extracted over the total linear advance in a borehole, expressed as a percentage), even in glacial till, clay, sand and gravel and bedrock. Sonic drilling can collect samples up to 300 mm in diameter and can drill down to 250 m in a vertical or angled hole. The environmental impact of sonic drilling is typically less than that of other drilling methods. Thus, having a small footprint and lacking the need to introduce fluid into the hole, this is an ideal drilling method where contamination is potentially a problem.

◘ **Fig. 4.26** A sonic drill bit (image courtesy of Sonic Drilling Ltd.)

- Reverse Circulation Drilling

The reverse circulation (RC) drilling technique (◘ Fig. 4.27) has been used in searching for mineral deposits since the early 1970s in Australia, and is actually a quick and economical technique for exploration drilling. It can be used in unconsolidated sediments or for drilling rocks. In reverse circulation drilling, the cuttings from the hole are transported to the surface where they are collected in plastic bags. This method offers higher productivity than diamond core drilling, but, as commented previously, the quality of the samples is lower. The RC sample content ranges from dust to 25 mm chips.

The RC method uses dual wall drill rods that include an outer drill rod with an inner tube situated inside the drill rod (◘ Fig. 4.28). The inner tube affords a continuous pathway with sealed characteristics for the drill cuttings to be translated to the surface. Thus, circulation is upward inside the drill stem and downward outside it. High-pressure air is the common way to define the drill flushing medium. Water can also be injected to reduce dust and to assist in transporting cuttings to the surface. At the surface, the cuttings are derivated to a cyclone for collection and bagging; the material can be further split in the drill location for assay, geological, and mineralogical studies. A good cyclone will gather more than 99% of the sample, with a sample interval normally 1 or 2 m from the drillhole. As one sample has been collected, another is being drilled and incorporated into the cyclone. The sample from a 1 m drillhole is approximately 50 kg.

- Rotary Drilling

Fig. 4.27 Reverse circulation drilling machine (image courtesy of Atlas Copco)

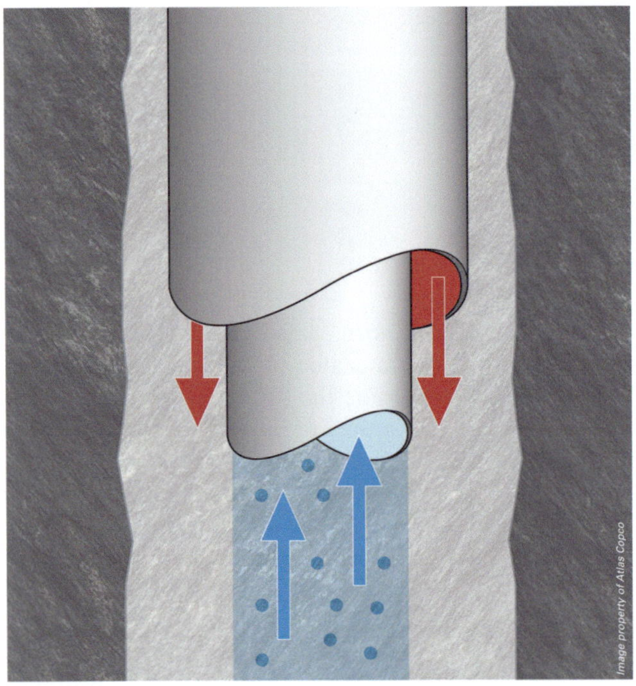

Fig. 4.28 RC system (illustration courtesy of Atlas Copco)

Rotary drilling with tricone bits is a noncoring method used for drilling through soft to medium-hard rocks such as limestone, chalk, or mudstone. Rotary drilling uses a different type of rotary bits although the most typical rotary bit is probably the tricone or roller rock bit (Fig. 4.29), which is made with tungsten carbide insets. As the drill string is rotated, the bit cones roll along the bottom of the borehole and the rock chips are flushed to the surface by the drilling fluid for examination. It requires minimal air volume, and downhole costs are low. For this reason, it is a very economical method of drilling.

- Auger Drilling

Auger drilling is probably the simplest form of drilling. It is a cheap and fast method to evaluate a site's viability. It involves the manual rotation of a helical screw into the ground. When the auger (a helical rotary tool for drilling a cylindrical hole) is deep enough, the earth is lifted by the blades of the screw to the surface. Auger drilling (Fig. 4.30) is a technique employed for identifying overburden thickness or sand and gravel thickness. This method works adequately in areas where quality and gradation of deposits are not very

Fig. 4.29 Tricone bit (image courtesy of Atlas Copco)

Fig. 4.30 Auger drilling (image courtesy of JS Drilling)

important. The rate of penetration depends on the type of formation being drilled. Auger drilling can reach depths of approximately 20 m. Great augers can be utilized for obtaining samples of sand and gravel deposits. The only limitation of this type of sampling is that undisturbed samples are never obtained. Auger drilling can give misleading results "when boulders are present, when deep weathering has occurred along planes of weakness of the rock, and when deep layers of partially weathered rock are present" (NSSGA 2013). This drilling method cannot be utilized to penetrate hard or consolidated rocks.

- Diamond Core Drilling

Diamond core drilling (Fig. 4.31) is the most expensive drilling method, but it is the most utilized technique for obtaining information about subsurface materials. In diamond core drilling, a cylinder of solid rock, the core, is extracted from depth. It is usually 27–85 mm in diameter (Fig. 4.32), but larger diameters (up to 200 mm) are most useful but much more costly. Due to the common hardness of the rocks and the time involved in translating the core from depth, the penetration in diamond core drilling is much slower than in other drilling methods. However, if core recovery is good, it has the benefit of carrying undamaged rock to the surface. Therefore, diamond drilling is considered to offer the best quality of sample. This technique collects rock samples for rock-type confirmation, weathering thickness, mineralogy, geologic structure, fracture patterns, rock-quality testing, recovery percentage, and other important quality features.

The diamond drill bit comprizes a cutting head using diamonds as the cutting medium. Water is used in diamond core drilling as a lubricant fluid and to remove crushed and ground rock fragments from the bit surface. In conventional rock coring, the entire drill stem and core barrel must be removed after each core run, which is a time-consuming process. For this reason, the wireline system is now a standard practice. The wireline system is designed to recover the rock core without removing the drill stem from the borehole. To obtain a core after the core bit is in place, the core barrel assembly is forced down the inside of the drill pipe using drilling mud pressure. When the core barrel assembly reaches the lower end of the drill stem, a locking device holds the barrel in place. As the overshot is lowered over the upper end of the assembly, the locking devices are released, permitting the removal of the entire assembly. The inner barrel containing the rock core is rapidly brought to the surface, leaving the outer core barrel and drill rods sill in position within the borehole. Figure 4.33 shows the head assembly and the attached overshot.

- Logging

Considering the high costs of drilling, a maximum of information must be extracted. Intense geological logging of core and drill cuttings is a common practice. Logging sand and gravel samples and rock cores provide geological features for defining geometry and estimating the volume of aggregates as well as the main characteristics of the rock and sand and gravel samples. For

Fig. 4.31 Diamond core drilling (image courtesy of Atlas Copco)

Fig. 4.32 Different core diameter sizes

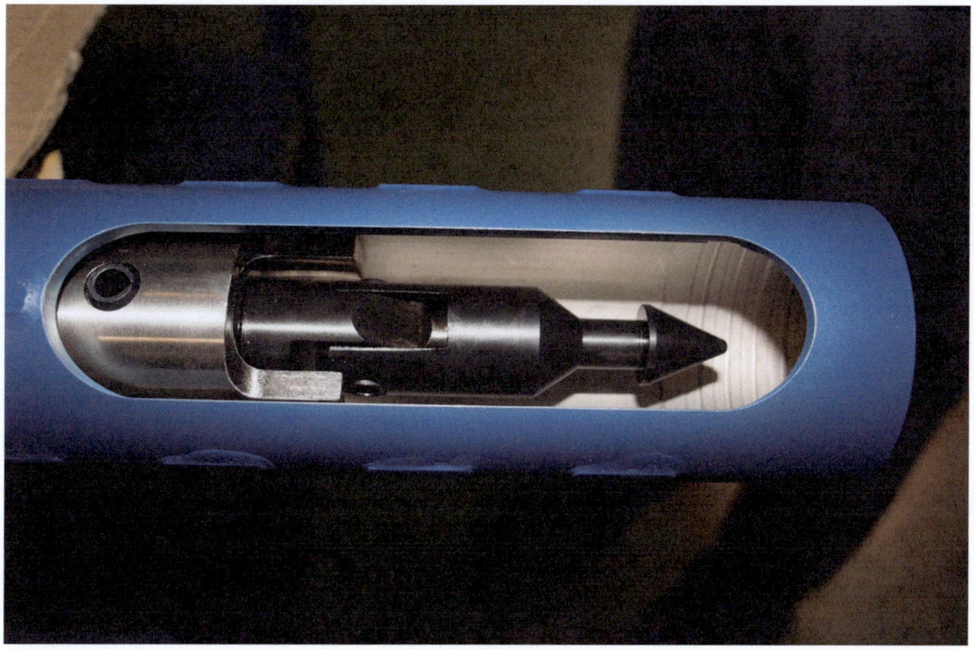

Fig. 4.33 Head assembly inside the overshot (image courtesy of TECSO)

instance, rock cores with "strong decomposition, weak strengths, considerable fracturing, low hardness or low rock quality designation (RQD) tend to make poor aggregates" (NSSGA 2013). A geological or engineering description of samples taken is always needed prior to the selection of samples for engineering or other analyses. In this description process, it is common to use standard terminology, standard symbols, and standard colors for minimizing subjectivity. A clear, accurate, and standardized logging procedure is very important to promote the uniformity of data through what is a long data-gathering period. Core logging in most cases should be carried out using a core that has been split along the main axis (Fig. 4.34). The use of split cores has two advantages (1) fresh surfaces can be observed and (2) half of the split core may be sent for testing whereas the other half can be retained for further works.

Although a great number of different logging methods are utilized in the industry, graphical scale logging (Fig. 4.35) includes several mapping columns along with extra columns for recording digital data, sketches, verbal comments, etc. The most important feature of all such logs is that they assemble many different types of geological observations on one form linked by a single down page scale. The graphical scale logging form is separated into columns. They will be referred to in numbered order from left to right, for example, column 1 (hole depth), column 2 (core recovery), column 3 (sample no.), column 4 (assay results—it will be necessary to devote several columns to insert all assay results), and so on.

- Exploration of Marine Aggregates

The exploration of marine sand and gravel is carried out using a combination of regional-scale paleogeographical reconstruction of the continental shelf and geophysical methods. Geophysical methods provide a high-resolution shallow seismic profiles and multibeam bathymetric data from specific seabed sites. Thus, "geological surveys for marine aggregates rely on a combination of multibeam bathymetry, high resolution shallow seismic profile data combined with grab and vibrocore (borehole) sampling" (Bellamy 2018). It is important to note that prospecting sites are chosen not only for their geological potential but also for other specific factors such as water depth, distance from the aggregate market, and environmental constraints (i.e., the existence of nature conservation areas).

Investigations of marine aggregates are assisted by the utilization of multibeam echo sounders (they resolve surface features with a resolution of under 0.3 m) to obtain complete coverage of the seafloor bathymetry. Consequently, the resulting seabed maps are more detailed than those obtained using single-beam mapping. Multibeam data provide a sound baseline for dredging depth and resource usage, highlighting the most important features such as sandbanks and sandwaves (Fig. 4.36) that constitute high-quality marine aggregate resources.

Nevertheless, this method does not reveal what is beneath the seabed so high-resolution shallow seismic profiles are needed. The seismic data provide an acoustically generated cross-section of the shallow geology (Fig. 4.37), "with a resolution of less than 0.5 m and a penetration for aggregate investigations of typically less

Fig. 4.34 Half and quarter core as samples (image courtesy of Pedro Rodríguez)

than 20 m below the seabed" (Bellamy 2018). Shallow seismic data interpretation requires complete knowledge of the geological context of a given area as well as the understanding of the methods of data acquisition. This allows obtaining the "true" geological features that can be differentiated from artifacts produced by the seismic system.

The data interpretation is assisted employing classical vibrocore sampling, with sample locations adequately selected from a previous seismic interpretation. Information about particle size is routinely undertaken on several subsamples per core. Core samples are also used to obtain sediment color, composition, lithology, and grading, providing the thickness of sediments overlying bedrock. "Using both seismic and core sample data, deposit margins and sediment spot thicknesses above bedrock can be plotted" (Bellamy 2018). If sufficient density of spot thicknesses is recorded, equal sediment thickness (isopachytes) can be plotted, producing charts that show the extent and thickness of aggregate deposits (Fig. 4.38). All these datasets combined can assist not only to assess geological potential for sand and gravel deposits but also in drawing up dredging extraction plans, monitoring resource utilization, and to establish mineral reserve volumes.

Bellamy (2018) states the main factors for determining when a company will decide to define a survey to search for marine aggregates "(a) geological context: key in determining the likelihood of aggregates occurring in a given site, (b) water depth: this must be neither too shallow nor too deep for dredgers to access the site, (c) steaming distances: the transport distance from the aggregate deposit location to the aggregate market is fundamental in whether an area of seabed might be commercially viable. This factor is considered in relation to the cargo capacity of the dredger and the value of the aggregate, with a larger capacity ship able to travel further on a longer cycle time than a smaller ship, (d) environmental and spatial constraints: Marine Protected Areas (MPAs), designated archaeological sites and other developments such as wind farms and cable routes can preclude marine aggregate extraction. These factors can reduce the area available for dredging or can cause uncertainty over free access to a given area, with the possibility of seasonal restrictions on dredging and/or exclusion zones being required, limiting the area of resource available, and (e) proximity to other aggregate license areas: if a new prospecting site is adjacent to existing license areas, there might be a greater chance of locating additional resources, if geological boundaries extend over license areas".

4.1.4 Sampling

Sampling is among the most fundamental activities in mining operations. The sampling of aggregate deposits is carried out for many reasons and at different steps in their evaluation (systematic sampling of the aggregate source is basic to determine the resources and reserves of the deposit) and exploitation. There are many definitions of sampling, but the concept is quite elementary. Examples include "to obtain samples that will show the nature and condition of the materials which they represent" (ASTM D75) and "to obtain samples that are representative of the products being evaluated" (NSSGA 2013).

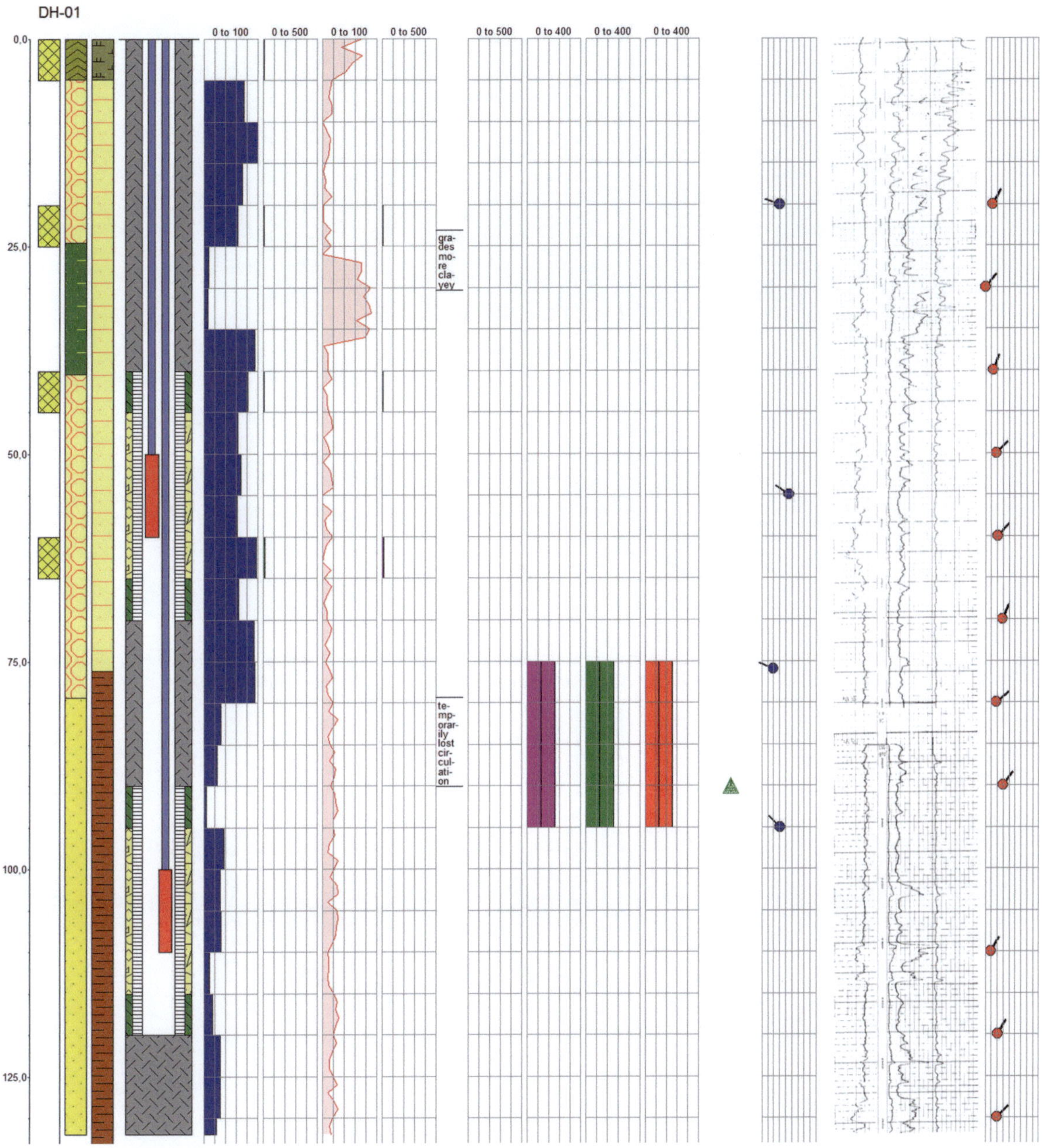

Fig. 4.35 Graphical scale log using Rockworks software

It is necessary to resort to what is known as sample and population. From a practical viewpoint, it is not possible to collect all items in a population unless the population itself is very small. In aggregate production, a population is defined as "any quantity of material produced at a given aggregate production plant with the same feed material and without change in plant setup or operating parameters" (NSSGA (2013) and a sample is "any subset of a population" (NSSGA 2013) or "an individual or group of individuals drawn from a large or infinite population" (Smith and Collis 2001). Although samples can represent any quantity of product, the aim of their utilization is for obtaining an estimate of the average and variability of a property for the population. The population estimate becomes better as the sample size increases.

Both definitions of sampling cited above are very similar as it is essential that the sample be representative. It is the key to a successful process of sampling. Geological interpretation and modeling cannot be carried out correctly if the initial data are wrong. Obtaining a representative sample is perhaps the most difficult control operation to carry out satisfactorily. To be representative, every portion of the population should

4

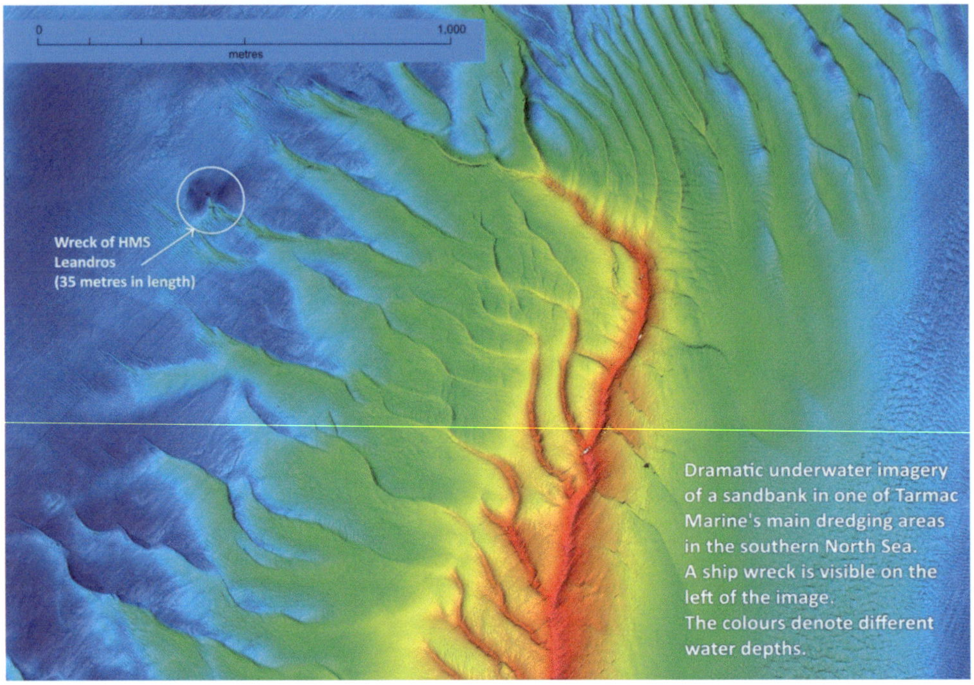

Fig. 4.36 Multibeam echo sounder image of the seabed in a licensed dredging area in the outer Thames Estuary showing the northern margin of the Kentish Knock sandbank with its associated sandwaves and a shipwreck to the west (Bellamy 2018)

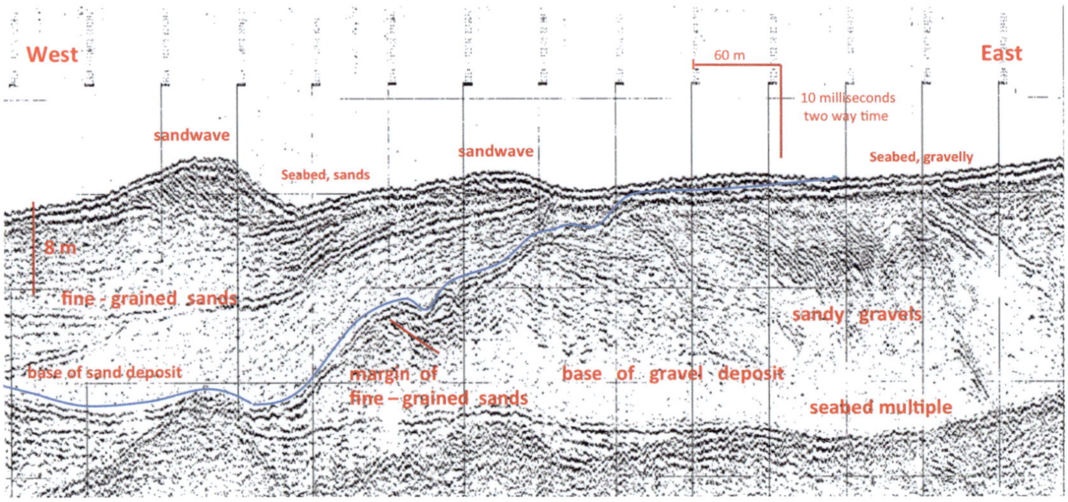

Fig. 4.37 Example of an interpreted high-resolution shallow seismic profile from sediments in the former Solent River floodplain 12 km east of the Isle of Wight. Vertical exaggeration × 9 (Bellamy 2018)

have an equal opportunity to be included in the sample. A small sample or a small number of samples can be highly representative of a population if observations show little variation and there has been no bias in collecting. On the contrary, when the variation is large, more and/or larger samples will be needed. The sampling process can be subdivided into three basic steps: (1) sample collection, that is, the collection of representative samples, (2) sample preparation, that is, sample reduction in size and amount, and (3) sample analysis.

Regarding the concept of bias, it is undoubtedly one of the most important items to avoid in sampling. The main problem of bias is that part of the population has more or less of a chance of to be sampled than the rest of the population. When this does not occur, the sample is said to be random. Consequently, if samples are not randomly obtained, they always have bias present. Unbiased sampling is performed exclusively when each component of the material being sampled has an equal possibility of being included in the sample.

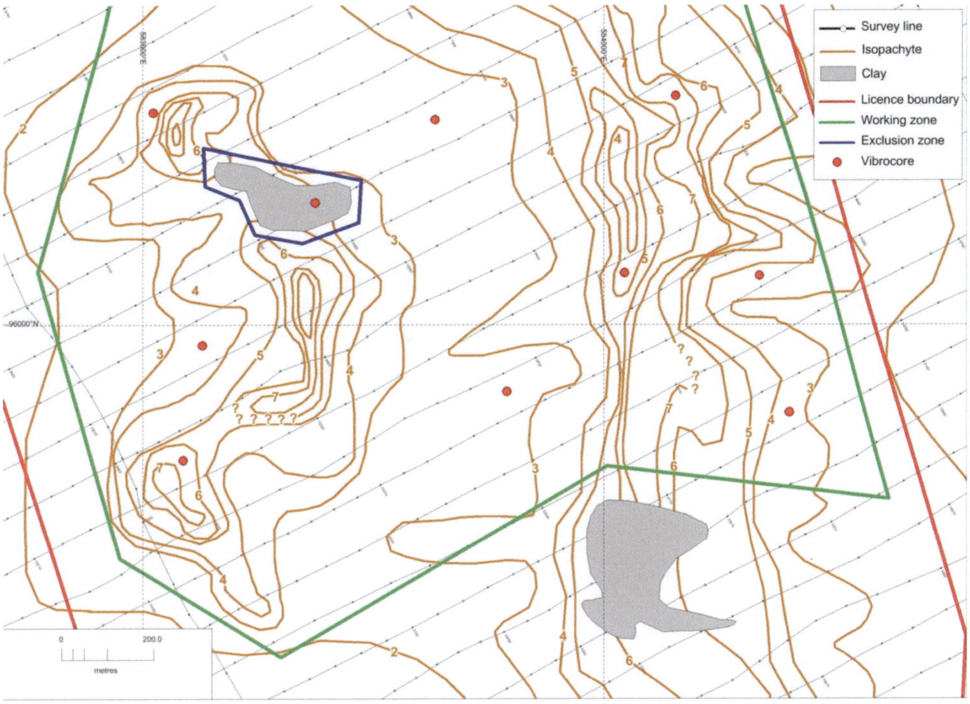

Fig. 4.38 Example of a seismic survey track chart with hypothetical sample sites, isopachytes and sediment boundaries (Bellamy 2018)

Different standards have been developed over the years by European (EN) and American (ASTM) organizations to serve as an aid in helping to define appropriate sampling methods as well as other topics such as minimum sample size or methodology to reduce laboratory sample size. The EN 932-1 standard, entitled Methods for Sampling, includes the following items: (a) bulk representative sample to be produced from sample increments in accordance with a sampling plan, (b) prescribed methods for sampling from various settings, (c) guidelines on minimum sample size and (d) guidelines on sample reduction on site. The EN 932-2 European standard, entitled Methods for Reducing Laboratory Samples, includes prescribed methods for obtaining laboratory test portions by sieving, splitting, and crushing bulk samples. For its part, the corresponding American standards are ASTM D75 (Sampling Aggregates) and ASTM C702 (Reducing Field Samples of Aggregates to Testing Size). Importantly, these procedures aim to obtain dependably representative samples for testing and assessment, variously for materials from a new source (initial type testing) or for routine production purposes (factory production control).

Sampling introduces sources of variation and error. The main items affecting the size of sampling errors are the sample amount, the consequences of splitting a sample for reducing the amount of sample, and the influence of the size of particles. The errors can be random and/or systematic. The former cancel each other out whereas the latter accumulates to create gross errors. Both types of errors are generated during the collection, preparation, and analysis of the samples. They must be recognized and accounted for. Examples of errors are improper choice of sampling method, inadequate collection of samples, improper location of samples, insufficient number of samples, inadequate sample preparation, and contamination of samples.

The type and number of samples collected depend on a range of factors that include "(1) the type of aggregate deposit and the distribution and grain size of the valuable phase, (2) the stage of the evaluation procedure, (3) whether direct accessibility exists to the material, (4) the ease of collection, which is related to the nature and condition of the materials, and (5) the cost of collection, funds available, and the value of the aggregate deposit" (Annels 1991). The cost of intense sampling of a low-value deposit (e.g., aggregates for construction) can be prohibitive.

Sampling patterns and methods as well as sample reduction schemes must be checked continuously and appropriately, including a complete analysis of the material with special attention to the particle size distribution and the nature of the particles in each size class. A good sample design must "(a) result in a truly representative sample, (b) lead to only a small sampling error, c) be cost-effective, (d) be one that controls systematic bias, and (e) be one such that the results of the sample study can be applied for the population with a reasonable degree of confidence" (Sarma 2009).

Several types of samples are taken based on the type of deposit to be sampled and the objective of sampling.

In the geological exploration of aggregate deposits, the main types of samples are grab samples, shot samples, pit samples, and core drilling samples. Regarding the locations where samples of aggregates can be taken for control of the production, the EN 932-1 standard establishes the following sampling procedures: "(a) sampling from stationary conveyor belts, (b) sampling at belt and chute discharge points, (c) sampling of pneumatically transported aggregates, (d) sampling of packed aggregates, (e) sampling of material in bucket conveyors, bucket loaders or grabs, (f) sampling for a silo, (g) sampling from stockpiles and (h) sampling from railway wagons, lorries and ships". This European standard notes that "aggregates should preferably be sampled from a stationary conveyor belt or from the stream of material". The ASTM D75 American standard more or less establishes the same sampling procedures.

4

For the sample to be representative, it is necessary to obtain a sufficient amount of sample. In this sense, EN 932-1 recommends the following minimum mass of a bulk sample:

$$M = 6 \times \sqrt{D} \times \rho_b$$

where

M is the mass of the sample, in kilograms,
D is the maximum grain size, in millimeters,
ρ_b is the loose bulk density, in megagrams per cubic meter.

Finally, the amount of sample must be reduced to an appropriate size for individual test requirements utilizing standard procedures. EN 932-2 European standard and ASTM C702 American standard take into account principally two types of sample reduction: using a riffle box and quartering (▶ Box 4.5: Sample Reduction Methods).

Box 4.5—Sample Reduction Methods

Splitting is a cost-effective method of reducing sample volume. Both methods of splitting (using a riffle box and by quartering) split the sample into two representative subsamples. In the riffle division method, the sample shall be mixed well and placed with a uniform thickness into the riffle tray and divided into almost two equal parts. Either of the two divided samples shall be selected at random each time the sample is reduced. The apparatus (◘ Fig. 4.39*) consists of a series of chutes that run in alternating directions and produce to randomly divided equal-sized fractions. The sample splitter normally has an even number of equal width chutes, which discharge alternately to each side of the splitter. It has to include two receptacles to hold the two halves of the sample following splitting. One of the fractions can be split again reintroducing the portion of the sample in one of the receptacles into the splitter; the process can be reiterated until a sample of the desired size is generated. The portion of material collected in the receptacle not used for further reductions can be reserved for other tests, when needed.*

The other typical method for sample reduction is coning and quartering (◘ Fig. 4.40*). In this method, the sample, after being placed in the working surface, shall be well mixed up and scooped into a cone-shaped pile where there will be neither loss of aggregate fraction nor the accidental addition of other materials. When the cone is formed, it must be flattened by pressing the cone top. The diameter should be approximately four to eight times the thickness. Then it is cut into quarters by two lines, which intersect at right angles at the center of the cone. The bulk of the sample is reduced by rejecting any two diagonally opposite quarters and reserving the other two, including all fine material. As in the case of the riffle division method, it is possible to consecutively mix and quarter the rest of the material until the aggregate portion is reduced to the desired size.*

◘ **Fig. 4.39** Riffle box for sampling (image courtesy of SAMCA)

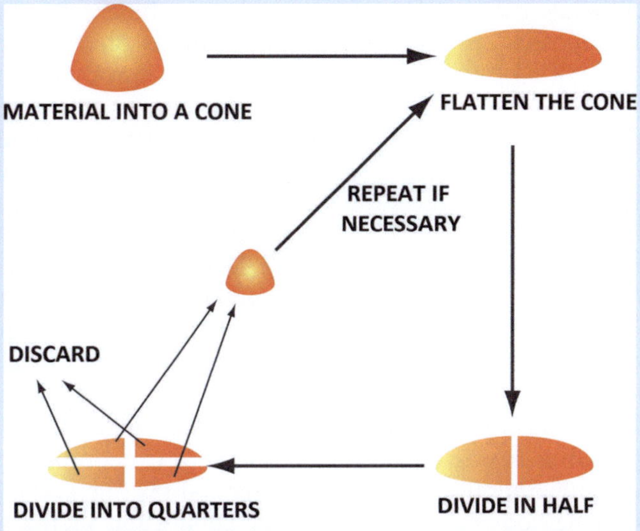

Fig. 4.40 Coning and quartering method

4.2 Evaluation Process

4.2.1 Introduction

To a large extent, the success of any mining project is based on knowledge of the deposit from a resources/reserves viewpoint. Although sampling and determining the properties of the material occupy a very important place in defining the overall characteristics of an aggregate deposit, the estimation of the resources/reserves, that is, establishing the amount of material, acquires a critical character. The results of the evaluation process will later influence the selection of the exploitation method. Trying to carry out this evaluation work with the minimum possible error is not only desirable but essential. Since the sampling is restricted to a very small volume of the entire investigated deposit, the techniques for evaluating the resources/reserves of the deposit are particularly relevant. In this sense, the use of computers and the spectacular development of computer applications specific for geological modeling in the recent decades have made it possible to continuously update the inventory of reserves by incorporating the last data from the different exploration and exploitation stages.

The estimation of the resources/reserves of an aggregate deposit can be carried out using two main methods (a) classical or geometric methods and (b) geostatistical methods. The latter appeared at the beginning of the sixties and have reached their maximum with the development of computers in the early eighties since extensive mathematical calculations are necessary for their realization. Nevertheless, the complexity of geostatistical methods makes them rarely used in the evaluation of aggregate deposits. Conversely, they are commonly used in metal mining projects (e.g., gold or copper—Bustillo 2018).

With the aim of never losing sight of the objectives of the resources/reserves evaluation process by any of the methods previously cited, it is necessary not to forget the fact that what is being carried out is a geometric process, that is, one or more geometric figures in which the amount of material (aggregates) must be calculated will be designed. The overall evaluation process always involves the following steps or stages: (a) definition of the surface of the aggregate deposit: it may be more or less complex but it will always be necessary to obtain the volume of aggregates, (b) estimation of the volume of aggregates that can be present in the deposit: it is estimated from the surface obtained in the previous step and with an estimated value of the thickness of the aggregate deposit. For it, the following formula will be applied (in a simplistic way):

$$V = S \times Th$$

where

V is the volume of the reservoir (in m^3)
S is the area of the aggregate deposit (in m^2)
Th is the thickness of the aggregate deposit (in m).

The last step in the estimation process is (c) the determination of the reserves of the aggregates in tons. knowing the volume of the deposit and obtaining the value of the average apparent density of the material, the product of both values will allow the quantity of tons of aggregates to exist in the deposit:

$$Q = V \times d$$

where
Q are the aggregate reserves (in tons)
d is the density of the material (in tons per m^3).

4.2.2 Classical Methods

Classical (also termed traditional, geometric, or conventional) estimation methods can be utilized to assign values to blocks (i.e., polygonal or inverse distance methods) and they are used for estimating the reserves of an aggregate deposit. Classical methods are not particularly accurate but can provide an order-of-magnitude of the reserves of the deposit. Although classical methods are now largely superseded by geostatistical techniques in mineral resource evaluation, they are still applicable in many situations and can produce an accurate end-result of the estimation process. Moreover, "if insufficient sampling does not allow the production of a mathematical model adequate to describe the nature of the aggregates in the deposit under evaluation using geostatistical methods, it is essential to utilize one of the classical methods" (Annels 1991). However, it is necessary to take into account that these methods do not quantify any estimation error, that is, they do not provide any objective way of measuring the reliability of estimates.

In classical methods, an aggregate deposit is converted to a geometric body composed of one or several solids that best express the size, shape, and distribution of the aggregate variables. The construction of these blocks depends on the method selected. Some methods offer two or more methods of block construction, introducing subjectivity. In such a case, a certain manner of construction is accepted as appropriate, preferably based on geology, mining, and economics. Numerous classical methods of reserve computations are described in the literature but some are only slight modifications of the most common ones. Depending on the criteria used in substituting the explored aggregate bodies by auxiliary blocks and on the manner of computing averages for variables, classical methods can be classified into six main types (1) sections or profiles method, (2) polygons method, (3) triangles method, (4) inverse distance weighting method, (5) contour methods, and (6) block matrix method (Fig. 4.41). Each of these methods has found application in the aggregate industry for estimation of the resources/reserves of the deposit. All are empirical and their use depends largely on the experience of the user. They proved useful and reliable where applied with close geologic control and adequate sample control.

The selection of the best classical method to use in the evaluation process is mainly based in the following variables: (a) the geology of the aggregate deposit, (b) the appraisal of geologic and exploration data, and (c) the accuracy needed. In the selection process, the time and cost of computations are significant considerations. The method (or combination of methods) selected should suit the purpose of computations and the required accuracy; it should also best reflect the character of the aggregate deposit and the performed exploration. In a complex or irregular deposit, it is advisable to use two or more methods for better accuracy and self-confidence. The average of these methods can be

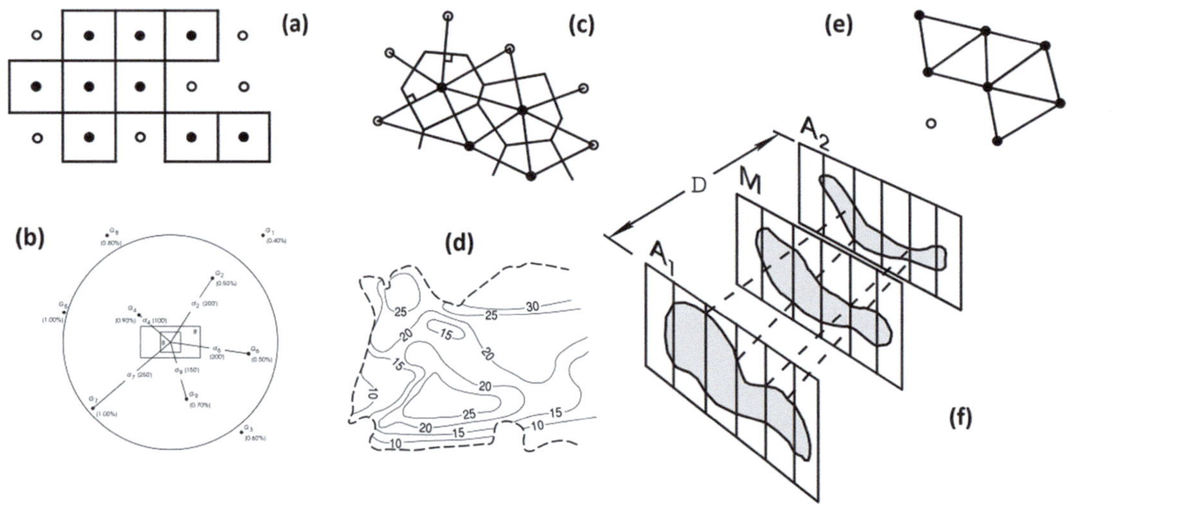

Fig. 4.41 Classical methods for resource/reserve calculations: **a** block matrices, **b** inverse distance weighting, **c** polygons, **d** contours, **e** triangles, and **f** sections

accepted as a final result, or the values of one method can be considered a control of others. As a consequence, utilization of two or more methods for computing the aggregate reserves for the same deposit is a common practice, so that no gross errors can occur. A common example of combined methods is where one method is applied to outline and divide the mineral body into blocks and another to determine the parameters (i.e., clay content) of each block.

- Cross-sectional or Profiles Method

Together with the polygonal method, the cross-sections or profiles method is one of the most used within the group of classical methods (▶ Box 4.6: Profiles Method).

Box 4.6—Profiles Method

The cross-sectional method is simple and rapid and is the most convenient method to compute the reserves of uniform aggregate deposits. Well-defined and large bodies of aggregate that are uniform in thickness (or show gradually changing values) can be estimated accurately by the cross-sectional method.

*The profiles method commonly applies to more or less irregular bodies of aggregates that have been systematically drilled on sections according to a regular grid, generating a reserve calculation based on cross-sections along these sections (*Fig. 4.42*). This method allows for careful consideration of the geology of the aggregate deposit and the development of a good geological model that is necessary for adequate reserve estimation. The previously cited requirement of a regular grid of investigation is precisely the most important restriction of the method since the aforementioned directionality, for various reasons, is not always achieved in the exploration of aggregate deposits for economic reasons. Nevertheless, the use of investigation methods (i.e., tomography) in granular deposits that generate information expressed in the form of profiles facilitates the use of this classical evaluation method. The distance between sections will define the accuracy of the calculation because widely separated sections can generate important errors due to the existence of changes in the aggregate properties not controlled by the successive sections. Although there are many geometric possibilities, in the traditional cross-section method the area of aggregate presence in a given cross-section is calculated (i.e., with a planimeter, counting squares, or through Simpson's rule) and the volume of the aggregate body is usually computed utilizing, as a solid figure, two consecutive cross-sections and the distance between them (*Fig. 4.42a*):*

$$V = [(A_1 \times A_2)/2] \times L$$

where

- V *is the volume, in* m^3
- A_1 *is the area of section* A_1*, in* m^2
- A_2 *is the area of section* A_2*, in* m^2
- L *is the distance between* A_1 *and* A_2*, in meters.*

*The interval between sections can be constant, for instance 50 or 100 m, or can vary to suit the geology and mining requirements. Another possibility is to compute the volume using half the distance to the two adjoining sections (*Fig. 4.42b*). Thus, the limits of the blocks defined lie exactly halfway between the drillholes. To increase the accuracy of computations, the number of blocks should be as large as possible. Care should be exercised to avoid arbitrary locations and construction of sections. Finally, the volume of each block multiplied by the bulk density of the aggregate material gives the tonnage of aggregates in the deposit. The sum of the tonnage of aggregates in each block generates the total tonnage of aggregates for the entire deposit.*

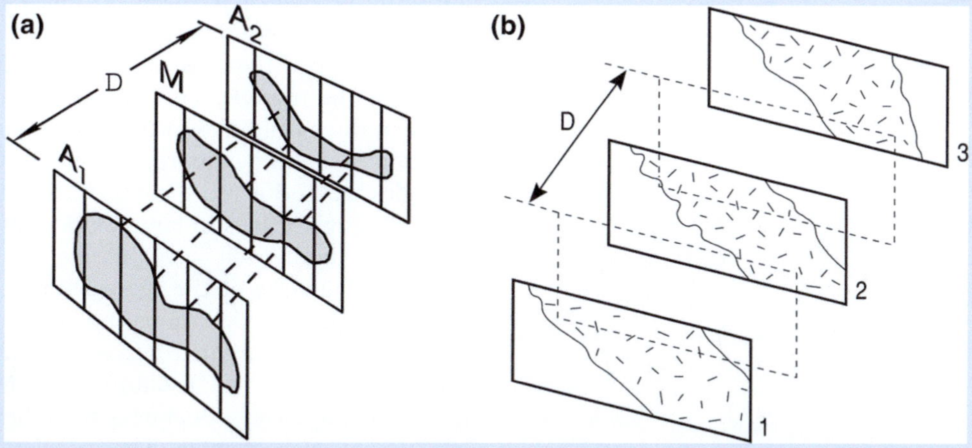

Fig. 4.42 Cross-sectional methods **a** solid figure formed by two consecutive cross sections and **b** solid figure obtained corresponding to half the distance to the two adjoining sections

4

- Polygons Method

Polygons method is used where drillholes are randomly distributed. It is based on the construction of a series of polygons whose center is a drillhole, assigning to each polygon the thickness of the aggregate body in the corresponding drillhole. It assumes that the thickness of the aggregate remains constant throughout the entire area of the polygon. This method of construction leads to a pattern of polygonal prisms. Because the polygonal estimate is based on assigning areas of influence around drillhole intercepts, this method corresponds with the intuitive idea that the quantity of information provided by each sample is proportional to its area or volume of influence. Polygons method is very popular although experience says that it is not very suitable. If the number of drillholes is large an excessive number of polygons are obtained, while if the number is small, the thickness of the aggregate body in each hole is assigned to an excessively large area (sometimes even square kilometers).

The drawing of polygons around drillhole data is established using a series of perpendicular bizectors of lines joining sample locations (Fig. 4.43). This is because the perpendicular bizector of a segment is the line for which points are at the same distance from either side of the segment. Each polygon contains a single sample location and every other point in the polygon is nearer to the contained datum than to any external datum. These concepts can be extended to three-dimensions, although polygons are typically drawn in two-dimensions and the volume of influence is defined perpendicular to the polygonal plane. In this method, the third dimension of the structure, that is, the height of the polygonal prism, depicts aggregate deposit thickness, which is perpendicular to the projection plane. The polygonal method is deficient in exposing the morphology of the aggregate body and the fluctuations of variables (i.e., thickness or clay content) within the individual blocks; although average thickness is computed, the pattern of its space distribution is not revealed.

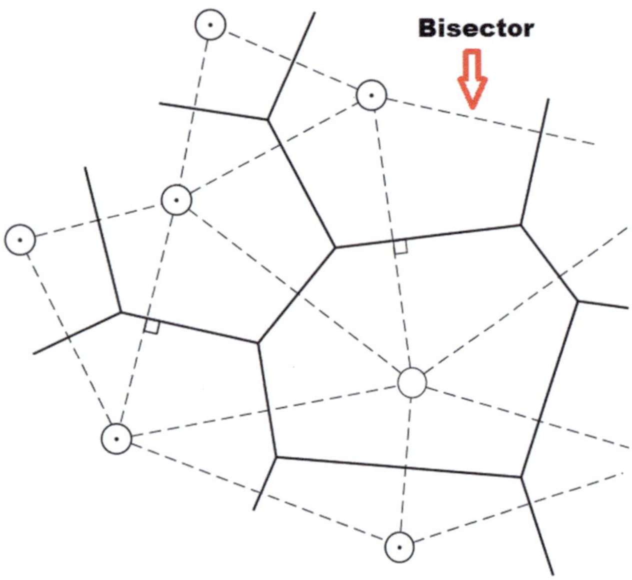

Fig. 4.43 Method of polygons using perpendicular bizectors

The mathematical procedure of estimation is somewhat similar to that used in the cross-section method. After the polygons have been drawn, the area of each polygon is computed, for example utilizing a planimeter. A polygonal prism is constructed using the thickness of the aggregate deposit as the height of the prism. The volume computed in the prism is later multiplied by the bulk density of the aggregate material to obtain the tonnage of aggregates. As in the previous method, the sum of aggregate quantity in each polygonal prism generates the amount of aggregates for the entire deposit. In aggregate deposits formed by several bodies, separate sets of polygons must be constructed for each one.

Favorable criteria for the use of the polygons method are the proven continuity of the aggregate body between drillholes and the gradual changes in thickness in the aggregate body. This method is successfully used in computing reserves of tabular deposits such as sedimentary bodies of granular aggregate deposits. The greater the number of polygonal prisms and the more regular the grid are, the more accurate the computations are. The utilization of this method is definitely incorrect if the bodies of aggregates cannot be satisfactorily correlated between drillholes, if these bodies are small and distributed erratically, or if intercalations of waste material (i.e., clay) are present.

- Triangles Method

The triangles method represents a modification of the polygon method. It consists of joining the adjacent boreholes to obtain triangles, in such a way that the final result is a set of triangles (Fig. 4.44) for each of which the corresponding thickness data are calculated as the average of the three values of the thickness at the apices of the triangle. This method has the advantage of using three drillholes in the calculation of the thickness for each triangular reserve block. Obviously, the triangles method is more conservative than the assignment of single values to large blocks, just as in the polygonal method.

The calculation of the aggregate reserves of the deposit includes the determination of the area of each triangle and its weighted aggregate thickness, obtaining these values by the usual methods. Multiplying the area by the aggregate thickness gives the volume of aggregates in each triangle. The calculation of tons of aggregate in each triangular prism and the entire reserves of the aggregate deposit is carried out using the

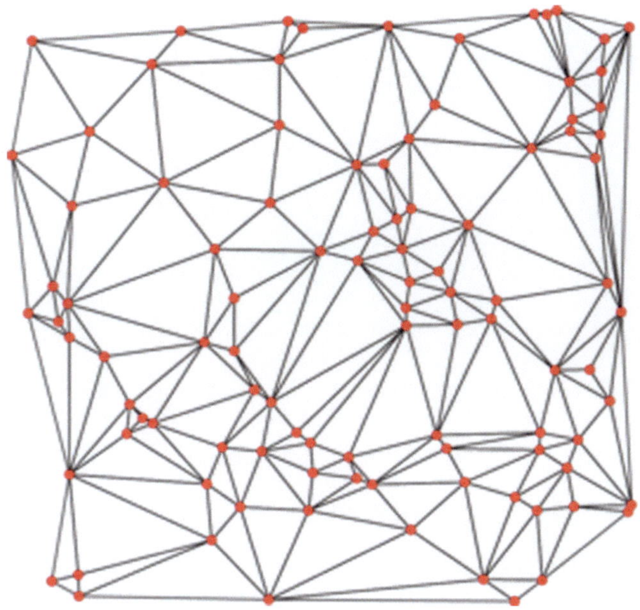

Fig. 4.44 Triangles method

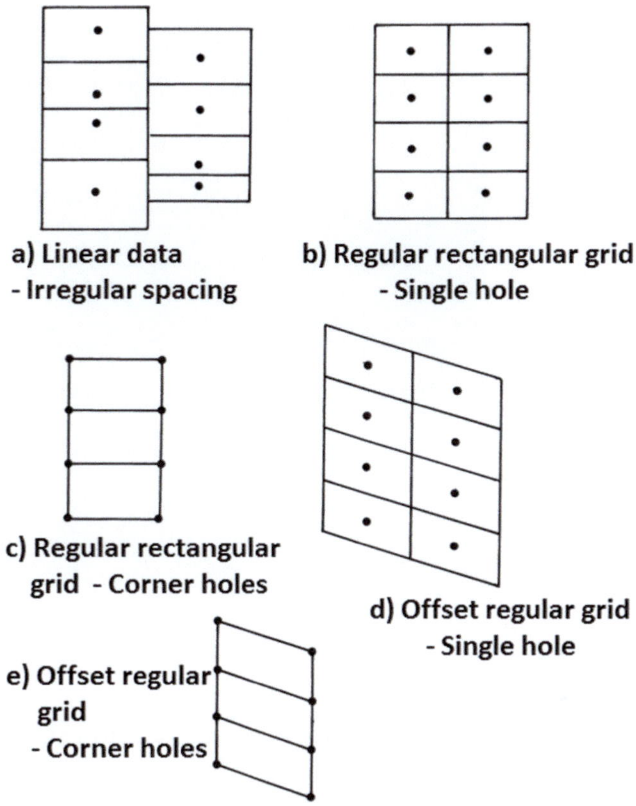

Fig. 4.45 Block matrices (Annels 1991)

procedures described above for polygonal or cross-section methods.

- Block Matrices, Inverse Distance Weighting and Contour Methods

These methods are not as widely used as the previous ones for the estimation of aggregate reserves in a deposit. In the block matrices method (Fig. 4.45), where the data are on rectangular or regular offset grids, regular blocks of square or rectangular shape can be fitted to the drillholes. The procedure of estimation of aggregate reserves is similar to that used in the polygon or triangles method. Contour methods (Fig. 4.41d) are very simple to use and produce good results, especially for aggregate bodies where there are certain natural regularities of the variations in thickness (Fig. 4.46). Finally, inverse distance methods (Fig. 4.41b) are a family of weighted average methods. The technique applies a weighting factor to each sample surrounding the central point of a block. The inverse distance is taken into consideration by assuming that the influence of a borehole over a point varies inversely with the distance. Although subjective, inverse distance weighting estimation procedures remain popular.

4.2.3 Geostatistical Methods

The term geostatistics arises from the combination of the random character presented by the variables to be studied (i.e., thickness of a sand and gravel body) with the geological character that, undoubtedly, they possess. These methods were developed by Matheron and his school in Paris from the beginning of the sixties and offer an optimal evaluation of the reserves of a deposit. Matheron (1963) stated that "geostatistics, in their most general acceptation, are concerned with the study of the distribution in space of useful values for mining engineers and geologists, such as grade or thickness, including a most important practical application to the problems arising in ore-deposit evaluation. Any ore deposit evaluation as well as proper decision of starting mining operations should be preceded by a geostatistical investigation that may avoid economic failures". In this sense, the classical methods mentioned above include little quantitative appreciation of the errors involved in estimation and this general concept of estimation error is fundamental to an appreciation of mineral resource estimates by any method; to solve this issue, geostatistics estimates the error involved in estimation.

Although geostatistics is not a technique broadly utilized in the evaluation of aggregate deposits, numerous authors have shown their interest as a complementary method capable of solving different problems associated with this type of deposit (i.e., distribution of clays within the sand and gravel deposit), especially those of a granular nature. In this sense, it is clear that the rock masses that can be used as raw material to obtain crushing aggregates do not present enough

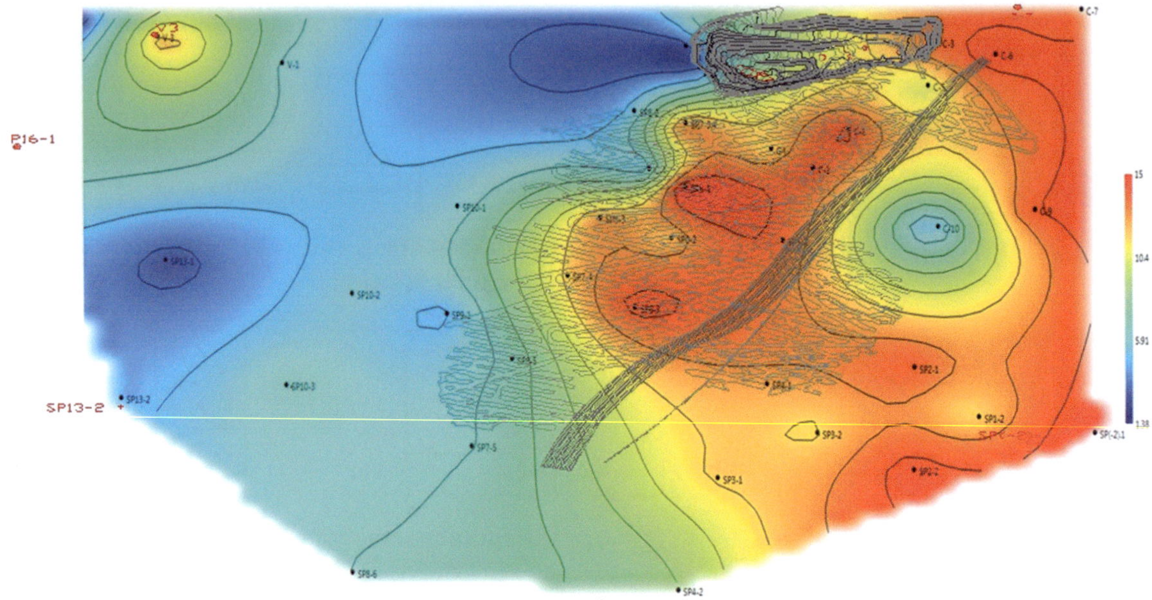

Fig. 4.46 Contour of aggregate average thickness in an aggregate deposit

problems to address the use of geostatistical methods in the evaluation of their reserves.

Formal geostatistical investigations of sand and gravel deposits are quite rare, and the primary focus of those studies is the spatial characterization of deposit thickness and its subsequent effect on aggregate reserves. Nevertheless, the aggregate sector will probably require methodologies of aggregate deposit evaluations comparable to those in other mining industry sectors due to the need of megaquarries to produce rock not only for aggregate but also for other products suitable for use such as chemicals, agricultural lime, filler, and cement. For this reason, acceptance of formal geostatistics can require changes in typical exploration and sampling protocols.

In the bibliography, Arthur (1994) presented an example of the potential application of geostatistical studies in sand and gravel deposit evaluation. This author describes the sampling methodologies that have never been implemented or even taken seriously by the aggregate sector. Jerosch (2013) showed the importance of geostatistics to evaluate materials likely to be used as aggregates and Mee et al. (2019) presented the feasibility of modeling sand and gravel resources using modern geostatistical techniques. Figure 4.47 is an example of the results obtained by these authors.

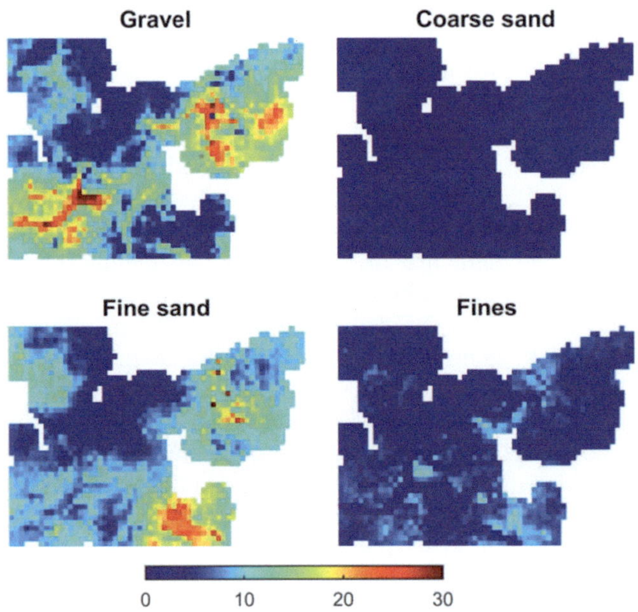

Fig. 4.47 Maps of the estimated percentages of gravel, coarse sands, fine sands and fines to a depth of 5 m (Mee et al. 2019)

4.2.4 Economic Evaluation

Project economic evaluation is "the process of identifying the economic feasibility of a project that requires a capital investment and making the investment decision" (Torries 1998). Many times, for commodities such as sand and gravel aggregates, economic and legal issues are far more important than geological aspects. Like other mineral investments, aggregate investment shows specific characteristics such as (1) the depletable and many times unique nature of the aggregate resources, (2) the specific location and features of the aggregate deposits, and (3) the existing of geological uncertainties. Thus, aggregate deposits can be mined only where they are found. This "decrease in flexibility increases the risk of aggregate mining ventures compared to other types of investment opportunities" (Torries 1998). Consequently, the utilization of adequate project economic evaluation techniques is more important in the aggregate mining industry than in many other industries.

Actually, the predominant economic evaluation technique, from prefeasibility study to mine operation, is the so-called discounted cash flow (DCF) method. It is the most used valuation instrument. In this method, a cash flow is designed to obtain all cash inputs and outputs over the whole life of a project. The cash flow model must recognize the time value of money discounting at an appropriate discount rate to obtain its present value, which is calculated yearly.

The main DCF criteria values are the following: (a) net present value (NPV): is a value that indicates of how much money a project adds to the company, indicateing the maximum value that the company may offer for a project at year 0; (b) internal rate of return (IRR): is a financial index utilized for measuring and comparing the profitability of an investment and (c) payback period (PP): is the period needed to pay back the initial investment from cashflow in future years; PP is an useful evaluation parameter because it indicates how long the company has to wait to get its return on investment. Nevertheless, payback period index does not account for the time value of money. The NPV, IRR, and PP methodologies is accepted by the industry, the financial community, and regulatory bodies.

All modern texts on project evaluation conclude that the preferred methods of evaluation, where sufficient data are available, are those that incorporate annual cash flow projections and that recognize the time value of money, the so-called dynamic methods. They include particularly the net present value and the internal rate of return, as opposed to those (i.e., PP) not considering the time value of money (named static methods). On an international level, economic assessments of aggregate mining projects are performed using NPV and IRR and sometimes PP.

- Net Present Value (NPV)

The net present value of an aggregate mining project is simply the difference between cash outflows and cash inflows on a present value basis. It is the backbone of a project evaluation process. The formula to calculate NPV is as follows:

$$\mathrm{NPV} = (R_0 - C_0) + \frac{R_1 - C_1}{(1+i)} + \frac{R_2 - C_2}{(1+i)^2} + \cdots + \frac{R_n - C_n}{(1+i)^n}$$

where

R is the expected revenues each year

C is the expected costs each year

I is the discount rate for the project.

A positive NPV means that the expected income is higher than the projected expenses and a negative NPV indicates a nonprofit or loss situation so that the project should be rejected. The larger the NPV, the richer the investors become by undertaking the project. On the other hand, the higher the discount rate is, the lower the NPV of the project. The selection of an appropriate discount rate is critical in the application of NPV as a valuation and ranking technique since this value gradually discounts the cash flow values and finally calculates the NPV of the project. The discount rate ranges between 5 and 15% above the interest rate of the required initial capital investment.

- Internal Rate of Return (IRR)

The internal rate of return (IRR) method is one of the most used investment analysis methods and, in addition to NPV, is probably the most common evaluation technique in the aggregate mineral industry. In the IRR method, the objective is to find the interest rate at which the present sum and future sum are equivalent. In other words, the present or future sum of all cash flows is equal to zero if the IRR value is used as the interest rate. It is clear that as the discount rate increases for a specific cash flow, the NPV of the cash flow decreases. The relationship between IRR and NPV can be written as:

$$\mathrm{NPV} = 0 = \left[\sum_{t=1}^{n} \frac{\mathrm{CF}_t}{(1+\mathrm{IRR})^t}\right] - I_0$$

where

CF_t is the cash flow in year t

I_0 is the initial investment (CF_0)

IRR is the discount rate that makes NPV = 0

n is the total number of years for the project.

IRR value is the interest rate at which the investor recovers the investment. The higher a project's internal rate of return, the more desirable it is to undertake the project because of the better the return on capital.

There are several reasons to explain the widespread popularity of IRR as an evaluation criterion. Most importantly, IRR is expressed as a percentage value and most managers and engineers prefer to think of economic decisions in terms of percentages compared with absolute values provided by the present, future, and annual value calculations. The acceptance or rejection of a project based on the IRR criterion is made by comparing the calculated rate with the required rate of return: if the IRR exceeds the required rate, the project should be accepted, but if not, it should be rejected.

- Payback Period (PP)

The payback (or payout) period falls under the heading static methods but it is one of the most simple and common evaluation criteria used by engineering and resource companies. It is the number of years required for a project to generate cash flow or profits equal to the initial capital investment. It is important to note that cash flow in the first year or even more years of a mining operation will be negative since these years are used to pay the previous investments (i.e., exploration). The PP method is a useful evaluation parameter because it indicates how long the company has to wait to obtain its return on investment. Payback periods lie between three and five years and shorter payback periods are needed in high-risk countries than in stable countries. The payback period is very useful in areas of political instability where the recovery of the initial investment within a short period is particularly important.

- Risk Analysis

The term risk has many meanings in the mining world. A broad definition of risk is "the effect of uncertainty on objectives" (ISO 31000) although it is very much a personal assessment. Risk analysis can be utilized by any organization regardless of its size, activity, or sector because mineral valuation comprizes many uncertain decisions and the risk associated to these decisions being wrong must be evaluated. The magnitude of uncertainties in mine development projects is larger than in most other industries used for comparison. For this reason, the key to successful evaluation is the incorporation and estimation of risk and its consequences. The inclusion or exclusion of risk in economic evaluation is essential. To evaluate the risk in a mining project (i.e., aggregates), which can be subdivided into technical, economic, and political, different methodologies can be used, with modifications to net present value analysis being the most commonly used (i.e., sensitivity analysis, risk-adjusted discount rates, scenario analysis and Monte Carlo simulation—Box 4.7: Monte Carlo Simulation).

Box 4.7—Monte Carlo Simulation

Where investments in mining projects are taken into account, the evaluation of the overall risk associated with a project can be very difficult. This is because there are too many influential risk factors with different types of uncertainties at once. A more quantitative risk assessment approach must also incorporate mathematical and statistical methods to assess the risk associated with a project. In the Monte Carlo method, a random number generator is used to calculate probability for each combination of events. The randomized calculation is repeated for many iterations so that an estimate of the overall probability for each outcome can be estimated.

In the same way that sensitivity analysis changes one variable at a time, the Monte Carlo simulation changes two or more input variables at the same time. Obviously, the overall impact on the project value will be much greater. There is an enormous number of combinations of different variables and different amounts of variation that we will have to address. For this reason, and the huge amount of calculations that result, Monte Carlo analyses are nearly always carried out using specific software packages that can model different combinations very quickly. The number of iterations is determined regarding the project size and the importance of risks (1000, 2000, 5000, and so on). A higher number of runs gives more accurate results. In most cases, to make the calculation easier, the variables are assumed to be independent from one another, although most of the variables are correlated. Regarding the presentation of Monte Carlo simulation results, for every combination of input variables a project value is calculated. After repeating the calculation for every combination (number of iterations), all the project values are plotted on a histogram and statistical parameters such as median, mean, mode, percentiles, etc., are taken into account. ▫ Figure 4.48 *shows the distribution of NPV values in a Monte Carlo simulation.*

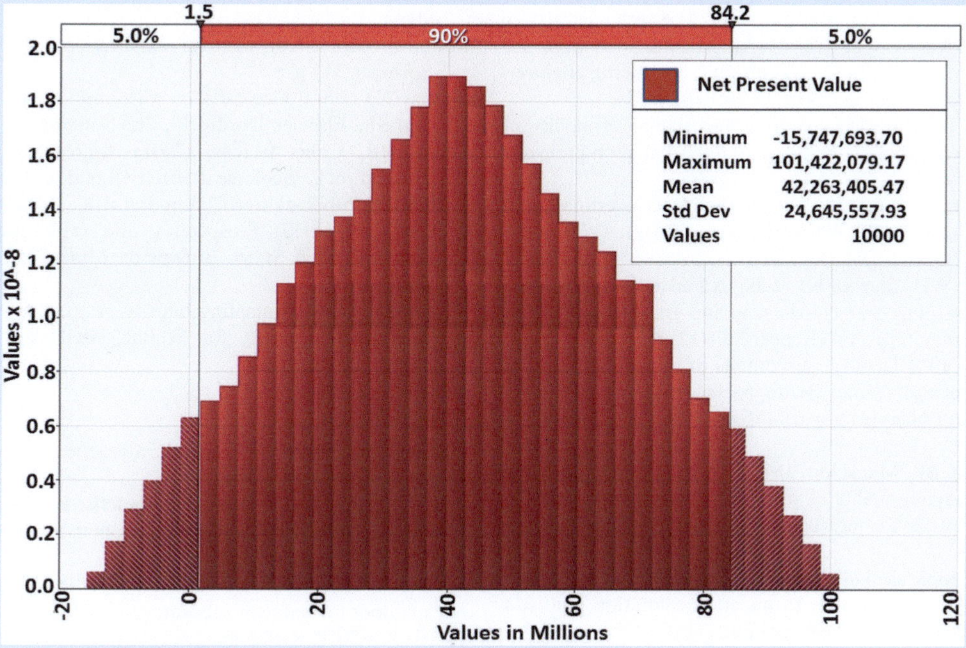

Fig. 4.48 Distribution of NPV values in a Monte Carlo simulation

4.3 Questions

Short Questions

- Enumerate the phases of exploration
- What are the main techniques used in the second phase of reconnaissance exploration?
- Explain in short the remote sensing technology
- Enumerate the main geophysical methods used in the exploration of aggregate deposits
- What is the ground penetrating radar technique?
- Define borehole geophysical logging
- What is the main difference between reverse circulating drilling and diamond core drilling?
- Definition of sampling
- Define the concept of geostatistics. What is the most important difference between classical methods and geostatistical methods regarding error estimation?
- What is the net present value of an aggregate mining project?

Long Questions

- Explain the sand and gravel mineral deposit model
- Explain the electrical resistivity tomography technique

References

Annels AE (1991) Mineral deposit evaluation: a practical approach. Chapman & Hall, London, p 436

Arthur J (1994) Computerised sedimentological modelling and resource evaluation of sand and gravel deposits. Cardiff, U.K., University of Wales, Ph.D. Thesis, 275 p

Bellamy AG (2018) Marine sand and gravel prospecting and seabed investigations for aggregate extraction. In: Hunger E, Brown TJ, Smith G, Anderson P (eds) Proceedings of the 19th extractive industry geology conference 2016 and technical meeting 2017. EIG Conferences Ltd., pp 205–215

Bliss JD (1993) Modeling sand and gravel deposits—initial strategy and preliminary examples. U.S. Geological Survey, Open-File Report 93-200, 31 p

Bustillo M (2018) Mineral resources—from exploration to sustainability assessment. In: Springer textbooks in earth sciences, Geography and environmental sciences. Springer International Publishing AG, Cham, 653 p

Chambers JE, Wilkinson PB, Weller A, Meldrum PI, Kuras O, Ogilvy RD, Aumonier J, Bailey E, Griffiths N, Matthews B, Penn S, Wardrop D (2011) Characterising sand and gravel deposits using electrical resistivity tomography (ERT): case histories from England and Wales. In: Hunger E, Walton G (eds) Proceedings of the 16th extractive industry geology conference. EIG Conferences Ltd., pp 166–172

Cox DP, Singer D (eds) (1986) Mineral deposit models. U.S. Geol Surv Bull 1693, 379 pp

Dirik K (2006) Advanced photogeology—lecture notes. Geological Engineering Department, Hacettepe University, 45 p

4

Gandhi SM, Sarkar BC (2016) Essentials of mineral exploration and evaluation. Elsevier Inc., Amsterdam, Netherlands, p 398

Gustaffson A (2010) Trends in exploration. In: M Smith (ed), Exploration drilling, Atlas Copco Craelius AB, Märsta, Sweden, pp 3–7

Hartman HL, Mutmansky JM (2002) Introductory mining engineering, 2nd edn. John Wiley and Sons, New York, p 584

Jerosch K (2013) Geostatistical mapping and spatial variability of surficial sediment types on the Beaufort Shelf based on grain size data. J Mar Syst 127:5–13

Laxton S, Coates J (2015) Geophysical and borehole investigation of aggregate resources in the Whitehorse area, Yukon. Yukon Geological Survey, Open File 2015-1, 26 p

Lucius JE, Langer WH, Ellefsen KJ (2006) An introduction to using surface geophysics to characterize sand and gravel deposits. U.S. Geological Survey, Open-File Report 2006-1257, 51 p

Marjoribanks R (2010) Geological methods in mineral exploration and mining. Springer-Verlag, Berlin, p 238

Matheron G (1963) Principles of geostatistics. Econ Geol 58(8):1246–1266

Mee K, Marchant BP, Mankelow JM, Bide TP (2019) Modelling the distribution and quality of sand and gravel resources in 3D: a case study in the Thames Basin, UK. Environ Model Assess 24:585–603

Milson J (2006) Geophysical methods. In: Moon CJ, Whateley MKG, Evans AM (eds) Introduction to mineral exploration, 2nd edn. Blackwell Publishing Ltd., Oxford, pp 127–154

NSSGA (2013) The aggregates handbook, 2nd edition. The National Stone, Sand and Gravel Association, Alexandria, Virginia, USA, 906 p

Roonwal GS (2018) Mineral exploration: practical exploration. Springer, 312 p

Sarma DD (2009) Geostatistics with applications in earth sciences, 2nd edn. Elsevier, Dordrecht, The Netherlands, p 205

Smith MR, Collis L (eds) (2001) Aggregates: sand, gravel and crushed rock. Geological Society, London, Engineering Geology Special Publication v. 17, London, UK, 360 p

Thoenen JR (1932) Prospecting and exploration for sand and gravel. United States Bureau of Mines, Information Circular 6668, 68 p

Torries TF (1998) Evaluating mineral projects: applications and misconceptions. Society for Mining, Metallurgy and Exploration Inc, Littleton, 153 p

Standards

ASTM C702/C702M—18. Standard practice for reducing samples of aggregate to testing size

ASTM D75/D75M—19. Standard practice for sampling aggregates

EN 932-1: 1997. Tests for general properties of aggregates—part 1: methods for sampling.

EN 932-2: 2015. Tests for general properties of aggregates—part 2: methods for reducing laboratory samples

ISO 31000: 2018. Risk management—principles and guidelines

Extraction Methods

Contents

5.1 Introduction – 144

5.2 Crushed Rock – 148
5.2.1 Surface Quarrying – 148
5.2.2 Underground Quarrying – 157
5.2.3 Super Quarries – 161

5.3 Sand and Gravel – 163
5.3.1 Dry Mining – 164
5.3.2 Wet Mining – 166
5.3.3 Marine Aggregate Extraction – 167

5.4 Questions – 169

References – 170

M. Bustillo Revuelta, *The Basics of Aggregates*,
Springer Textbooks in Earth Sciences, Geography and Environment,
https://doi.org/10.1007/978-3-031-42961-3_5

Abstract

This chapter reviews the main topics related to aggregate extraction from surface and underground methods. Although aggregate sources can be widely variable, the methods of extraction are very similar throughout the world, with crushed rocks mainly extracted from quarries (surface and underground mining) and sand and gravel from gravel pits. In the latter, the aggregate can be extracted above or below the waterline. These extraction methods are fairly well described together with loading and hauling equipment. Working hard rock quarries is a more complicated and intensive process than quarrying sand and gravel. Drilling and blasting in hard rock quarries are also briefly explained in this chapter because they are often necessary to extract the required rock. To illustrate the development of each mining method, several case studies are included in the text. The last heading of the chapter is devoted to marine aggregate extraction.

5

5.1 Introduction

Extraction, as used in this book, refers to all methods of removal of aggregates from their natural position. In general, the term extraction is used interchangeably with the terms mining and quarrying. The method to extract aggregates mainly depends on whether the material is rock or sand and gravel, the conditions at the site, and the end product. Although aggregate sources can be widely variable, the methods of extraction are very similar throughout the world, with crushed rocks mainly extracted from quarries (surface—Fig. 5.1—and underground mining) and sand and gravel from gravel pits. In the latter, aggregates can be extracted above the waterline, for instance, in benches 5–10 m high, or it can be extracted below the waterline. This separation is caused by the geological environment in which the two materials are found and the degree of consolidation or induration of the deposit. In several countries, such as the United Kingdom, aggregates are also obtained from sea-dredged materials (◘ Fig. 5.2) (marine aggregates—see ▸ Chap. 3). The mining system and excavating equipment are different depending on the type of aggregate.

A large proportion of world aggregate production is obtained using quarrying methods because sand and gravel are not always readily available. Quarrying is a term commonly used to describe the production of construction and building materials (i.e., dimension stone—◘ Fig. 5.3—and crushed rocks for aggregate). Thus, in quarrying the target substance is the very rock itself. Quarrying can be further subdivided into surface and underground methods, with surface mining being predominant in aggregate extraction.

Sand and gravel do not require crushing because they are unconsolidated or weakly consolidated sediments but still need to be screened to obtain the sizes

◘ **Fig. 5.1** Surface quarry for extracting dolomitic aggregates

Fig. 5.2 Sea-dredged materials (image courtesy of EMOdnet)

Fig. 5.3 Monte Coto Pinoso dimension stone quarry (Alicante, Spain) (image courtesy of Levantina)

needed. For its part, crushed stone is produced from solid rock formations of sedimentary, igneous and metamorphic origin (see ► Chap. 3) and must be drilled (Fig. 5.4) and blasted prior to extraction. It is further alternatively crushed and screened. One notable difference between crushed stone quarries and sand and gravel pits lies in the amount of land used because sand and gravel pits often require a higher rate of land occupation. This is because "sand and gravel deposits are generally very shallow in nature and bigger areas of land require disturbance for a similar quantity of production" (Bustillo 2018) (Fig. 5.5).

Fig. 5.4 Drilling blastholes (image courtesy of Atlas Copco)

Fig. 5.5 Large area mined in a sand and gravel deposit

In both sand and gravel extraction and surface quarrying, the first step is site preparation, which involves the cleaning of any trees and vegetation to expose the sand, gravel or rock. The topsoil and overburden are then stripped (Fig. 5.6) and relocated before extraction of aggregates using bulldozers, conventional loaders and haul trucks. Overburden, which may include soil, subsoil, unconsolidated material, contaminating materials, and/or weathered materials, can be utilized to construct mounds or walls termed berms or can be adequately stockpiled.

Often required by regulation, organic soil is stripped separately from the rest of the overburden since this material will be further used in environmental restoration works (Fig. 5.7). In this case, organic soil requires careful handling and storage if its physical structure and chemical features must be preserved. Nevertheless, where possible it will be directly relo-

Fig. 5.6 Overburden stripping in an aggregate quarry

Fig. 5.7 Overburden disposal close to the mine for environmental restoration works

cated onto previously worked out areas as part of progressive restoration. In any case, topsoil and overburden should not be stockpiled over parts of the deposit where later mining can be developed. Sometimes consolidated overburden can need blasting before it can be removed. In summary, overburden removal processes are established for minimizing the quantity of handling and transport distance. All aspects related to site preparation must conform to the approved mine planning. Site preparation also includes the construction of roads, berms, processing and maintenance facilities as well as other plant infrastructure.

It is important to note that the exploitation processes are dynamic and change throughout the life of the mining operations. This means that the original extraction plans are continuously revised and updated to reflect the changes that affect the system. Some examples of these changes are the following: (a) irregulari-

Fig. 5.8 Underground aggregate mine (image courtesy of John Deere)

ties in geology (i.e., unknown faults or watercourse), (b) unexpected turns in market demand (i.e., product specifications), (c) regulation changes generating significant differences in the production methods, and (d) technology advances producing lower-cost production methods.

Finally, because the geology of each deposit is different, no two extraction systems are the same, thus needing a unique design. Even the characteristics of the rock within the same deposit can vary, changing the quality and thickness of the material from quarry to quarry and commonly from one side of a quarry to the other. In this sense, as commented below, underground mining (Fig. 5.8) provides to mine selectively and going straight to the desired material. To develop a correct extraction planning, all information that affects the last mining design should be obtained and utilized. Computer-aided mine design is currently a necessary tool in quarry planning and development.

5.2 Crushed Rock

Working hard rock quarries is a more complicated and intensive process than quarrying sand and gravel. Crushed stone is produced from quarries that are much deeper than sand and gravel quarries. Hard rock quarries "have outputs ranging from 100,000 t/y to approximately 5 Mt/y" (Langer 2006). Therefore, the investment in plant and equipment can be very large. As a rule, mining crushed stone differs from mining sand and gravel because the material is almost always first drilled and blasted (Fig. 5.9). As commented before, most of the quarries that extract hard rocks are surface quarries but the presence of underground quarries is becoming more common in many countries.

5.2.1 Surface Quarrying

When the quarry design is finished, the sequence of mining development should be established with the aim of assure the production of saleable products at the earliest possible time. The development of most modern surface quarries involves stripping the overlying soil and weathered rock prior to the extraction of the hard rock underneath. This is because the establishment of a production face remains the essential first element. The detailed planning of overburden and soil extraction as well as availability of storage space without interrupting mining operations, sterilizing reserves or incurring

Fig. 5.9 Blasting in an aggregate quarry

extraordinary costs is a significant topic previously to the development of a quarry.

The thickness of the overburden, ranging from loose soil to partially weathered rock or a combination of the two, varies considerably from almost nil to over 30 m in some operations, being removed by a combination of hydraulic excavators, ripping, and blasting using explosives. Soil and partially weathered rock can be removed using a bulldozer or a hydraulic excavator whereas harder or more consolidated materials need drilling and blasting, which are required to fracture the rock to enable mechanical excavations. Bulldozers (◘ Fig. 5.10) are utilized extensively in surface mining and are very useful under difficult conditions. Ripping or mechanical breakage, using a dozer fitted with a tooth at the rear (◘ Fig. 5.11), is possible where the rock mass is already fractured extensively. A ripper is basically constituted by a large tooth attached to the rear of the bulldozer. When working, they are pulled through the material as the machine moves forward.

Quarry design depends on several factors such as (a) pre-existing topography, (b) intended output, (c) infrastructure, and (d) environmental impact. As a rule, four main types of quarries can be established (◘ Fig. 5.12): (a) hillside, (b) hilltop, (c) coastal, and (d) valley bottom. The quarry is worked using a bench system, that is, extracting the rock in layers or benches, which are formed by drilling, blasting, and excavating rock. The quarry becomes deeper with each subsequent bench.

The aggregates can be extracted by successive horizontal benches in full width or by simultaneous horizontal benches. In the former, a working face of a height determined is developed at one end of the deposit over its full width, which will advance parallel to itself. This is the most common method and is applied to all types of massive rock deposits (sedimentary, igneous, or metamorphic). If the deposit is thin in depth,

◘ **Fig. 5.11** Ripping dozer

◘ **Fig. 5.10** Bulldozer (image courtesy of Eduardo Revuelta)

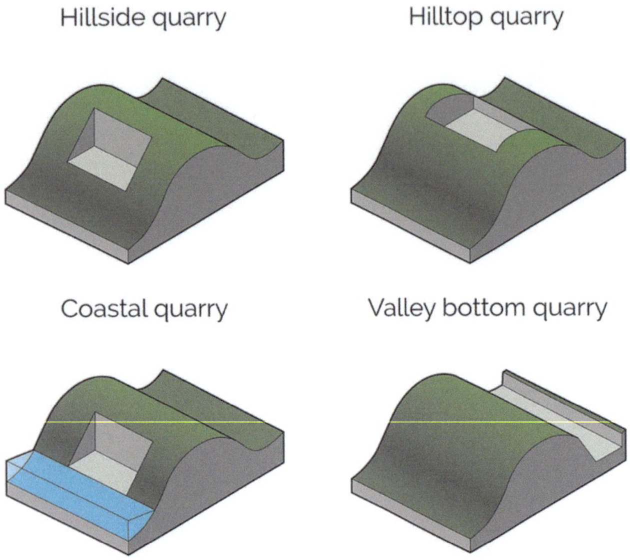

Fig. 5.12 Types of surface quarries (image courtesy of Epiroc)

a single bench can be sufficient. A second bench following the same system must be necessary where the deposit is thick. In some cases, a third is obligatory and so on (Fig. 5.13). Sometimes two can be worked simultaneously for homogenization. In this method, overburden stripping works are generally carried out simultaneously over the whole quarrying site, and "reclamation is possible only at the last stage and the boundary slopes alone can be considered the ultimate pit" (Lot and de la Rupelle 2000).

In the case of simultaneous horizontal benches, a first working face is opened and immediately a second one is also opened at a lower level. Both the working faces come under operation almost simultaneously.

Fig. 5.13 Aggregate quarry with three benches

Depending on the thickness of the deposit, more faces can be further opened. This method is largely recommended in deposits of low thickness compared to their area (i.e., basalt or limestone deposits). The works of overburden stripping and extraction can be developed at the same time. Where applicable, "this method affords the possibility of a more immediate reclamation on the heels of extraction operations (case of restoring cultivation)" (Lot and de la Rupelle 2000).

Three principal and elementary phases can be defined in the aggregate extraction process: (a) drilling and blasting, (b) loading and (c) transport (haulage).

- Drilling and Blasting

Drilling (Fig. 5.4) and blasting (Fig. 5.9) are necessary to extract the required rock, being actually the technology of rock blasting highly developed and regulated in each country. 'They are developed in benches (quarrying progresses downward on several levels or benches (Fig. 5.13)'. Drilling and blasting are carried out to reduce the size of the rock mass.

Different drills and explosives are utilized in surface mining market due to the variability of rocks extracted. The principal factors affecting the type of drill utilized are the following: "(a) deposit geology and thickness, (b) rock hardness and chemical composition, (c) climatic conditions, and (d) regulatory, geographical and local restrictions" (NSSGA 2013). In general, the hardness of the rock is the main factor determining the basic type of drill employed. In soft rocks (i.e., limestone), a rotary drill using a drag bit is utilized while a down-the-hole or in-the-hole hammer drill is normally used in hard rocks (i.e., andesite).

In blasting, a line of shot holes is drilled (to depths ranging from a few meters to 20 m or more) at a distance back from the quarry face (burden) (Fig. 5.14), and explosives (i.e., gelatine dynamites—Fig. 5.15—can be used as the main explosive charge where high density and energy is required or as a primer for ANFO—see ▶ Box 5.1) charged and detonated (Fig. 5.9). The bench height is two to five times the burden distance whereas 1:1.25–1:2 is often the ratio between the burden and the spacing. The drill pattern is designed based on the hole spacing in the bench and burden for a given hole diameter (Fig. 5.16). When detonated, the rock is broken into manageable fragments for further crushing processes. The most widely used explosive for aggregate quarry blasting is a mixture of fuel oil (FO—6%) and ammonium nitrate (fertilizer) (AN—94%); this mixture is termed ANFO (▶ Box 5.1: ANFO).

Fig. 5.14 Line of shot holes drilled at a distance from the quarry face (image courtesy of Octavio de Lera)

Fig. 5.15 Cartridges of gelatine dynamite (image courtesy of Octavio de Lera)

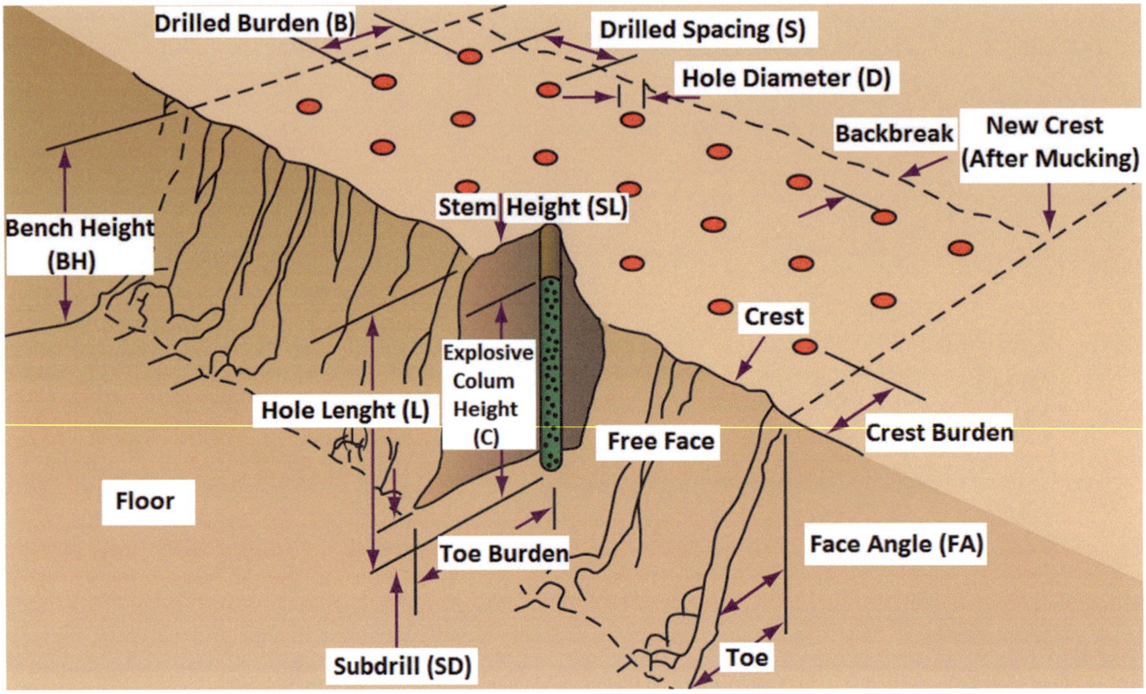

Fig. 5.16 Main blast pattern parameters in surface bench blasting (Bustillo 2018)

Box 5.1—ANFO

ANFO gives the maximum energy and velocity of detonation (approximately 3660 m/s). Even after mixing the two components, the final product remains fairly dry since the percentage of fuel oil in the mix is very small, and it is absorbed in the pores of the small granules (prills) of ammonium nitrate. Because of their insensitivity, ANFO cannot be detonated by a detonator, and it should be detonated by a primer of high explosive (e.g. one or two cartridges of dynamite with detonator).

ANFO is supplied basically as poured or packaged. Where the quantity of explosives is high (i.e., large open pit mines), ANFO is supplied in separate component containers on a truck, mechanically mixed at the worksite, and poured into blastholes. Hence, it is also called bulk ANFO. Where the requirement of the mine is low, for example, in small quarries, ANFO is usually supplied in nylon bags (Fig. 5.17). Figure 5.18 *shows the ANFO loading operation using nylon bags. Poured ANFO "proves more effective than the packaged form as it fills the entire cross section of the blasthole whereas the package leaves a gap between the walls of the blasthole and external diameter of the package"* (Gokhale 2011). *If a blasthole has a significant quantity of groundwater seeping into it, it cannot be charged by poured ANFO. In such cases, ANFO is premixed and packed into thick cylindrical plastic bags sealed at both ends. The process of charging a blasthole with packaged ANFO is very tedious and time-consuming.*

The advantages of ANFO are its safety, ease of loading, and low price. In the free-flowing form, they have a great advantage over cartridge explosives because they completely fill the borehole. Regarding disadvantages, there are two disturbing aspects about the use of ANFO in large surface mines. The first is the quick evaporation of diesel oil; the second is the high solubility of ammonium nitrate in water. Where atmospheric temperature and humidity are high, it becomes essential to add extra fuel in the ANFO mix to address the degree of evaporation likely up to the time of detonation. With respect to solubility in water, if the humidity of the atmosphere is high, ammonium nitrate, being highly hygroscopic, absorbs a large quantity of water and becomes less effective in the process. In this context, it is worth noting that after approximately 9% water content, the ANFO mix becomes insensitive and fails to detonate. If the ANFO mix contains aluminum as a sensitizer, such a mixture is called ALANFO. It is particularly useful "for blasting hard rock masses, being 10–15% the most commonly used percentage of aluminum in ALANFO" (Bustillo 2018).

■ **Fig. 5.17** Packages of ANFO (image courtesy of Pedro Rodríguez)

■ **Fig. 5.18** Loading operation of ANFO nylon bags (image courtesy of Octavio de Lera)

The top portion of the hole is further filled with a non-explosive material, sand or crushed stone, which is referred to as stemming (■ Fig. 5.19). The explosive in each hole is initiated with detonators. There are three types of detonators (■ Fig. 5.20) based on the source of energy used for starting detonation in the detonator: electric, nonelectric, and electronic; the electronic detonator is similar to the nonelectric detonator but with the pyrotechnic delay element substituted by a microchip. In turn, they can be instantaneous or with a delay element built into them. Finally, an initiating device placed at a very great distance for the sake of safety always detonates a primer cartridge. The most used initiation transmission line in blasting is probably the detonating cord (■ Fig. 5.21), which transmits a detonation wave. It is made of a plastic tube with a 3–5 mm outside diameter filled with penthrite in uniform density (10–15 g/m). When a detonation is initiated, it travels at a velocity of 6500 m/s with very high shock energy. Another option of initiation is a signal or shock tube (■ Fig. 5.22).

Blasting lasts less than a second, being formed by many smaller individual blasts. These individual blasts are separated by delay periods between charges of a few thousandths of a second. The delay periods assure that each hole will have to break only the rock immediately in front of it (■ Fig. 5.23). When the explosive is detonated, large volumes of gases are rapidly generated causing crushing of the rock. The stemming materials keep the gases in the rock to maximize the effect.

The pattern of blast-holes, the amount of charge used, and the timing of the blast must be carefully de-

Fig. 5.19 Stemming of a borehole on top

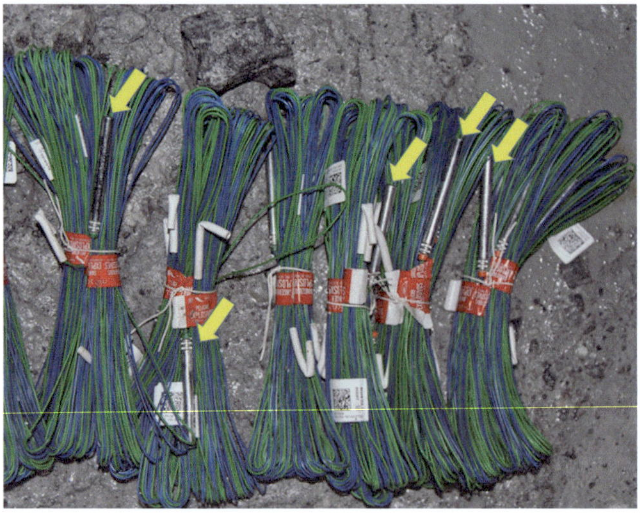

Fig. 5.20 Detonators (image courtesy of Magnesitas de Rubián, S.A.)

Fig. 5.21 Detonating cord

Fig. 5.22 Signal or shock tubes

fined "to ensure the maximum efficiency of the explosive, the production of appropriate-sized blocks, and the creation of a new, stable quarry face" (Bustillo 2018). The pattern of blast-holes and the delay periods are designed to reduce good vibrations from the blast. The pressure waves from each blast interfere with each other and decrease vibration as well as reduce the risk or fly rock (pieces of rock thrown clear of the blast site). Currently, laser profiling is sometimes utilized for characterizing the quarry face before blasting so that the correct amount of explosive is used. It is necessary to remember that a poor blasting significantly increases production costs. After blasting, if very large pieces of rock blocks are obtained, it is necessary to reduce them to obtain smaller blocks compatible with the

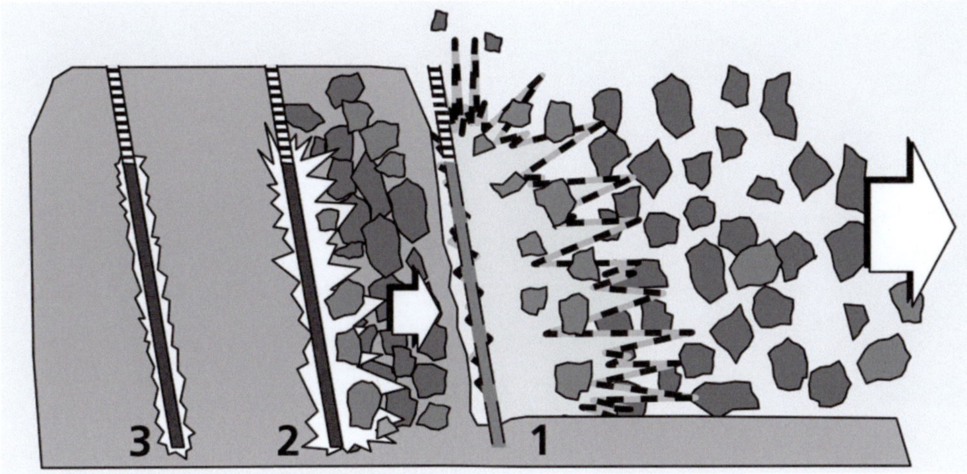

■ **Fig. 5.23** Effect of blast timing (illustration courtesy of Atlas Copco)

■ **Fig. 5.24** Hydraulic hammer for crushing large pieces of rocks (image courtesy of Benito Arnó e Hijos, S.A.U.)

■ **Fig. 5.25** Wheel front loader

crusher dimension. This process can take place using either secondary blasting or with a hydraulic hammer (■ Fig. 5.24).

- Loading and Transport

The blasted material is dry and can be extracted and loaded utilizing bulldozers, front loaders (■ Fig. 5.25), backhoes or hydraulic excavators (■ Fig. 5.26). These machines load the rocks into off-road dump trucks (■ Fig. 5.27) or conveyors to transport the material to a fixed primary crusher frequently located in the processing plant or to a mobile primary crusher on the quarry floor (■ Fig. 5.28). In this case, the machine can be brought up to the face where it can be directly fed by the excavator. The numbers and capacity (i.e., from 15 to 100 tons) of dump trucks will usually be established based on the needed production rate. Because hauling is a major cost in quarrying operations, much attention is given to the design of the access ramps and the distances traveled. Unlike sand and gravel deposits, where particle size distributions are relatively narrow, the crushed rock in the quarry tends to have a broad variety of particle sizes, which precludes the utilization of conveyors as the primary transporting system. They can operate on a much steeper gradient than dump trucks, and are an important alternative. ▶ Box 5.2 (El Aljibe Quarry) shows an example of a surface aggregate quarry located near Toledo, Spain.

Fig. 5.26 Hydraulic excavator

Fig. 5.27 Dump trucks for transporting the aggregate

Fig. 5.28 Mobile crusher on the quarry floor (image courtesy of Octavio de Lera)

Box 5.2—El Aljibe Quarry—Text and Images Courtesy of Benito Arnó e Hijos

El Aljibe quarry (Fig. 5.29) *is a crushed stone quarry located near Toledo (Spain). It is an open-pit quarry formed by descending benches developed in mylonite metamorphic rocks. The thickness of the overburden material ranges from centimeters to 1.5 m. From a geological point of view, the quarry is placed in the axial part of the Central Hercynian zone of Spain, within the Toledo Shear Zone that separates the Toledo Migmatite Complex unit to the north and the Paleozoic metasediments to the south. The quarry mines these materials to produce track ballast for high-speed and conventional trains, high strength concrete and asphalt mixtures and subbase and base courses in road and rail track construction. They have a density of approximately 2700 kg/m³ and an average uniaxial compressive strength of 131 MPa. Ballast rock is qualitatively described as tough rock with good abrasion resistance.*

Drilling and blasting are used to mine approximately 0.5 Mt/y of run of mine (ROM) although this quantity can change depending on the market demand. Only one blast is loaded at a time during one working shift. Each blasting process is carried out with three rows of blastholes drilled in a staggered pattern of 2.5 m × 3 m. The blastholes were 89 mm in diameter and up to 19 m deep with an inclination from the vertical of 9°. The burden of about 30 times the blasthole diameter d is within the range of 25d–35d recommended for vertical blastholes loaded with light and heavy explosives.

The charging pattern is adapted to the amount of water in the holes; two 60 mm × *620 mm gelatin cartridges (2.5 kg)* *are charged at the bottom of the hole as primers and smaller gelatin cartridges of 50 mm × 380 mm (approximately 1 kg) are loaded up to the water level. A total of 75 kg of ANFO was further loaded in the dry part of the hole up to an uncharged length of 2.5 m (*◘ Fig. 5.30*). Thus, the total explosive consumption per hole is approximately 82.5 kg. Finally, the 16.5 m column of explosive stems with 2.5 kg of granular materials (drill cuttings). Nonelectric detonators are used in all holes. The total explosive consumption in each blasting process is approximately 4000 kg.*

After the blasting process, a hydraulic excavator with a bucket of 4.5 m³ mucks the fragmented rock into 60 t average rigid dumper trucks that haul the material from 250 to 850 m to the bin of the primary crusher.

◘ **Fig. 5.29** El Aljibe crushed stone quarry (Toledo, Spain)

◘ **Fig. 5.30** Bench blast (image also courtesy of Antonio Durán)

5.2.2 Underground Quarrying

In the future, "the provision of hard rock resources suitable for aggregates may give rise to increasing levels of conflict, particularly where they coincide with attractive landscapes or other forms of land use" (Millar et al. 2011). Consequently, licenses to operate new quarries are likely to become progressively complicated to obtain. To solve this troubling situation, the underground mining of aggregates is an important strategic component to accomplish the intense aggregate demand in many countries while eliminating most of the physical and visual impacts of a surface mining within urban communities. Aggregate underground mining may be an economic necessity to maintain the supply. This alternative can be integrated into an efficient land use planning system to capture the social-economic benefits of underground space after depletion of aggregate resource. Different reasons can explain the increasing trend of underground aggregate production: "(a) permits are difficult to obtain for new quarries, (b) underground mining maximizes reserves on a permitted property and (c) selective mining of rock is possible" (NSSGA 2013). Nevertheless, it must be taken into account that underground mining is more expensive than surface mining and is not as common. In general, underground mining of aggregates is approximately 50–60% more expensive than mining aggregates from surface quarries with similar rock properties.

According to Brown et al. (2010), the main potential advantages of underground mining of aggregates in sustainable development terms are the following: "(a) the ability to extract hard rock from areas without disturbing the surface landscape would lead to a reduction in visual intrusion, habitat disturbance and land-use changes in comparison to surface quarrying, (b) there is potential for enhanced revenues for the aggregates producer by putting the land surface above an underground aggregates operation to economical use during extraction operations, (c) improved security of supply arising from an increase of the total aggregate resource available through the additional consideration of underground resources, (d) increased provision of underground space (after extraction) in areas that are close to high population density areas (that coincide with high aggregates demand centers), (e) the containment of some noise, dust and vibration (from air-over pressure) associated with operations of drilling, blasting, loading and hauling of rock, and (f) with underground mining, there is no need to strip and store soils

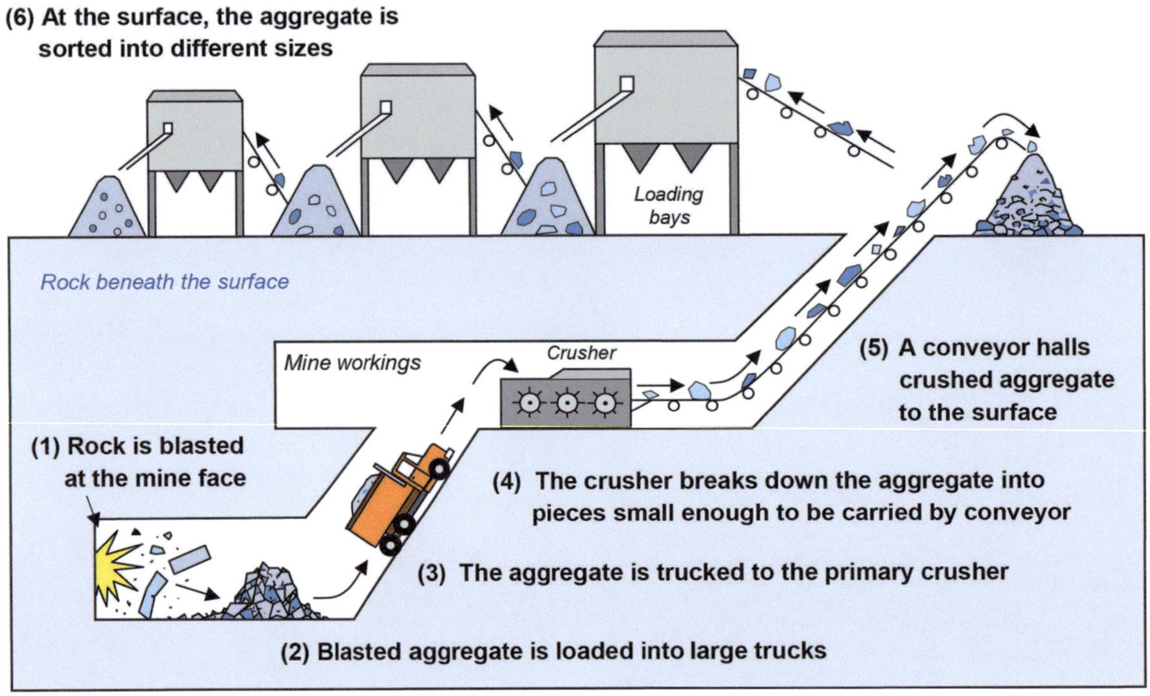

Fig. 5.31 Example of how an underground aggregate mine works (illustration courtesy of Kentucky Geological Survey)

and overburden from above hard rock resources, which provides further potential environmental benefits by removing the dust, noise and fuel usage caused by soil stripping activities; it is also expected that underground mines will offer a reduction in land restoration activities and costs". Another added benefit of underground aggregate mining is to work in a constant underground climate rather than a variable surface climate. Thus, working conditions are not affected by rain, snow, and wind.

In some cases, underground aggregate mines are opened from the floors of existing quarries, facilitating the quick production of the commodity. The depth of the mining level below the land surface can vary from 20 to 100 m and the mine can go into production when the entrance road or ramp has reached the proper level. Figure 5.31 shows an example of how an underground aggregate mine works.

Surface expression of the mining operation through subsidence must be considered an inadmissible result generated by underground mining of aggregates in populated areas. For this reason, the selection of an adequate underground mining method is essential, specifically the room and pillar method. It is one of the most classical underground mining methods (Figs. 5.32 and 5.33) and "recovers the material as completely as possible in open stopes, called rooms, leaving pillars of rock to support the hanging wall (hence the name room and pillar) but without jeopardizing working conditions and personal safety" (Bustillo 2018). Having selected this mining method, determination of the opening dimensions and the supporting pillar dimensions are the next items to be taken into account. In this sense, the most important factor when considering the development of an underground quarry is the stability of the ground overlying the deposit to be mined. For this purpose, it is necessary to conduct adequate studies of rock mechanics. They must be utilized as an aid to analyze if the ground is competent to stay open without other ground support that stabilizes the material surrounding the excavation. ▶ Box 5.3 (Kanpanzar Quarry Project) shows an example of an underground aggregate quarry project located in the north of Spain.

Fig. 5.32 Two rooms and one pillar in a room-and-pillar underground system (Spain) (image courtesy of Pedro Rodríguez)

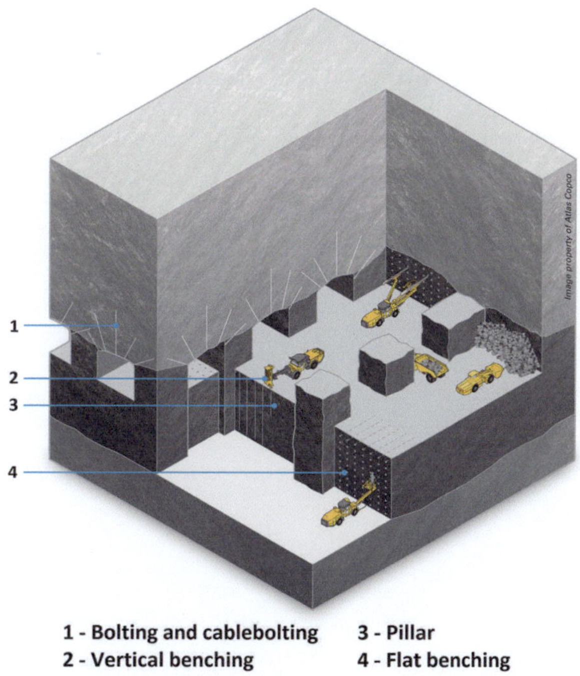

Fig. 5.33 Room and pillar method (illustration courtesy of Atlas Copco)

5

Box 5.3—Kanpanzar Quarry Project—Text and Images Courtesy of Calcinor

Cantera de Kanpanzar is a company founded in 1958 by the Etxebarria family, which later joined the Calcinor Group. The quarry is located in the massif of Udalaitz (Guipuzcoa, Spain) and mines Cretacic limestones in a surface quarry. The new project in the company is the transition of surface exploitation to an underground quarry (Fig. 5.34) *because the surface reserves will be depleted. The underground quarry will be working in approximately 2024, although the worldwide economic crisis delayed the final project, which started in 2002.*

Oviedo University began its studies in the limestone massif in 2003 and the first galleries of the underground quarry were excavated in 2004. There were more than 4 km of galleries in 2011, with the access to the underground quarry located in the restored benches of the surface quarry. There are actually four entries to the underground quarries but the project will finally have up to 14 entries. The feasibility study was developed through pilot galleries and the conclusions established a room and pillar method as the best underground method for the extraction of the limestone rock.

The Kanpanzar project comprises an area of 400 m wide and 300 m deep on four levels, with six rows of pillars (Fig. 5.35) *for each level. The opening of the room will take place in three phases. In the first one, several galleries with an approximate section of 8 m × 7 m are developed to study the geotechnical and hydrological behavior of the rock mass. Based on the results of this study, the location of the pillars will be established. They will have sizes between 22 m × 22 m and 32 m × 32 m. The design of the mine will be carried out in such a way that the areas with the greatest fracturing and water circulation will be included within the pillars. Once the galleries have been prepared and the pillar model defined, the second phase of the project is the opening of the room, widening the gallery to the location of the pillar. The third and final phase of the project will be the lowering process for reaching the final height of the room. The possibility of creating rooms up to 20 m wide and 25 m high makes possible competitive prices for the aggregate product. The economic viability of underground quarries is also because the high quality of the rock makes it possible to dispense with any type of support within the mine (i.e., gunite or bolts).*

The chosen method does not allow large recovery rates, although the abundance of limestone resources guarantees the extraction of sufficient material to be a cost-effective underground quarry. The initial estimates establish a recovery rate (ratio between the extracted material and that which remains in the pillars) of approximately 65%. If four extraction levels are finally developed, separated by horizontal rock masses, the overall recovery rate will be about 32%. This underground mining method, unlike other methods, allows the recuperation of the pillars once the exploitation is finished.

Fig. 5.34 Underground Kanpanzar quarry (image courtesy of Calcinor)

Fig. 5.35 Pillars of the underground quarry (image courtesy of Calcinor)

■ Fig. 5.36 Standard drill jumbo

Further aspects to be considered in the development of an aggregate underground mine are ventilation (it must be supplied to all underground workings), removal of underground water by pumping, access (using ramp and tunnel since shafts normally need very high expenditures per ton of production) including its location and size, drilling and blasting methods (■ Fig. 5.36), and loading and hauling (the number of loaders and trucks—Fig. 5.37—required is high because space is always limited resulting in increased time to load trucks).

Nevertheless, notwithstanding its environmental superiority, this alternative must overcome a broadly held perception that queries the economic viability of this option. Defining the economic feasibility of underground mining for crushed rock aggregates is essential for determining whether the aggregates can be produced and delivered to the market from an underground aggregate mine at a competitive cost. Obviously, this cost must be comparable with that from classical surface mines located further afield. In this sense, "underground crushed rock aggregate mines may be able to compete for a share in the overall market by replacing/displacing aggregate produced in the quarries" (Brown et al. 2010).

5.2.3 Super Quarries

Although the location of an aggregate quarry is needed close to the market of consumption, as noted in ► Chap. 1, the required reduction of environmental impacts and/or social conflicts related to aggregate mining operations has lead in the last four decades to a decrease in the number of aggregate quarries and the increasing volumes extracted from the remaining quarries. All these factors generated the development of the super quarry concept (i.e., Barton 1996). These super quarries are mostly developed in less inhabited coastal regions, and are linked to the trend of globalization of aggregate production and its trade. In this sense, different factors promoted the international transport of aggregates: "(a) geological (physical lack of suitable aggregates or presence of aggregates of poor quality), (b) societal/environmental (permitting problems related to mining licenses, etc.) and/or logistical/economical (interstate land shipping is often much more expen-

■ Fig. 5.37 Loading in an underground quarry (image courtesy of Kerford)

sive than international sea transport)" (Langer 1995). Moreover, "super quarries are identified as a possible solution to long-term problems of economic marginality and population decline in the remotest rural areas" (Cowell 2000). Foster Yeoman's Glensanda coastal super quarry is probably the most famous aggregate super quarry in Europe (▶ Box 5.4: Super Quarry Glensanda). The development of this super quarry in the 1980s was due to "the growing consensus of opinion that large-scale coastal super quarries will be required to meet the future demand for construction aggregates in Northern Europe" (Kirk 1993).

Box 5.4—Super Quarry Glensanda—Text and Images Courtesy of Aggregate Industries

August 2017 marked the 30th anniversary of the first shipment made from super quarry Glensanda (◘ Fig. 5.38*), located in the center of a peak in Morvern Peninsula, near Oban on the West Coast of Scotland. Glensanda is one of the largest single quarry operations in the world, producing more than 7 million tons of crushed rock granite aggregates per year (aggregate sizes vary from fine sand up to 180 mm) although it has a current capability to excavate up to 9.5 million tons per year. It is the flagship site of construction materials producer and supplier Aggregate Industries, and the largest granite quarry in Europe, exporting to markets all across Northern Europe. Super quarry Glensanda has a 760-million-ton reserve of granite. To minimize visual impact from the coast, the quarry is sited 1.6 km inland.*

Glensanda was first discovered by John Yeoman on a boating holiday. The site was named as a potential location in a report commissioned by the government in the 1970s to identify coastal super quarries. Extraction began in 1986, and it undertook its first shipment on 5th August 1986 to Houston, Texas, delivering the granite to the US Gulf market just 10 days after setting sail. Since then, over 5500 shipments have been made to projects throughout Europe including the Channel Tunnel, the TGV rail line from Bordeaux to Bayonne and the Pont de Normandie in France, Deurganckdok in Antwerp, Schipol airport in Amsterdam, Moorburg Powerstation in Hamburg, the A1 motorway Gdansk to Torun, Riga airport, the Olympic park for the London 2012 Olympic Games, and the new London Gateway container port.

The site extracts Strontian granite, which occurs on the northwest side of the Great Glen Fault. It is assigned to the Argyll and Northern Highlands Suite of late Caledonian granitic intrusions. The pluton extends over an area of approximately 200 km^2 *in a north- to south-trending outcrop from the northwest shores of Loch Linnhe to the southern slopes of Meall a' Ghruith.*

Glensanda is one of the UK's top ten tonnage ports, and is only accessible by sea due to the surrounding rugged mountainous terrain. All vehicles, personnel, and materials are transported to/from the quarry by barge, boat and other vessels. It is renowned for being one of the most efficient mineral extraction operations in the world, whereby granite is crushed at the top of the mountain, fed into a 'glory hole' taking the rock 300 m down and out of sight into the core of the mountain, then through a 1.8 km horizontal tunnel to the foreshore where the rock will be washed, screened, and stored ready for ship loading. The ships are capable of transporting almost 100,000 tons of high-quality granite on each shipment.

Fig. 5.38 Super quarry Glensanda (Scotland, UK) (image courtesy of Aggregate Industries)

■ **Fig. 5.39** Low degree of consolidation of sand and gravel deposits (image courtesy of Marcos Endrina)

■ **Fig. 5.40** Dewatering processes in a sand and gravel deposit (image courtesy of Cemex)

5.3 Sand and Gravel

Sand and gravel have been extracted from rivers and streams worldwide since there has been a construction aggregate market. This is because this type of aggregate is the only available option for aggregate resources in many countries. It is important to bear in mind that sand and gravel extraction from rivers can originate far-reaching environmental impacts. In this sense, "each river or stream has its own unique set of geologic, hydrologic, climatic, and anthropogenic characteristics, and associated list of potential environmental impacts" (Langer 2003).

There are numerous methods to extract sand and gravel. Many of the methods used in hard rock quarries are also employed in this type of deposit. Unlike the rock materials that are used as aggregate resources, sand and gravel deposits are in a loose state. Their low degree of consolidation (■ Fig. 5.39) allows the aggregates to be extracted utilizing conventional mechanical methods, which means that drilling and blasting using explosives is not necessary. This advantage of sand and gravel extraction over crushed stone quarries is counteracted by the grading factor: while in quarries, the size of the material to be treated in the plant can be defined through blasting processes, in sand and gravel deposits this characteristic is previously imposed by transport and sedimentation processes.

The "excavation technique and equipment employed in sand and gravel extraction depends on the type of deposit, the general topography of the site, the conditions of planning and restoration and on the scale of operation" (Smith and Collis 2001). From an extraction viewpoint, sand and gravel deposits can be subdivided into two main groups: (a) dry mining and (b) wet mining, in which the water table is so near the surface that almost all the deposit is underwater. In wet mining, again, two options are also considered: (a) dewatering the site by pumping (■ Fig. 5.40), or (b) working the site wet. Dry working is preferred because it enables a more selective extraction and is more cost-effective. Total or partial dewatering is developed to allow the materials to be more easily mined. Nevertheless "the final decision to dewater or not depends mainly on deposit thickness, permeability of the materials and restoration requirements" (Bustillo 2021). Some extraction methods are applicable only to wet or dry mining but some are common to both. Sand and gravel extraction can occur in several in-stream or near-stream environments (■ Fig. 5.41).

Prax (2000) delineates three types of exploitation based on the relative positions on the water table, the

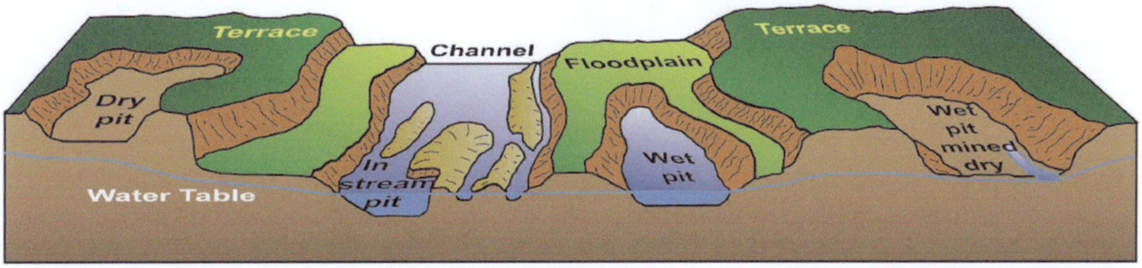

■ **Fig. 5.41** River deposits and methods of extraction (Langer 2003)

top and bottom (bedrock of the deposit): "(a) 30%, surface deposit: the level of the water table is below the bottom of the deposit (can be worked dry); (b) 70%, submerged deposit: the level of the water table is above the top of the deposit (can be worked wet), and (c) partially submerged deposit: the level of the water table is midway between those of the top and bottom of the deposit (generally workable in two distinct phases, dry, then wet)". Many types of equipment utilized in mining hard rocks (i.e., front-end loaders, power shovels and haul trucks) can also be used to extract this type of loose sediments from either dry deposits or those that can be dewatered by pumping. However, different machines and techniques are adapted to suit the particular situation in wet mining. The intrinsic features of the different sand and gravel deposits clearly exert a decisive influence on the extraction operation.

5.3.1 Dry Mining

Extraction of aggregates with dry mining, whether they are naturally dry or as a result of dewatering, is more effective in terms of reserve recovery. In upland areas, such as "alluvial fans, high terraces and some glacial meltwater deposits" (Langer et al. 2004), sand and gravel mining does not penetrate the water table; that is, these aggregates are naturally dry and can be extracted by employing classical earth-moving equipment, typical of any open-pit mining. Shovels, loaders, bulldozers or similar equipment load the sand and gravel into dump trucks and/or articulated dump trucks (Fig. 5.42) or onto conveyor belts to transfer to the processing plant located close to the extraction site. On well-maintained haul roads, the rigid body dump truck may reach higher speeds. In the articulated dump truck, "the articulation of the body and the multiple wheel drive generate a dump truck that can successfully on very poor and constricted haul roads as often found when removing overburden, soils, and sand and gravels" (Smith and Collis 2001). Figure 5.43 shows the main differences between the bodies of both types of trucks. However, long-distance transport by conveyor belts is a widespread method because the maximum particle size is small enough. If the sand and gravel materials are too consolidated, digging with a front-end loader or power shovel is almost impossible, and in this case a bulldozer with a ripper is required to loosen the material. Mechanical rippers allow the selective extraction of materials and have less environmental impact than blasting.

Fig. 5.42 Articulated dump truck

Dry mining is developed as a three-dimensional excavation until reaching the bed of the sand and gravel deposit. Depending on the depth, the advance is made with a single front constituted frequently by several benches (Fig. 5.44) (if the exploitable thickness exceeds 8–10 m), each bench having one extraction face. Extraction options can be (a) digging from above or (b) scraping from below. Since dry mining is located on the upper terraces of the river, it has the advantage of greater visual accessibility for the materials extracted as well as better working conditions of the mining equipment. ▶ Box 5.5 (Asunción Sand and Gravel Quarry) shows an example of a dry mining extraction located near Segovia, Spain.

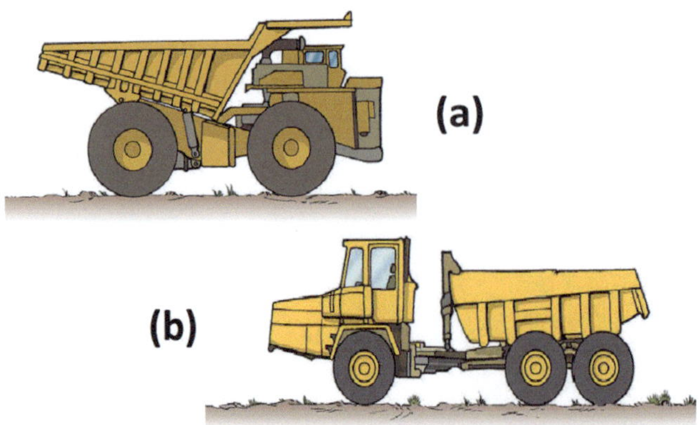

Fig. 5.43 Illustration of a rigid body dump truck (**a**) and an articulated body dump truck (**b**)

Fig. 5.44 Two benches sand and gravel exploitation (image courtesy of Marcos Endrina)

Box 5.5—Asunción Sand and Gravel Quarry—Text and Images Courtesy of CONORSA

The Asunción sand and gravel quarry is located near Segovia (Spain). It is a sand and gravel pit developed in arkosic sandstone originating from the weathering of igneous and metamorphic rocks. The structures of the sediment (Fig. 5.45) *indicate eolian and fluvial origins. The quarry mines these sediments to produce mainly concrete, both ready-mixed concrete, and precast concrete. The thickness of the overburden material (mainly clays) ranges from 0.3 to 1.5 m.*

Since the characteristics of the sand and gravel materials vary according to the area in which they are found, different analyses are performed prior to the extraction processes. Samples are studied to obtain data about grain size distribution or sand equivalent. Based on the information obtained, the most appropriate areas for extraction are established in each case. All this process is coordinated with the processing plant, which generates, among other things, significant energy savings.

Obviously, drilling and blasting are not used in the quarry, and the extraction process (in a bench of 8 m) is carried out using conventional earth moving equipment (hydraulic excavator with 3.3 tons capacity bucket) that loads the sand and gravel into trucks of 33 tons for transferring to the processing plant (Fig. 5.46). *The extraction area is approximately 10,000 m² per year. Processed materials in the processing plant are loaded using wheel loaders with a 7.5 tons capacity bucket.*

Fig. 5.45 Structures in the sediments

Fig. 5.46 Hydraulic excavator loading the aggregates into a truck

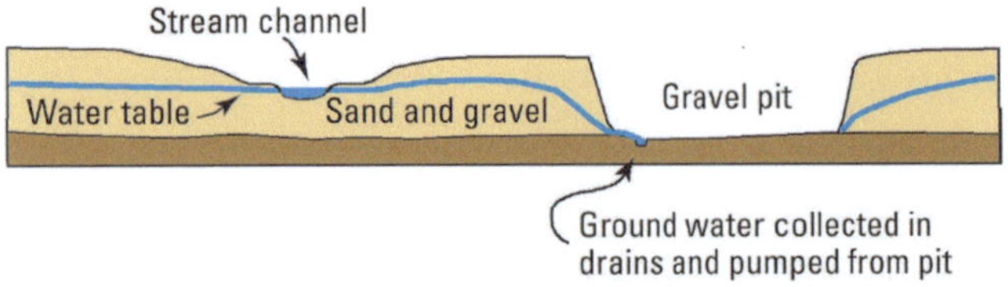

Fig. 5.47 Groundwater can be collected in drains in the floor of gravel pits and pumped from the pit (Langer 2003)

Fig. 5.48 Pumping out the water to draw down the water table (image courtesy of CEMEX)

If sand and gravel pits penetrate the water table (i.e., in low terraces and some glaciofluvial deposits), in some geologic settings, wet pits may be converted to dry pits by collecting the groundwater in drain excavations. They can also be developed by collecting groundwater in drains on the floor of the pit (Fig. 5.47) and pumping out the water (Fig. 5.48), that is, with a total evacuation of the water body from the site. For this purpose, mobile pumps are installed in the initial excavation to pump with the objective of drawing down the water table and holding the level below the needed digging level. In some cases, slurry walls or other barriers to groundwater flow around the pit can be needed (Fig. 5.49). The amount of water that must be pumped can be remarkable since the hydraulic conductivity is high and the radii of influence are several hundred or thousands of meters. After the groundwater drains from the deposit, sand and gravel can be mined like a surficial deposit, with all the advantages of using dry mining techniques.

5.3.2 Wet Mining

Classical wet sand and gravel deposits include those that occur beneath water bodies. These deposits cannot be dewatered, so the materials are extracted using specialized equipment. The extraction of aggregates located at shallow depths can be carried out using draglines with cables whereas the utilization of dredges is imperative if sand and gravel are located deeper. Stream channels can also be occasionally or constantly diverted to enable sand and gravel extraction utilizing typical earth-moving equipment in the natural streambed.

Draglines (Fig. 5.50) are employed in wet mining to extract the aggregate although other equipment such as slackline excavators can also be used. A dragline is a land-based crane that casts its bucket into the water above the area to be mined (Fig. 5.51). “The bucket is allowed to sink to the bottom and is then dragged back to the crane; as the bucket is dragged along the deposit, it becomes filled with material; the

Fig. 5.50 Dragline (image courtesy of Aggregates Manager)

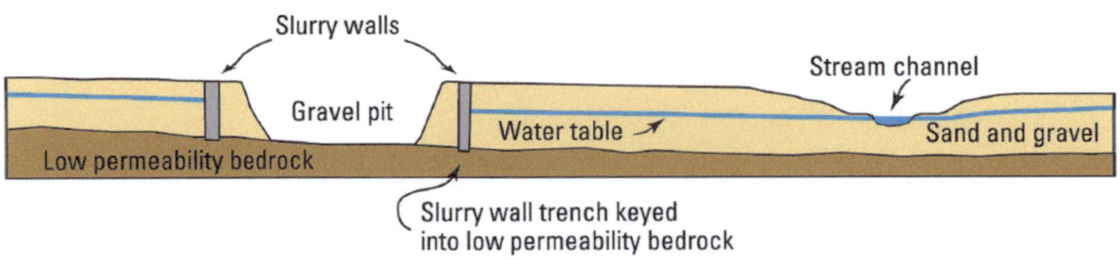

Fig. 5.49 Slurry walls to divert the groundwater (Langer 2003)

■ **Fig. 5.51** Dragline casting its bucket into the water

■ **Fig. 5.52** Belt conveyor (image courtesy of canning conveyor)

bucket is raised out of the water, swung over the land and dumped" (NSSGA 2013). The dragline works on a bench on the highwall or on top of the highwall itself. The excavated sand and gravel will be stockpiled to drain, taking as long as 12 h, and must be rehandled to the processing plant. This process utilizing excavators or shovels clearly increases the final operational costs of the quarry. Slackline excavators are used in large and deep alluvial or glacial deposits since they are routinely mined over wide areas and a long period.

Finally, for the hauling of the aggregates from the extraction area to the processing plant it is easier to use belt conveyors (■ Fig. 5.52) due to the small particle size of the extracted material. Nevertheless, hauling using dump trucks is still generally utilized. The selection of conveyor or truck is carried out according to economic and environmental/social aspects.

Regarding dredging techniques, they are often used due to their ability to target specific sections of the deposit. Dredging equipment is typically situated on floating platforms that allow it to move freely above underwater deposits. In this position, the extraction of the materials is accomplished either with large suction pumps (■ Fig. 5.53) that pull the material to the platform level or by one or more bucket excavators that carry the material up to the same level. The dredger would discharge the aggregates to auxiliary barges for delivery to the plant, although, at locations where dredging is performed near shorelines, the use of pumps and pipelines to transport material to onshore storage locations is very common. For its part, grab dredgers "can work in very deep water and can extract deposits containing large cobbles and boulders that the suction dredgers are unable to extract" (Smith and Collis 2001).

In some areas such as floodplains and low terraces, another possibility is to fill the pit with water because it may not be practical to drain the excavation. In these cases, the operator may extract the material by using typical wet mining techniques.

5.3.3 Marine Aggregate Extraction

Most marine aggregate extraction is developed at short distances from landing ports close to the consumer site because this material has a low value. They are developed in discrete geographical locations dictated by the spatial extent of the resource. Marine aggregate extraction is mainly undertaken by purpose-built (specific) dredging vessels (■ Fig. 5.54), avoiding the requirement of a permanent exploitation platform at sea. To locate the dredging area, vessels use satellite navigation systems. Dredgers employ powerful pumps to draw seawater and a mixture of sand and gravel from the seabed, which are transferred into the ship's cargo hold. Some dredgers with cargo capacities of up to 9000 tons work using pumps capable of drawing aggregate at almost 3000 tons of material per hour, at water depths up to 50 m (even 90 m, for Japanese vessels). In this sense,

■ **Fig. 5.53** Floating suction dredge (image courtesy of Tronox)

5

Fig. 5.54 Dredging vessel (image courtesy of British Marine Aggregate Producers Association)

it is important to note that the ability of dredging vessels to access aggregate deposits is strongly controlled by the water depth, with most dredgers incapacitated to work further down water depths of 70 m. The high capital cost of these specializing dredgers requires maximum use and minimum travel times.

The type of dredging vessel is determined "by the existing regulatory regime, economic considerations, contactor requirements and site conditions" (Velegrakis et al. 2010). There are two principal types of dredging vessels used for offshore sand and gravel extraction: (a) static or anchor suction dredging and (b) trailer suction dredging (Figs. 5.55 and 5.56). Both techniques use centrifugal pumps to draw up the sediment through pipes of up to 1 m diameter. Static suction dredging involves vessel anchoring or remaining stationary over a deposit. This method is suitable in small spatially restricted or thick, localized aggregate deposits; its main disadvantage is the low efficiency of recovery.

Trailer suction dredging "requires the dredger to move over the seabed, permitting large areas with thin and more evenly distributed deposits to be worked; this method is also utilized for thicker deposits to limit the environmental impacts to the superficial layers" (Garel et al. 2009). Unlike anchor dredging, trailer suction dredging may be easily controlled to exclusively recover the payable materials without penetrating the substrate.

The total dredged cargo commonly includes a mixture of different grain sizes. This material can be retained on board for further discharge and processing ashore or can be processed on board using screening techniques, spilling back to the sea the unwanted material. Screening marine sediments can remove specific sizes of dredged material to meet the requirements of a particular customer or market. This method enables only loading and transporting cargo of the needed type. Screening can remove either fine sediments to pro-

Fig. 5.55 Trailer suction dredger (atozhub.weebly.com/mechanical-engineers/dredging)

■ **Fig. 5.56** A trailer suction hopper dredger working at the entrance of Santander harbor (image courtesy of María Ángeles Morán)

vide basic gravel material or remove coarser sediments to load sand. Properly managed screening may expand the life of a dredging area to a large extent by making fuller use of the natural deposit. Nevertheless, it is not permitted many times since screening is considered likely to have unacceptable environmental impacts. The amounts of material returned to the seabed through screening can be very significant and reach up to 60% of the total dredged sediments.

After a cargo has been loaded, the vessel transports the aggregates to a wharf for discharging. Aggregate dredgers are designed to self-discharge the cargo at the wharf. The wharves are located in major ports, close to large population cities. This is because these cities require large quantities of aggregates for construction activities. A cargo will be discharged at the wharf to a stockpile in a holding area using a range of machineries such as bucket wheels, scrapers (■ Fig. 5.57) or wire-hoisted grabs that place the aggregate onto conveyor belts to deliver the material ashore at rates as high as 2000 tons per hour. Alternatively, hydraulic discharge is also utilized principally for beach-nourishment processes.

Dredging activities "are subject to environmental impact assessments and should be controlled by the competent regulatory authorities. In several EU Member States, dredging vessels are required to be equipped with Electronic Monitoring Systems (EMS), which allow the monitoring of their activities and their compliance with the licensing terms" (Velegrakis et al. 2010)

■ **Fig. 5.57** Dredger unloading its cargo of marine aggregates close to the center of London using scraper buckets (image courtesy of Mark Russel—BMAPA)

(■ Fig. 5.58). The EMS records the position of the vessel while dredging operations are developed, thus assuring that activity is only carried out in license areas.

5.4 Questions

Short Questions

- Define extraction of aggregates
- What is the first step in aggregate extraction?
- What is the main difference between mining crushed rock and mining sand and gravel?
- Enumerate the factors defining the type of drill to use in crushed stone surface quarrying
- Define the most widely used explosive for aggregate quarry blasting
- Explain the most common underground mining method utilized in aggregate extraction
- What are the main factors defining the excavation technique and equipment employed in sand and gravel extraction?
- What is the main difference between dry mining and wet mining in sand and gravel extraction?
- Explain what a dragline is and how it works
- What is the Electronic Monitoring System (EMS) in marine aggregate extraction?

Long Questions

- Explain the wet mining method in sand and gravel extraction
- Explain the different processes involved in marine aggregate extraction

5

Fig. 5.58 Dredging control system in a trailer suction dredger (image courtesy of Royal IHC)

References

Barton H (1996) The Isle of Harris superquarry: concepts of the environment and sustainability. Environ Values 5:97–122

Brown TJ, Coggan JS, Evans DJ, Foster PJ, Hewitt J, Kruyswijk, JB, Millar DL, Smith N, Steadman EJ (2010) Underground mining of aggregates, ASRP project No. 7 assess the feasibility of the underground mining of aggregates, 322 p

Bustillo M (2018) Mineral resources—from exploration to sustainability assessment. In: Springer textbooks in earth sciences, geography and environmental sciences. Springer, International Publishing AG, Cham, 653 p

Bustillo M (2021) Construction materials—geology, production and applications. In: Springer textbooks in earth sciences, geography and environmental sciences. Springer International Publishing AG, Cham, Switzerland, 603 p

Cowell R (2000) Localities and the international trade in aggregates: coastal superquarries on the horizon? Geography 85:134–144

Garel E, Bonne W, Collins MB (2009) Offshore sand and gravel mining. In: Encyclopedia of ocean sciences, 2nd edn. Academic Press, pp 182–190

Gokhale BV (2011) Rotary drilling and blasting in large surface mines. CRC Press, Taylor and Francis Group, London, 744 p

Kirk M (1993) Coastal quarry or superquarry? A review of strategy for hard-rock coastal quarries. Quarry Manag 20:19–23

Langer WH (1995) Geologic and societal factors affecting the international oceanic transport of aggregate. Nonrenew Resour 4:303–309

Langer WH (2003) A general overview of the technology of in-stream mining of sand and gravel resources, associated potential environmental impacts, and methods to control potential impacts. US Geological Survey, Open-File Report OF-02-153, 44 p

Langer W (2006) Crushed stone. In: Kogel JE, Trivedi NC, Barker JM, Krukowski ST (eds) Industrial minerals and rocks—commodities, markets and uses. Society for Mining, Metallurgy and Exploration, Inc., Colorado, pp 171–180

Langer WH, Drew LJ, Sachs JS (2004) Aggregate and the environment. American Geological Institute, US Geological Survey, AGI Environmental Awareness Series, 68 p

Lot J, de la Rupelle G (2000) Stripping, extraction, loading and transport in quarrying massive rocks. In: Primel L, Tourenq C (eds) Aggregates. A.A. Balkema, Rotterdam, pp 164–199

Millar DL, Brown TJ, Kruyswijk JB, Smith N, Coggan JS, Foster PJ, Stradman EJ, Evans DJ, Hewitt J (2011) Assessing the feasibility of underground mining of aggregates in southern and eastern England. In: Hunger E, Walton G (eds) Proceedings of the 16th extractive industry geology conference. EIG Conferences Ltd., pp 54–70

NSSGA (2013) The aggregates handbook, 2nd edn. The National Stone, Sand and Gravel Association, Alexandria, Virginia, 906 p

Prax (2000) Stripping, extraction, loading and transport in the exploitation of sand and gravels deposits. In: Primel L, Tourenq C (eds) Aggregates. A.A. Balkema, Rotterdam, pp 200–227

Smith MR, Collis L (eds) (2001) Aggregates: sand, gravel and crushed rock. Geological Society, London, Engineering Geology Special Publication v. 17, London, 360 p

Velegrakis AF, Ballay A, Poulos S, Radzevicius R, Bellec V, Manso F (2010) European marine aggregates resources: origins, usage, prospecting and dredging techniques. J Coastal Res 51:1–14

Processing Techniques

Contents

6.1 Introduction – 172

6.2 Sand and Gravel – 175
6.2.1 Washing – 178
6.2.2 Screening – 187
6.2.3 Stockpiling and Loadout – 197
6.2.4 Dewatering – 197
6.2.5 Materials Handling – 206

6.3 Hard Rock Aggregate – 208
6.3.1 Crushing – 208
6.3.2 Screening, Stockpiling, and Loadout – 221

6.4 Questions – 223

References – 223

M. Bustillo Revuelta, *The Basics of Aggregates*,
Springer Textbooks in Earth Sciences, Geography and Environment,
https://doi.org/10.1007/978-3-031-42961-3_6

Abstract

Processing techniques basically involve crushing and screening in crushed stone aggregates while sand and gravel materials are processed using screening and washing methods. These processes occur in an aggregate processing plant, whose objective is to prepare the materials in an adequate form for their final utilization as aggregates. Both types of aggregates finally need stockpiling and load out for transport. Upon sale, the final product is loaded on trucks, trains, or barges for transport to the market. Sand and gravel processing for most applications is first accomplished through washing to extract clayey components. Sand fraction separation, grading of the gravel into several, sand classification, and dewatering are the following processes until the final product is obtained. For crushed rock aggregates, after blasting in the quarry, trucks or conveyors move material to the processing plant where crushing (most operations use jaw and gyratory type crushers for primary crushing whereas secondary crushers are usually cone crushers or horizontal-shaft impact crushers) and screening are the main processes for producing fractions with the needed requirements. The processing methods require efficiency since the additional cost is added to the final product each time the aggregate is handled and/or processed.

6.1 Introduction

Aggregate processing consists of transporting material extracted in the quarry (rock or sand and gravel) to the processing plant. This is because, depending on the characteristics of the aggregate deposit, some processing is necessary before the product is suitable for use. The aim of the processing plant is to prepare the materials in an adequate form for their use as aggregates. It is established in key parameters such as particle size, shape and distribution, physical and mechanical properties, and lack of contaminant materials. In this sense, some properties (i.e., particle shape and rock type) will influence the type and range of plant utilized.

Aggregates are processed using either a mobile device or at a fixed plant. The latter is constituted by a large quantity of equipment interconnected by belt conveyors (◘ Figs. 6.1 and 6.2). In general, this type of plant tends to be associated with larger deposits providing extended life ("ideally 30 years or more for hard rocks; ten years or so for sand and gravel"—Smith and Collis 2001) to supply a great number of markets in terms of aggregate types, tonnage, and locations. In contrast, mobile plants (◘ Figs. 6.3 and 6.4) are utilized in specific short-term projects (i.e., quick recovery of a road) and where moving from site to site is essential. They can also be used to temporarily increase the capacity of a fixed plant. In these cases, the life of the operation and volume of the reserves are often established entirely by the single market. Mobile plants can be either rubber-tired or track-mounted. In some cases, a third type of plant must be considered: semifixed plant. It is associated with both crushed stone and sand and gravel where it is expected to live for less than ten years.

In modern fixed processing plants, the equipment is controlled using a computer situated in a centrally located control tower (◘ Fig. 6.5). Moreover, monitoring instrumentation, sampling points, and control measures are almost always provided to ensure the quality requirements of aggregate utilization.

Processing techniques basically involve crushing and screening in crushed stone aggregates whereas sand and gravel materials are processed using screening and washing methods. Both types of aggregates finally need to be stockpiled (◘ Fig. 6.6) and loaded out for transport. After screening, sorting, and washing, belt conveyors transport the material to the stockpile. Mineral separation is utilized for rejecting materials with undesirable physical and/or chemical properties. Upon sale, the final product is loaded on trucks (◘ Fig. 6.7), trains (◘ Fig. 6.8) or barges (◘ Fig. 6.9) for transport to the market.

The selection criteria for the design of an aggregate processing plant are based on a great variety of factors such as mineral reserves, mining permits, the investment needed, quality of the materials destined to the market (products, volume and price sale), production costs, production capacity, flexibility of the plant, and environmental concerns. For instance, the characteristics of the material extracted in the quarry are a key factor because the final applications of the aggregate will depend on the physical and chemical features of the rock (see ▸ Chap. 2). The flexibility of the plant is another key factor since it must be designed according to the possible changes in the demand for the products and variations in a specific sector (i.e., granulometry and volumes). In this sense, the existence of regulation stocks (◘ Fig. 6.10) in the plant can resolve specific problems in the demand for products.

Fig. 6.1 Belt conveyors in a limestone processing plant (Image courtesy of Calcinor)

Fig. 6.2 Belt conveyors in a sand and gravel processing plant

Fig. 6.3 Mobile plant (image courtesy of VAEX Maquinaria)

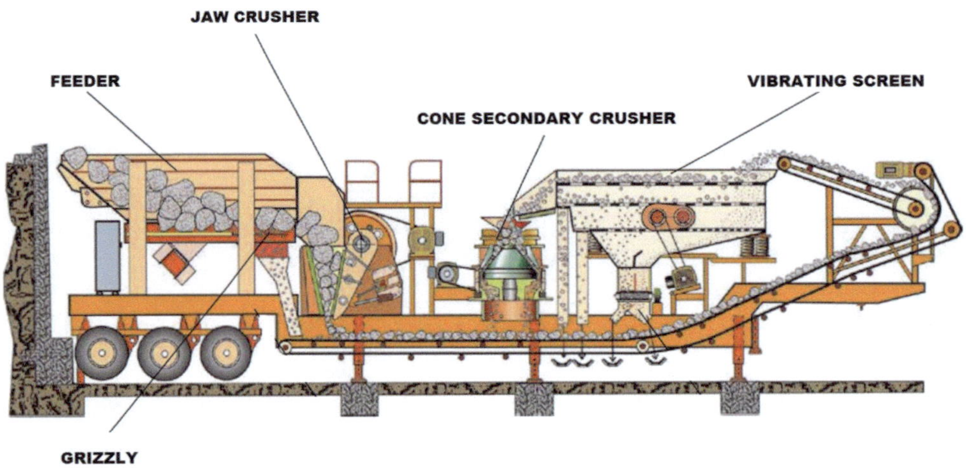

Fig. 6.4 Portable plant (illustration courtesy of Metso)

Some of the processes mentioned (i.e., washing and sometimes screening) require substantial amounts of water that must be available on site. The utilization of water and wet processing techniques facilitates the classification of particles and the rejection of fine particles (i.e., clays). The processing plant must often include equipment for pumping and slurry handling as well as solid–liquid separation and final dewatering equipment. In addition, it is important to note that land must be available near the site to accommodate the complete processing plant. The processing methods require efficiency since the additional cost is added to the final product each time the aggregate

■ **Fig. 6.5** Control tower in a fixed plant (image courtesy of Benito Arnó e Hijos, S.A.U.)

is handled and/or processed. This is because aggregates have a very low-cost margin due to their very low price in the market. A detailed description of the processing plant can be developed based on the type of aggregate, that is, sand and gravel or crushed stone.

6.2 Sand and Gravel

After extraction, sand and gravel to be processed in the plant are transported from the pit by conveyor or haul truck. The material is further stored in a stockpile and a gate at the bottom of the stockpile releases a controlled amount of sand and gravel to a screen where sand is separated from gravel. These materials in some circumstances are utilized as-is, which is called pit-run gravel. However, it is most processed before utilization. There are a great number of ways to configure the plant design based on the characteristics of the sand and gravel deposit and its final application, but it is principally based on screening and washing processes. Sand and gravel plants (■ Fig. 6.11) produce 10,000 to 1 million tons per year, with most in the range of 100,000–300,000 tons per year.

The processing of sand and gravel for most applications is carried out through washing for removing clay, separation of the sand fraction, grading of the

■ **Fig. 6.6** Storage of different aggregate fractions (image courtesy of Benito Arnó e Hijos, S.A.U.)

Fig. 6.7 Transporting aggregates by truck

Fig. 6.8 Transporting aggregates by train (image courtesy of Texas Crushed Stone Company)

Fig. 6.9 Transporting aggregates by barge (image courtesy of CalPortland)

Fig. 6.10 Regulation stocks in an andesite quarry (image courtesy of Benito Arnó e Hijos, S.A.U.)

Fig. 6.11 Sand and gravel processing plant

gravel into different sizes, sand classification and dewatering, and in some cases crushing of oversize gravel (Fig. 6.12) to produce smaller more saleable material (Fig. 6.13). Processing provides scope for adjusting the grain size distribution to match market requirements for the final product. It is also common to blend materials from the same and different sites with the aim of adjusting grain size to meet user requirements and to maximize the utilization of resources. Although the final application controls the size of the aggregate to use, the greatest demand is concentrated in 4/10 mm and 10/20 mm sizes.

6.2.1 Washing

Washing removes silt and clay (material finer than 0.063 mm) that are present in the aggregates, either adhered to the particles or present as clay-bound agglomerates, as well as other undesirable components such as roots and organic matter (Fig. 6.14). They are removed after extraction, but further processing is required to improve the cleanliness of the aggregate fractions. Silt and clay content can be so high as to produce difficulty in handling, screening, and crushing while clay material contaminates the final product (some aggregate applications—i.e., concrete production—request a very low content of clay). In summary, aggregate is washed to make it meet specifications.

The amount of silt and clay in a sand and gravel deposit is a significant feature for defining its economic viability. The silt content is important, but the clay content is much more important because it is harder to remove. In general, the washing process needs to remove 96–98% of material below 0.063 mm and 100% of material below 0.150 mm. Detailed planning of any washing treatment process should consider the kind of pollution and its magnitude as well as the amount of material to be processed. These factors will decide the kind of treatment required and its scale.

Based on the degree and nature of the undesirable components of aggregates, washing processes can be incorporated at different stages of the treatment of materials, starting from the initial up to the final level of classification of sands. These processes can be very simple, such as a brief washing on a screen, or complex, with many treatment operations (i.e., bucket wheels and hydrocyclones). On the other hand, it is important

Fig. 6.12 Oversize gravel

Fig. 6.13 Crushing stage of oversized gravel

Fig. 6.14 Roots and other contaminants eliminated by washing processes (image courtesy of SAMCA)

to note that washing processes use huge amounts of water that become polluted mainly by fines. Therefore, recycling and clarification of water are carried out as complementary operations. Consequently, removal of the silt and clay from the wash water is needed prior the water may be reused or discharged into the local water environment. The reuse of water is vital as its control becomes very strict. The process of washing incorporates several devices utilized in series or parallel as appropriate to a particular requirement of operation. The following are the most utilized in an aggregate washing plant.

- Bars with Spray Nozzles

In coarser fractions, washing can be performed by applying water jets (Fig. 6.15) through spray nozzles directed as a water curtain and under pressure at the material being classified, aiming to remove the impure particles adhering to the material. This device enables the adjustment of water flow to wash the needed tonnage of sand and gravel with maximum efficiency. Nozzles are commonly manufactured in polyurethane (Fig. 6.16) because they are economical and abrasion and corrosion resistant. They are installed in metallic pipes placed cross-sectionally in relation to the material flow. Nozzles must be oriented at 45°-back toward the feed end of the screen. Thus, a water dam is generated, which slows the material, increases the retention time and helps turn the material for effective washing. Washing materials by this method is effective "only if the equipment assures dispersion of the aggregates on the grids and uniform advance of the grains by short throws, thereby exposing the entire surface of the grains to the action of the water jets" (Exbrayat and Mishellany 2000).

- Barrels

Barrels have been used to wash aggregates since the very early days of washing plants. Two types of barrels are available in the market: washing barrels and scrubber barrels. Washing barrels turn at a relatively slow speed and are usually incorporated where the feed material does not need intensive attrition to clean it.

Scrubber barrels (Fig. 6.17) are typically used in aggregate processing plants for beginning the process of removing silt and clay from the raw feed. They turn at a more rapid 50% of critical speed, creating more material-on-material attrition that causes the fines and clays to be almost eliminated in the aggregate. Scrubber

Fig. 6.15 Washing process in a vibrating screen (image courtesy of SAMCA)

Fig. 6.16 Polyurethane nozzles for washing placed cross-sectionally to the material flow (image courtesy of SAMCA)

6

Fig. 6.17 Scrubber barrel (image courtesy of Quarry King)

barrels are horizontal drums in which feed material is cascaded in water, employing impact and attrition to remove silt and clay. They are autogenous and the residence time of the aggregate particles is relatively short, approximately two minutes. Throughput is based on the size of the unit and ranges from 100 to 500 tons per hour, but larger throughputs are performed with the utilization of several units in parallel. Where scrubber barrels do not adequately remove silt and clay, log washers can be used.

- Log Washers

Log washers (Fig. 6.18) are used to remove clays from sand and gravel and crushed materials. The amount, type, and percentage of undesirable components to be extracted are considered when selecting the type of log washer. In some cases, two or more units in a series can be needed. When in operation, they sit at a slope between zero to 14° (with 8° being typical), depending on the severity of the washing application. As the amount of clays increases, the retention time is increased raising the slope. It is recommended to eliminate the sand-size material before the feeding to the log washer. This is because the finer material tends to cushion the washing action.

Feed is introduced at the lower end of the unit whereas water is introduced at the upper end. Sand and gravel are transported upward by the paddles against the water flow (Fig. 6.19), maximizing attrition and allowing the washers to remove clay and other undesirable components. Later, an overflowing weir device located at the lower end of the washer removes the undesirable components whereas the cleaned aggregates are discharged at the upper end.

- Spiral Classifiers

In spiral classifiers (Fig. 6.20), a continuously revolving spiral (one or two) moves the sands up the slope. The shape of the spiral classifier tank is usually rectangular and its slope ranges from 14° to 18°. The size of particles to feed the spiral classifiers is in the region of 150 mm and coarser. The pulp agitation degree and the tonnage rate of sand extraction are determined by the speed of the spiral. "For coarse separations, a high degree of agitation may be necessary to keep the coarse particles in suspension in the pool, whereas for finer

Fig. 6.18 Log washer (image courtesy of McLanahan)

Fig. 6.19 Paddles and flow of water (image courtesy of McLanahan)

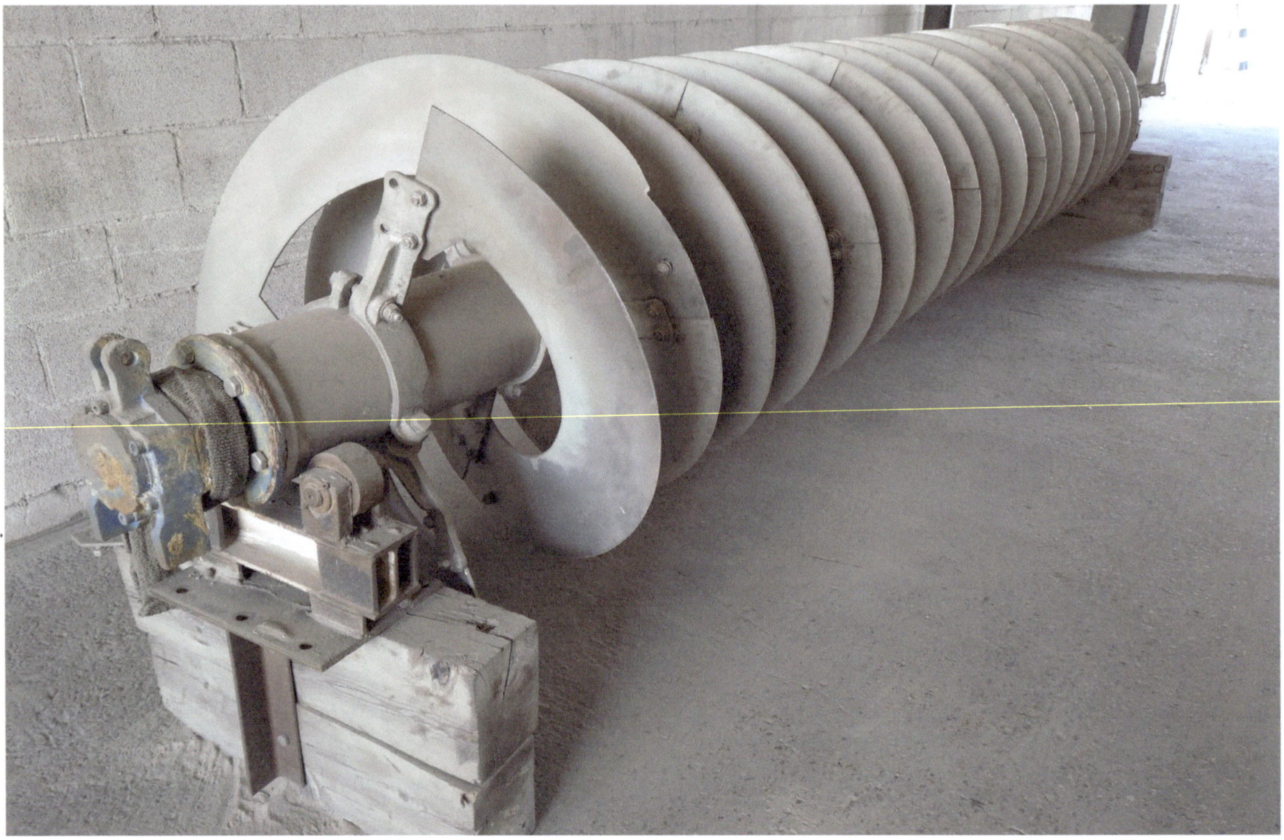

Fig. 6.20 Spiral classifier (image courtesy of SAMCA)

separations, less agitation, and lower raking speeds, are required" (Wills and Finch 2016). However, it is very important that the speed is high enough to transport the sand up the slope. In the operation of these devices, the dilution of the pulp is very important. Spiral classifiers never operate at more than approximately 50% solids.

- Bucket Wheels

For fine-grained material (coarse-grained material is already cleaned during the screening process), a washing bucket wheel (Fig. 6.21) is utilized, which is a wheel of large diameter incorporating a series of perforated buckets. The dewatering wheel principle originated more than a century ago. In this device, a suction dredger or a compressed-air dredger feeds the feeding mixture of sand, gravel, and water. The solids are deposited in the bottom of a tank and are excavated by the bucket wheel and the material is either dumped onto a conveyor or into a stockpile. The fines and clays are evacuated by the overflow from the tank. According to performance, the bucket wheel rotates only at 0.4–2 rpm because the wheel speed should not give rise to substantial movement. The wheel diameter and bucket width determine the solid handling capacities. For its part, the surface area of the slurry bath determines the hydraulic capacities. Particles ranging from 0.1 to 0.15 mm can be separated with little pollution of the water. Bucket wheels are very popular due to their high water capacity and they use very low horsepower.

- Hydrocyclones

Most sand washing plants involve the dispersion of sand in water and the separation of fines in a hydrocyclone (▶ Box 6.1: Hydrocyclone), which delivers a partially dewatered coarser product.

Fig. 6.21 Dewatering bucket wheel

Box 6.1—Hydrocyclone

*In Hydrocyclones, rapid settling and classification are carried out by incrementing the force acting on the particles by substituting the gravitational force with centrifugal forces. Hydrocyclones (commonly abbreviated as just cyclones) use the forces of centrifugal flow to accelerate the decantation of particles suspended in a fluid by weakening the viscosity and density effects. Hydrocyclone (*Fig. 6.22*) has grown into one of the most interesting and broadly used classifiers, capable of sizing particles as fine as 0.005 mm (the removal of silt and clay is accomplished by operating with a cut point in the 0.040–0.070 range). It is usually applied in closed circuits within crushing systems. The main advantage of hydrocyclones is that they have large capacities in comparison to their size and can split at finer sizes than most other screening and classification devices. Hydrocyclone units can be installed on simple supports as single units or in clusters. The simplicity of installation and maintenance are further advantages of hydrocyclones.*

*A typical hydrocyclone is formed by a conically shaped vessel, open at its apex, or underflow joined to a cylindrical section, which has a tangential feed inlet (*Fig. 6.23*). Particles entrained in the fluid under pressure through the tangential entry experience a strong centrifugal force because of the spiraling motion of the carried fluid. This generates a vortex in the hydrocyclone, with a low-pressure zone along the vertical axis. As a result, the particles settle through the fluid toward the outside of the cyclone. Sometimes, hydrocyclones are provided with nozzles just above the apex to inject water. This is because the nozzles help to compensate for water loss and loss of fines.*

The largest particles settle more quickly and migrate to the apex opening. Due to the action of the drag force, smaller particles may not settle quickly enough to reach the envelope and move toward the zone of low pressure, being carried upward to the overflow. The greater the speed of the fluid is, the lower the particle separation size. In this sense, the hydrocyclone works as a centrifugal sedimentation classifier. Hydrocyclones are forthcoming in a wide range of sizes, in accordance with their applications. They range from 2.5 m in diameter down to 10 mm, which corresponds to cut sizes of 300 µm down to 1.5 µm, respectively. These devices cannot be utilized where coarse separations are needed.

Fig. 6.22 Hydrocyclone for separation of fines

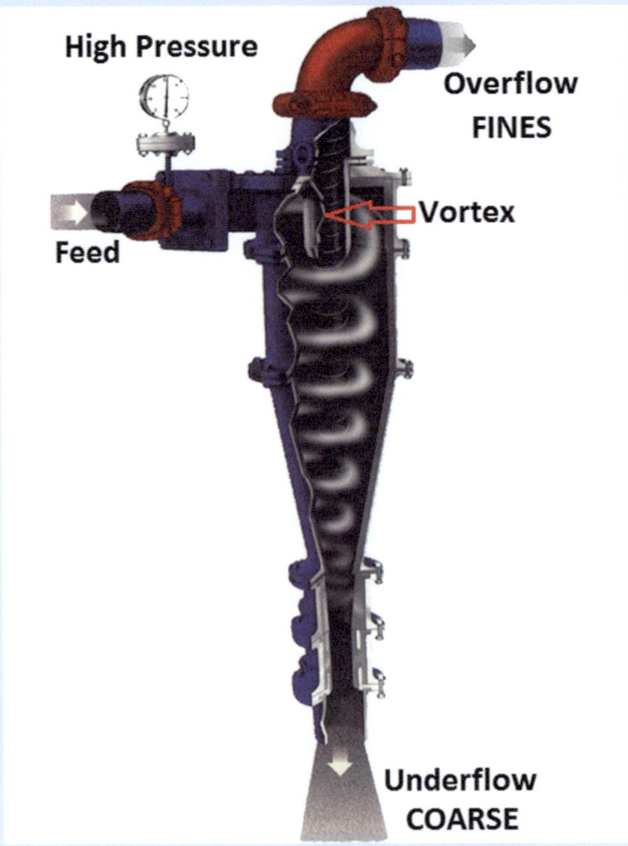

Fig. 6.23 Schematic representation of a hydrocyclone

6.2.2 Screening

Two distinctly types of separation methods based on size can be taken into account: screening and classification. Screening is the mechanical separation of aggregate particles into various sizes while classification is a method of separating particles on the basis of the speed with which the particles fall through a fluid medium (i.e., water—Fig. 6.24). Selection between screening and classification is influenced by the fact that finer separation demands large areas of screening surface and can be expensive compared to classification for high-throughput applications. Unfortunately, ideal separation is never achieved because some undersized particles will be reported to the oversized stream, and some oversized particles will be reported to the undersized stream. As a rule, in aggregate processing plants, screening is used while classification is only used in very specific cases.

Screening is one of the oldest methods of separation and is utilized in many industries worldwide. Industrial screening is "extensively applied to size separation from 300 mm down to 40 μm, although the efficiency decreases rapidly with fineness" (Wills and Finch 2016). For its part, classification is almost the only possibility to separate particles smaller than approximately 0.1 mm. Thus, "the selection of an appropriate type and size of screen is essential because an aggregate plant with inadequate screening capacity limits the plant production rate and increases the difficulty of producing consistent products" (NSSGA 2013). Equally important is the selection of the screening media.

Screening uses a geometrical pattern for size control: the screen. In its simplest configuration, a screen is a hard perforated and rigid surface that is uniformly perforated with apertures (Fig. 6.25), which are normally all the same size. The size and shape of the apertures (an aperture is an individual opening in the screening surface), the proportion of open area (it is the surface percentage that is aperture) and the material properties of the screening surface are typically critical to the performance of a screen device. In screening, the material is presented to the screen surface so that the material finer than the apertures falls through the screen (they are called undersize, passing, minus, or lower product) and the oversize is conveyed to the discharge end of the screen (termed oversize, retained, plus, or upper product) (Fig. 6.26). The desired separation size is called the cut size. The extent to

Fig. 6.24 Hydraulic classifier (image courtesy of SAMCA)

Fig. 6.25 Screen

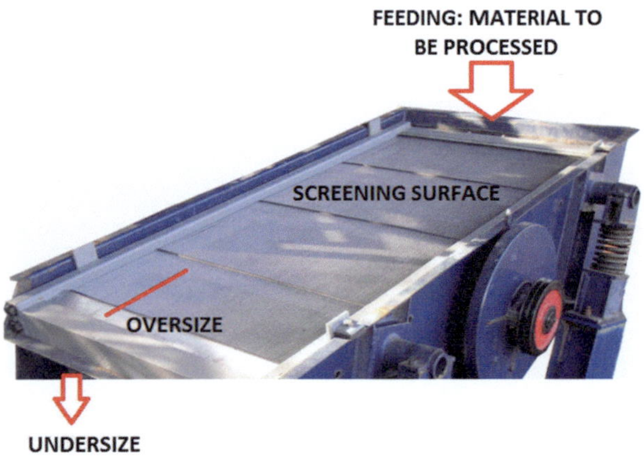

Fig. 6.26 Screening process

which particles are located in the wrong product stream indicates the ineffectiveness of the separation process. Screens are associated with all types of crushing units at the feed and discharge stages. Each screen provides two products; to obtain more products, it is necessary to use additional screens.

Three principal constraints limit the utilization of screening in industrial-scale size separations. First, the screens tend to become blinded by particles that get stuck in the apertures (plugging—Fig. 6.27—and blinding); this trend tends to increase quickly as the size of the apertures decreases. Below an aperture size of approximately 0.5 mm, it becomes a very serious problem. Second, every particle must be presented to an aperture at least once to achieve a reasonable separation. Near-size particles, those that are only slightly smaller than the screen aperture, can need to be presented many times before they fall through. Thus, "to screen very fine particles efficiently requires repeated presentations of a very large number of particles to a limited number of apertures; this is because the number of apertures is limited and the number of particles per unit mass increases very rapidly as the particle size decreases" (Bustillo 2018). Consequently, the separation rates decrease very significantly as the separation size decreases.

The third constraint is related to the screening surface and the number of apertures. The screening surface "must be strong enough to support the weight of the material, be flexible enough to withstand the vibration, and provide enough open area to allow the desired throughput of aggregates" (NSSGA 2013). With decreasing cut size, a reduced proportion of the screening surface is occupied by apertures and the percentage of the open area of the screen surface decreases. The reason for this is that the rigidity and mechanical integrity of a surface demands a certain minimum amount of material between the apertures. As the number of apertures increases and their size decreases, the proportion of the open surface decreases as well.

There are many types of screening surfaces available in the aggregate market. The two most important factors for the selection of the most adequate in each case are the screen panel life expectancy and open area. Screening surfaces can be placed into one of three general categories: woven wire screen (Fig. 6.28),

Fig. 6.27 Plugging in a screen

Fig. 6.28 Woven wire screen

Fig. 6.29 Perforated plate

perforated screen plate, and profile wire or bar. Woven wire cloth is the most utilized for screening aggregates. Common options for this type of surface are high-carbon, oil-tempered and stainless steel. Perforated plates (Fig. 6.29) are used in very intensive utilization and/or secondary screening and are available in various steel types and hardnesses. Plates are more expensive, but they resist wear and have a long life, are less blinding, have higher efficiency, and have a high degree of accuracy in sizing. The final selection depends on the size range, nature of the application, required capacity and efficiency of the screen.

Screening surfaces are manufactured from steel, rubber, or polyurethane (Fig. 6.30). Rubber and polyurethane are classically used in high wear areas to increase service life, reduce noise, and decrease blinding by flexing the aperture that can be tapered outward from the top. The rubber surface is ideal in dry, high-impact applications while polyurethane is utilized in cases where water is added or the feed has a slurry form. Rubber displays the longest wear life of any screen media in the most aggressive conditions.

Regarding the shape of the apertures (Fig. 6.31), several configurations are available: square, round, and rectangular (with the flow or across flow). Square configuration is the most used in aggregate industry because it provides the most accurate separation and good wear life with a reasonable open area. Round apertures are utilized in some heavy-duty applications but the open area is lower than for square apertures. Rectangular apertures with the flow are designed for screening regular shaped particles whereas across flows are used where pegging is likely to occur.

Screen performance is controlled by screen capacity and efficiency. The former can be measured in tons per hour of aggregates being fed to the screen. For its

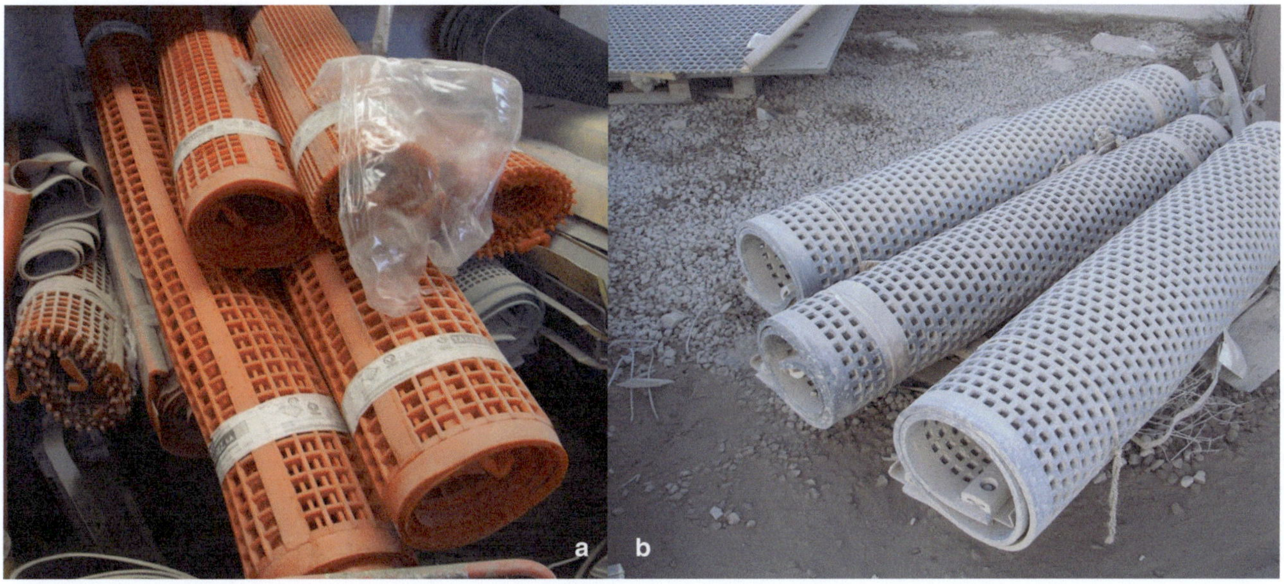

Fig. 6.30 Screening surfaces of polyurethane (**a**) and rubber (**b**)

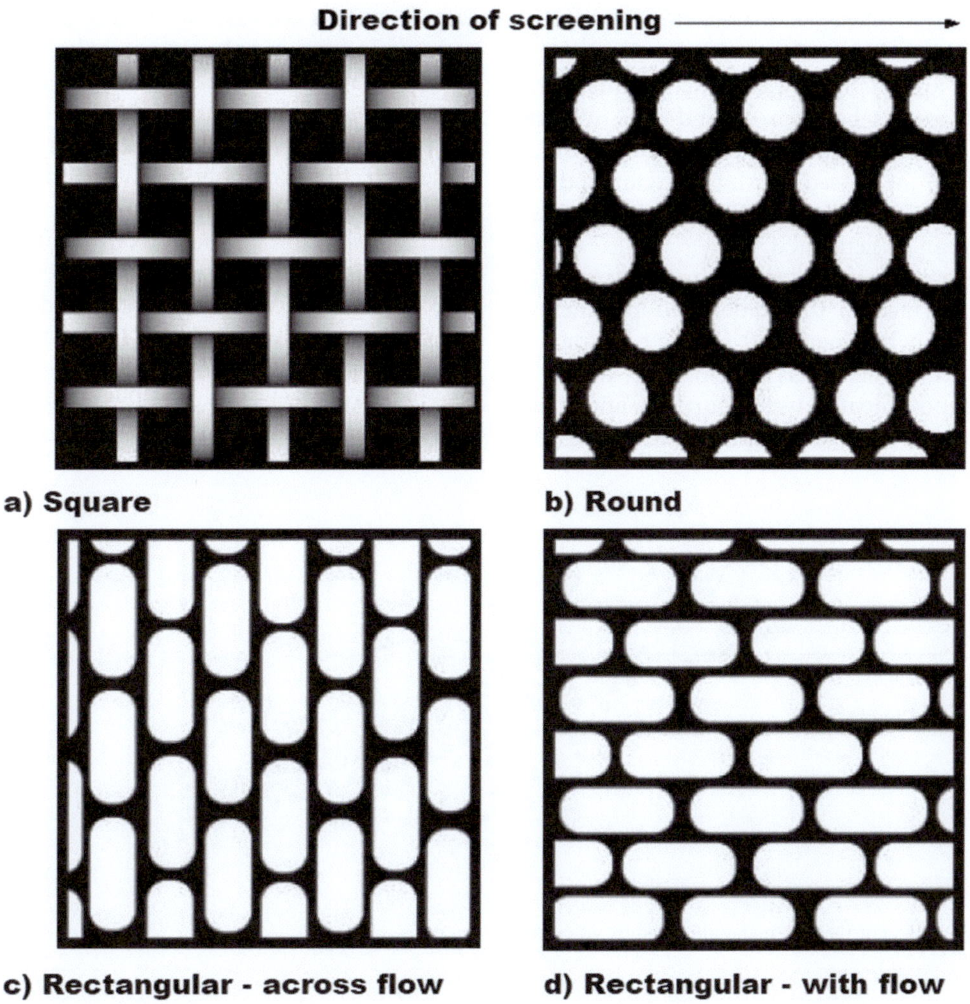

Fig. 6.31 Common aperture shapes

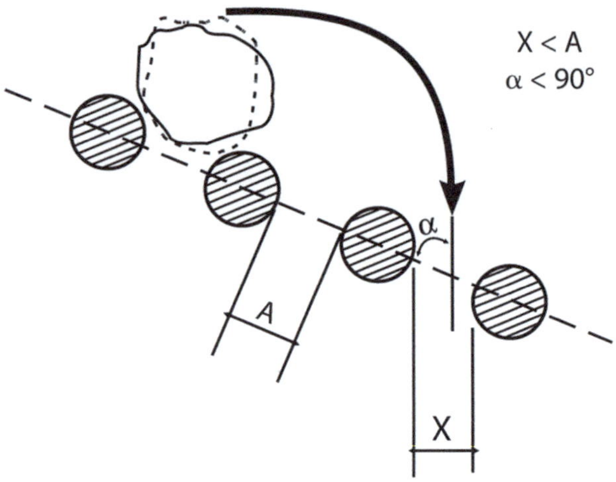

Fig. 6.32 Screen angle or slope (illustration courtesy of Metso)

part, the efficiency of screening is calculated as the percentage of the undersize particles passing through the apertures in any time interval divided by the percentage of lower product in the feed (this can be reversed if the oversize is the product to be recovered). For instance, 75% screen efficiency means that 25% of the material is being rejected with oversized material. Industrial screens are classically performed to offer screen efficiency with ranges between 90 and 95%. Where a screen operates at low efficiency, important problems (i.e., overload of the closed crushing circuit and contamination with particles of size noncompliant with specifications) can be generated.

Whether a particle passes through the apertures of the screening surface to the undersized product will rely principally on its shape and size but characteristics such as screen dimensions and loading, the nature of the relative movement between particles and screening surface, the percentage of open area or degree of perforation, and screen angle can also influence the separation process. Regarding the particle shape, fairly small plate-like particles can tend to bridge apertures and be misplaced with the oversized product. Screen effectiveness must always be coupled with capacity.

The screen angle or the slope of the screen deck also considerably affects the separation process because the slope of the screening surface influences the angle at which particles are presented to the screen apertures (Fig. 6.32). If particles approach to apertures at shallow angles, there will be a narrower efficient aperture dimension; thus, near-mesh particles have less opportunities to pass. Some screens utilize this effect to achieve separations significantly finer than the screen aperture. For example, banana screens incorporate a variable-angle slope that allows for increased throughput. Where the discharge end of a screen deck is inclined down from the horizontal, material cascades more rapidly down the slope and either passes through an opening or over the screen surface in accordance with some probability. Hence, the capacity of a screen must increase as the deck slope increases. Efficiency will remain constant or increase up to a critical slope and decrease as the slope increases. In crushing plants, screen decks are installed with angles between 20° and 30° below the horizontal.

Vibration is another factor that significantly affects screen performance. Screens are vibrated to throw particles off the screening surface and they are another time presented to the screen. The motion is in the horizontal plane of the screen, but it can also be circular or reciprocating, or with a vertical component. The right type of vibration also induces stratification of the feed material (Fig. 6.33), also termed trickle stratification, of the material. Consequently, the finer particles of aggregate pass through the bed and meet the deck surface giving a sharp separation whereas the

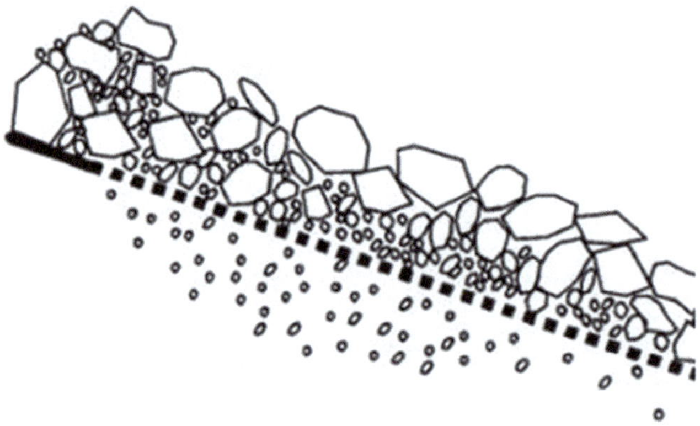

Fig. 6.33 Stratification of the feed material caused by vibration

Fig. 6.34 Grizzly for rough screening

coarser particles remain at the top. This is because the motion of the screen decreases the internal friction in the material.

Wet screening is classically taken into account when the feed materials have high moisture content (i.e., 40%), when fine sizes must be screened (below 5 mm) or when fine particle agglomerates are present. Dry screening is limited to the materials above 5 mm in size. Screen efficiency can be improved with wet screening by helping to move fines through the aperture. As a rule, it can be established that the more dilute the feed is, the more efficient the process.

- Screening Equipment

The types of screens most used to separate particles in the aggregate industry include the following.

Grizzlies

Grizzlies (Fig. 6.34) are used for rough screening of coarse to very coarse materials. They are most often found in crushing circuits and are installed for sizing the feed to the primary crusher. Deposits containing boulders may be fed through a grizzly screen to remove large boulders; sometimes, these boulders go to the crusher to produce crushed aggregates. Crushing is now a common feature at many sand and gravel processing plants, and is very necessary to maximize the saleable product. A grizzly is a group of heavy manganese steel rails, bars, or rods placed in parallel. The spacings in between are on the order of 5–200 mm. Rail grizzlies can be installed to operate in a horizontal flat plane, but they are often inclined in the direction of the material flow to aid the transport of material across the screen. The inclination ranges from 30° to 40°. Coarse material slides on the inclined surface of the bars and fine material falls through. Grizzlies are static but they can be vibrated to improve their performance, for instance where sticky materials are present. These devices are less prone to blockage, providing a positive control upon feed rate.

■ **Fig. 6.35** Trommel (**a**) and the screen inside the device (**b**)

Trommels

A trommel, revolving screen, or drum separator is a horizontal, or slightly inclined in the direction of the material flow, rotating cylindrical screen with perforated steel plates (■ Fig. 6.35). These machines are one of the oldest screening devices, comprizing a cylindrical screen rotating at between 35 and 45% critical speed. They can handle material from 55 mm down to 6 mm and even smaller sizes can be achieved under wet conditions. The material is fed at one end of the cylinder and moves down the inside of the inclined surface. Uundersized materials fall through the screening surface and the oversized is transported to the discharge end through the rotating motion down the incline. Trommels may handle both dry (i.e., in construction materials) and wet feed material.

Vibrating Screens

Vibrating screens are the most important and versatile screening machines in the aggregate industry (▶ Box 6.2: Vibrating Screens). They are classically used for applications above 2 mm and are located within crushing plants. Vibrating screens are specifically well-equipped to meet the fineness requirements of the aggregate market.

Box 6.2—Vibrating Screens

In the vibrating screens, the perforated surface is set into a frame that is agitated vigorously. The movement of the frame can be designed to facilitate the translation of the material bed along with the screen. The selection of the mechanism controls the type of screen motion causing the movement of the particles on the screen. The vibrating screens are easily adjusted to enhance screen efficiency. Changes in slope, speed, stroke, and direction of rotation provide the flexibility needed. The rotation direction can be contra-flow or in-flow. Contra-flow slows the particles more and allows more effective separation while in-flow permits a better performance.

Vibrating screens (■ Fig. 6.36) *have rectangular screening surfaces with feed and oversize discharge at the opposite end. They perform size separations from 300 mm down to 100 µm. One, two, or three screening surfaces can be set into a single frame in single-, double-, or triple-deck configurations, respectively. The screen decks are mounted one above the other deck; as a consequence, each deck has a smaller aperture than the deck above it. On the other hand, all the assemblies can be horizontal or inclined from feed to discharge; inclined screens are the most popular for aggregate applications. They use gravity, helping to move the material and requiring low energy consumption. The rapid relative movement engineered between the particles and the screening surface facilitates rapid rates of screening. Screen slopes can change from 15° to 30° in dry applications to slightly flatter slopes in wet separations. As mentioned above, water is sometimes used to enhance the separation process.*

The Mogensen sizer (■ Fig. 6.37) *is a classical vibrating screen, a probability device useful in the separation of particles in the size range of 5–0.1 mm. It utilizes the principle that particles smaller than the aperture statistically require a certain number of presentations to the screen to pass. The Mogensen sizer system is made up of several screening decks (up to six decks) that are arranged on top*

Fig. 6.36 Vibrating screen (image courtesy of Metso)

of one another and have different inclinations. This arrangement enables particles much finer than the screen apertures for passing easily whereas causing larger particles to be rejected by one of the screen surfaces.

The Banana (Fig. 6.38*) or multislope or multi-angle concept screening machine is a major innovation in screening technology. It is a nonconventional vibrating screen capable of achieving exceptional throughput per screening area. The screen includes as many as six inclinations varying from 45° to 10°. The screens are often designated for fitting modular polyurethane deck panels although woven wire or punched plates are sometimes utilized. Banana screens are popular in high-tonnage sizing applications where both efficiency and capacity are important. They are designed with a linear-stroke vibrator.*

Fig. 6.37 Mogensen sizer (image courtesy of AT Mineral Processing)

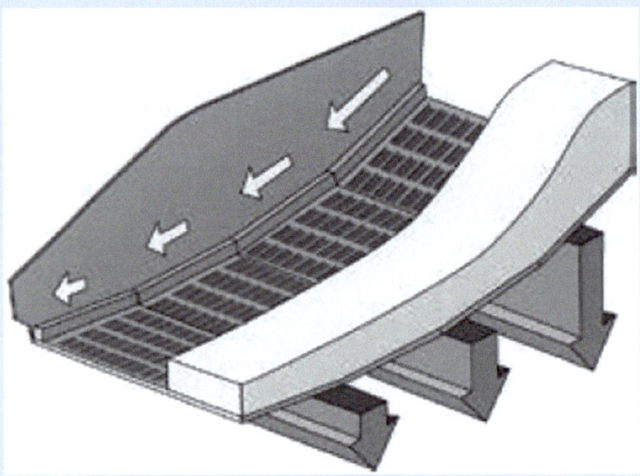

Fig. 6.38 Illustration of a banana screening machine

6.2.3 Stockpiling and Loadout

Finished aggregate materials are stockpiled by a belt conveyor or by mobile equipment. Many plants use some combination of the two methods. Upon sale, the stockpiled aggregates (◘ Fig. 6.39) may be sold as a single-size product or, as commented previously, two or more materials may be blended to make a new, graded product. For example, sand and properly sized gravel may be mixed in a defined proportion to be utilized with cement to produce concrete. Procedures must be followed carefully when stockpiling and handling the final product. It is important to bear in mind that when aggregate falls too far from conveyors onto coned stockpile (◘ Fig. 6.6) or when it is dumped from trucks down a slope, the material can be separated from a well-blended product into individual size fractions, originating an inadequate product.

Stockpile sites must be well-drained and have an adequate bearing capacity for supporting the weight of the materials. For this purpose, the base for stockpiles must be constituted by sound, tough, and durable rock fragments and sand. These materials must also to pass the maximum sieve size of the aggregates to be stockpiled and are free of deleterious substances. Improper handling can also result in contamination of the aggregate with foreign material from underneath the stockpile. The final product is loaded on trucks, trains (◘ Fig. 6.6) or barges (◘ Fig. 6.7) for transport to the market. Proper management of aggregate stockpiles needs several stages such as appropriate environmental regulation compliance, inventory control and accurate stockpiling, and load-out practices, with the aim of avoiding problems such as degradation, contamination, and segregation (◘ Fig. 6.40).

The loadout of aggregates is an important stage that needs to be well-designed and considered. Anything impacting loadout cycle times will be costly to an operation. Loadout can be carry out using tunnels, feeders, conveyors, bins, scales, or other devices. One of the most common methods employs a front-end loader to transport the aggregate fraction from the stockpiles to the trucks.

6.2.4 Dewatering

The unwanted fines and other contaminating materials generated in the washing processes are allowed to settle in a pond or can be further dewatered using thickening and filtration processes. Thus, the water can be

◘ **Fig. 6.39** Stockpiled aggregates

Fig. 6.40 Segregation process in a stockpile

recycled back to the plant. From a water utilization point of view, thickening and filtration are preferable to silt ponds as water losses (infiltration and evaporation) are largely avoided, decreasing the water needed.

- Settling Ponds

The use of settling ponds or silt lagoons is the most classical method for the disposal of fines to discharge the waters (Fig. 6.41). In these structures, the solids are allowed to settle under gravity. Nevertheless, the utilization of this approach is clearly diminishing. One advantage of settling ponds is that they can treat water from multiple sources (e.g., washing plants and contaminated run-off). Settling ponds are located in excavated areas below ground level and are all designed to slow the flow of surface waters to improve the settling and consolidation of suspended solids. A common design includes the construction of two or more ponds in such a way that the water becomes progressively cleaner as it passes from pond to pond. Sometimes, the utilization of these constructions is constrained by land availability and site topography. Once full, settling ponds can be restored and revegetated since the cost and time involved to clean the ponds with an excavator is excessive.

Settling processes can be assisted using selected chemical flocculants. Flocculants (Fig. 6.42) quickly settle out almost all the suspended solids in a dirty water stream (flocculation involves the attachment of particles to long-chain polymers). Regarding the flocculant composition, the development of synthetic polymers has been a major advancement in physical separation during the last 50 years, and they have largely replaced organic polymers.

As an alternative to the construction of a settling pond, the use of thickeners and filter presses is becoming more usual. The installation of thickeners decreases the requirement for four or five silt lagoons down to one lagoon. On the other hand, attaching a filter press can eliminate the need for a lagoon. The combination of a thickener and a press filter saves space and recovers a great amount of water for recycling back to the plant.

- Thickening

Thickening of sludges is carried out using thickeners (Fig. 6.43), consisting of shallow tanks from which the clear water is removed at the top whereas the thickened suspension is recuperated at the bottom. The basic objectives of these devices are the generation of a clarified overflow of water and a thickened underflow of the needed concentration. Clarified water (Fig. 6.44) overflows at the top of the tank and is recycled. The solids concentration in a thickener varies from that of the clear overflow to that of the thickened underflow being discharged. Since the process is continuous, there is no region of constant composition in the thickener. Consequently, different zones are maintained within the thickener that corresponds roughly to the different stages of settlement (Fig. 6.45). The degree

Fig. 6.41 Settling pond

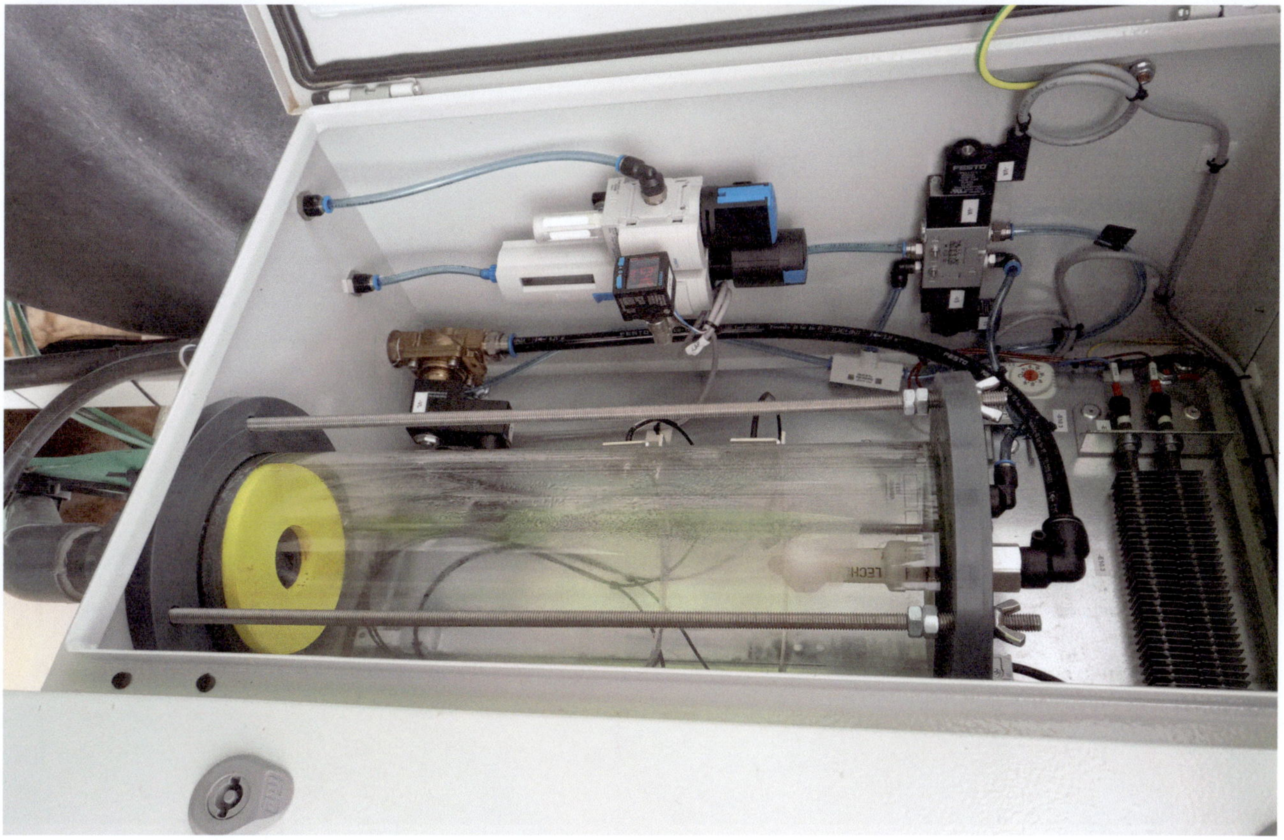

Fig. 6.42 Dosing system for flocculant addition (image courtesy of CONORSA)

6

Fig. 6.43 Thickener

Fig. 6.44 Clarified water of the thickener

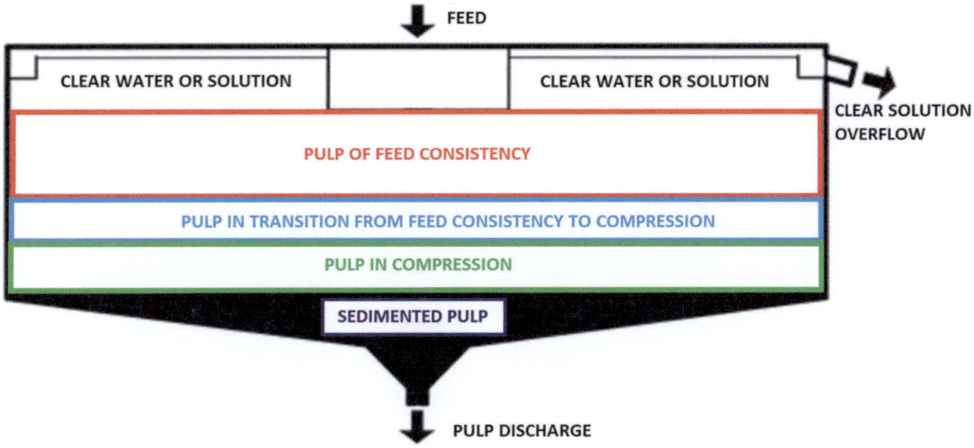

Fig. 6.45 Thickener zones

of thickening produced is controlled by the residence time of the particles and hence by the thickener depth.

The design of a continuous thickener consists of a large, shallow cylindrical tank with a conical bottom sloping toward the center, a diameter ranging from approximately 2–200 m, and a depth of 1–7 m. Thickeners considerably decrease the volume of slurry requiring subsequent handling. Thickeners are very useful when land availability is limited or when health and safety risks are unacceptable. The thickening process can also be assisted using automated dosing systems with chemical flocculants.

- Filtration

Rapid settling of solid particles in water originates clarified water that may be decanted. The resulting thickened slurry may need dewatering using filtration methods. Filtration is the process of separating solids from liquid by employing a porous medium, termed the filter. It retains the solid, termed the cake, although allows the liquid to pass. The filter medium (Fig. 6.46) causes a clear separation of particulates from the fluid but always with the minimum consumption of energy. Thus, "correct selection of the filter medium is an essential part of efficient filter operation, both in terms of operating cost and filtrate clarity" (Cox and Traczyk 2002). Filter media are manufactured from many different materials such as cotton, wool, silk, glass fiber, paper, or metals.

The filtration occurs under a varied group of conditions and the choice of the most suitable type of equipment will depend on many factors. Whatever type of equipment is used, "a filter cake gradually builds upon the medium and the resistance to flow progressively increases throughout the operation" (Bustillo 2018). There are five steps in the cake filtration process: cake formation, moisture reduction, cake washing, cake discharge, and medium washing. The filtering process is completed when nearly all the liquid has been eliminated from the pulp and the filter cake is removed from the filtering medium.

Fig. 6.46 Membrane for filtering (image courtesy of TEFSA)

In aggregate processing plants, filtration is carried out using a filter press. This device removes the silt that a thickener has collected and extracts the water forming the cake, which is later discharged. Finally, the cake is moved by a loader to a disposal area or many times they are utilized as backfill material. One of the most typical filter presses used in aggregate processing plants is the vertical plate filter press (► Box 6.3: Vertical Filter press).

6

Box 6.3—Vertical Plate Filter Press

*Vertical plate filter presses are great pieces of dewatering equipment, and are principally utilized with slurries made up of approximately 50–60% solids. They separate the liquids and solids using the pressure filtration method. Each device is designed according to the volume and type of slurry that must be dewatered. In this type of filter press (*Fig. 6.47*), filter plates are made from lightweight polymer and suspended vertically from a steel frame.*

*During the filling cycle, the filtering medium is situated against the sides of the plates and the slurry is pumped into the machine. The solids, which are distributed evenly during the filling cycle, fill each chamber. The slurry pressure presses (with pressures up to 8 bar) the pulp against the medium, forcing the liquid through the cloths. As a consequence, the solids formed a cake on both surfaces of the frame (the filter cloth prevents the solids from escaping from the plates). The chamber spaces are full of dewatered solids and the slurry feed is finished once required moisture level has been reached. The cycle is complete and the filter cakes release. When the press is opened, the plates separate and the cake falls by gravity (*Fig. 6.48*). To finish, the filter cloth was cleaned, the chambers were closed, and the cycle was repeated. The cycle time typically takes 10 min. The filtrate, basically clean water, exits the filter plates and discharges through the side of the plates (*Fig. 6.49*).*

The plates have square shapes with ribbed or studded surfaces although circular plates are available on the market. The plate sizes vary from around 500 mm × 500 mm to 2500 mm × 2500 mm and frames from 10 to 200 mm in thickness. The number of plates ranges between 25 and 50 although up to 100 plate filters can be present.

The selection of the thickness of the cake is critical to the adequate functioning of the press. The thickness can range from 15 to 50 mm. Obviously, the main problem is that the permeability of the cake decreases as the filter cake becomes thicker. On the other hand, if large quantities of clay are present, even small increases in the cake thickness can produce large increments in the cycle time; for instance, increments of a thickness of 10 mm (from 20 to 30 mm) can double or triple the cycle time on less permeable slurries. As a consequence, it is common to work with a thickness of 25 mm if clays are a concern.

The ability to select cycle times, cake thicknesses, feed pressures, and plate styles are what enable plate filter presses to achieve much drier filter cakes than other technologies. When used in combination with a clarifier, the both devices can recover 90–95% of water as clean water.

Fig. 6.47 Vertical plate filter in an aggregate processing plant

Fig. 6.48 Filter cake (image courtesy of McLanahan)

Fig. 6.49 Water exiting from the filter plates (image courtesy of SAMCA)

Fig. 6.50 Rotary kiln for thermal dewatering (image courtesy of SAMCA)

The level of dewatering that can be achieved through the methods considered previously is limited. If further upgrading is needed, it must be attained through thermal processing. This method is comparatively very expensive because fuel consumption is one of the largest and most important operational costs. Equipment for thermal processing includes direct or indirect heat rotary dryers (Fig. 6.50). However, this method of dewatering is very rare in aggregate processing plants.

▶ Box 6.4 is an example of a sand and gravel processing plant.

Box 6.4—Asunción Sand and Gravel Processing Plant—Text and Images Courtesy of CONORSA

The Asunción sand and gravel processing plant is located near Segovia (Spain). It processes sand and gravel materials extracted in the Asunción sand and gravel quarry (see ▶ Chap. 5)*, which is near the quarry to minimize the costs of transportation. The material is an arkosic sandstone originating from the weathering of igneous and metamorphic rocks and the structures of the sediment indicate eolian and fluvial origins. The plant processes approximately 130,000 tons per year to generate different size fractions (0/2 mm, 0/4 mm, 3/6 mm, 8/20 mm, and 20/100 mm) depending on the requirements of the market.*

After transported by truck to the processing plant, sand and gravel are discharged directly to a grizzly of parallel bars with apertures of 100 mm to screen out large cobbles and boulders (Fig. 6.51)*. This device separates the oversized material from the smaller, marketable sizes. The material that passes through the grizzly is fed using a vibrating feeder to a conveyor belt that sends the material to a multideck vibrating screen (maximum 3 screen surfaces). Screening separates the sand and gravel into different size ranges. The vibrating screen has water jets with 35 spray nozzles for washing the material. Water is sprayed onto the*

Fig. 6.51 Truck discharging the sand and gravel to the grizzly

material throughout the screening process. The screens are tilted at an angle of around 15° from the horizontal. After screening, sand is translated to a hydrocyclone to finish the separation between sand and clay/silt; the sand is later stockpiled.

*The water with silt and clay (dirty water stream) originating in the hydrocyclone is sent to the first clarifier where the silt and clay settle to the bottom and are removed; flocculant is added to facilitate the settling of solids. The exact amount of flocculant is calculated through measurements every 120 s. After passing through a log washer, silt and water are sent to a mined pit for filling; subsequently, the area will be restored with pine trees. The remaining water is translated to a second clarifier (*Fig. 6.52*) and later to a concrete deposit for further reuse in the vibrating screens.*

Fig. 6.52 Second clarifier

6.2.5 Materials Handling

In an aggregate processing system, materials handling equipment is needed for permitting the choice of many arrangements of material storage and recovery. This equipment includes different elements such as feeders, belt conveyors, and storage bins (Fig. 6.53). The first two constitute the backbone of the crushing and screening processes and they become an integral part of the system itself.

- Feeders

Feeding is basically an operation with a short distance traveled but with a close regulation of the rate of passage needed. Feeders are necessary to deliver a uniform flow of aggregate fractions among storage, handling, and processing equipment. They are vital where principal operations are interrupted by a storage step. Feeders provide a reduction in wear and tear, abrasion and segregation as well as the help in dust control. Feeder design must consider "desired flow rates, delivery of a stable flow rate, the feeding direction, and particle size range of feed to be handled" (Roberts 2001).

A typical feeder is formed by a small bin, which can constitute an integral part of a large bin, with a gate and a suitable conveyor. Feeders are basically designated according to their conveying element structure (i.e., apron) and their flow-actuating system (i.e., vibrating). Many types have been designed such as apron, belt (Fig. 6.54), chain (Fig. 6.55), roller, rotary, reciprocating plate, table, vibrating, and many others. Among them, apron and vibrating feeders are probably the most used in aggregate processing plants.

Apron feeders (Fig. 6.56) are the most robust units and offer a long service life with low maintenance. They are used with dry, wet, heavy, abrasive, coarse, or lumpy materials, especially jaw crusher feed. In this device, the aggregate is conveyed by overlapping metal plates or pans mounted on strands of conveyor chains. Vibrating feeders (Fig. 6.57) are designed for large-sized materials and are principally utilized for feeding primary crushers. They are robust and versatile and

Fig. 6.53 Aggregate storage bin

Fig. 6.54 Belt feeder

have a low purchase cost compared to apron feeders. Vibrating feeders are formed by a sloped trough with a flat having shallow sides and an open end. The vibrating motion is provided by two types of mechanisms: electromagnetic and electromechanical.

– Belt Conveyors

Belt conveyors (Figs. 6.1 and 6.2) are the arteries that keep material flowing through the aggregate plant and are of key importance to keep the process flow stable. They are reliable, efficient, and have relatively low operating costs. The use of belt conveyors is the most common and widely utilized system of handling aggregate fractions. It incorporates a wide variety of widths, lengths, accessories, and options. In this system, the belt is a flexible and flat loop mounted over two pulleys. One of these must be connected to a drive to provide motion in one direction. Proper selection of the belt conveyor includes three principal factors: capacity, length and lift, and weight and size of the materials. For its part, the belt width is influenced by the material gradation and the maximum particle size. Belt conveyors "can operate at a maximum angle of 18°, depending on the material being conveyed, although it can be increased to approximately a 20-degree inclination is

Fig. 6.55 Chain feeder (image courtesy of Metso)

Fig. 6.56 Apron feeder (illustration courtesy of Metso)

necessary" (NSSGA 2013). Beyond these values, the material can slide down the belt.

6.3 Hard Rock Aggregate

After blasting in the quarry, trucks (Fig. 6.58) or conveyors move material from the quarry to the processing plant, where crushing and screening are the main processes for obtaining the final product with the needed requirements.

6.3.1 Crushing

Crushing is carried out to produce particles of a given size distribution and particle shape. In this sense, the size distribution of the aggregate is essential, not just the degree of size reduction. Crushing is a dry operation and can be developed in two or three basic stages (i.e., primary, secondary, and tertiary). They are carried out to better control the product size and quality while

Fig. 6.57 Vibrating feeder

minimizing the generation of unwanted fine particles. This is because aggregate producers are very sensitive to excessive generation of fines. More than two size-reduction stages are considered when rocks are very hard or when it is necessary to minimize the production of fines. It means that the harder the rocks, the more crushing stages and/or larger equipment is needed.

The quantity of fines increased as the material progressed from primary to secondary and further phases. Currently, many quarries do not produce significant amounts of aggregate coarser than 20 mm or even less. This is because the trend for highly specified aggregate has meant that products have become increasingly finer. As a consequence, the decrease of the top size of aggregate produced generates an increase in the amount of fines produced. For instance, a top size of 40 mm originates from 5 to 10% of fines whereas a top size of 10 mm originates between 35 and 40% of fines.

The reason for two or three stages of crushing is because of the limited reduction ratio that it is possible to achieve with a single crusher. The reduction ratio (RR) of a crushing stage is defined as the ratio of the maximum particle size entering to the maximum particle size leaving the crusher, although other definitions are sometimes used. RR varies from 3:1 to 20:1 or even greater, although it is in the range of 4:1–8:1 for many crushing devices. Where minimum fine content is needed, crushers should be operated at a low-reduction ratio in a closed circuit. Experience has shown "that each crushing machine operates most effectively, as measured by capacity, maintenance and power costs, utilization of capital investment, etc., within a range of values of the reduction ratio" (Smith and Collis 2001).

The most common devices utilized for crushing in aggregate plants are jaw, gyratory, cone, and impact crushers. Table 6.1 shows the main criteria to select the most adequate machine in each crushing state. Jaw and gyratory crushers are both capable of handling hard, abrasive rocks whereas impact crushers are restricted to less abrasive rocks (i.e., limestone). Many crushers also incorporate abrasion and attrition processes, which lead to the production of fine materials. All three modes of fracture (compressive, tensile, and shear) (Fig. 6.59) "can be discerned depending on the rock mechanics and the type of loading" (Wills and Finch 2016). Obviously, the nature of the fracture products is different in each case. Comminution machines "are usually constructed in such a way that one

6

Fig. 6.58 Truck transporting material from the quarry to the processing plant

Table 6.1 Main criteria to select the crushing in each crushing state (Metso 2014)

Major crusher type	Typical process stage	Feed size up to (mm)	Typical max. endproduct size (mm)	Typical capacities (t/h)	Abrasiveness		Amount of fines produced
					Low	High	
Gyratory crusher (large)	Primary	1500	200–300	Over 1200		×	Low
Jaw crusher	Primary	1400	200–300	Up to 1600		×	Low
Horizontal impact crusher	Primary/ secondary	1300	200–300	Up to 1800	×		Medium/high
Cone gyratory crusher	Secondary	450	60–80	Up to 1200	×	×	Low
Cone gyratory crusher	Tertiary	300	0–30	Up to 1000	×	×	Low/medium
VSI Barmac, B series	Tertiary	40	0–30	Up to 600	×	(×)	High

kind of breakage prevails and the mechanism that will predominate depends on the size of the particles and the mechanical configuration of the comminution device used" (Bustillo 2018).

Most aggregate processing plants use jaw and gyratory type crushers for primary crushing. The main characteristics of these devices are the wide input side and narrow discharge, handling a large tonnage of rock. A common feature of both types of devices is that the rock pass through a fixed although adjustable opening prior to be discharged. "Power demand, volume of rock processed, and product control are all

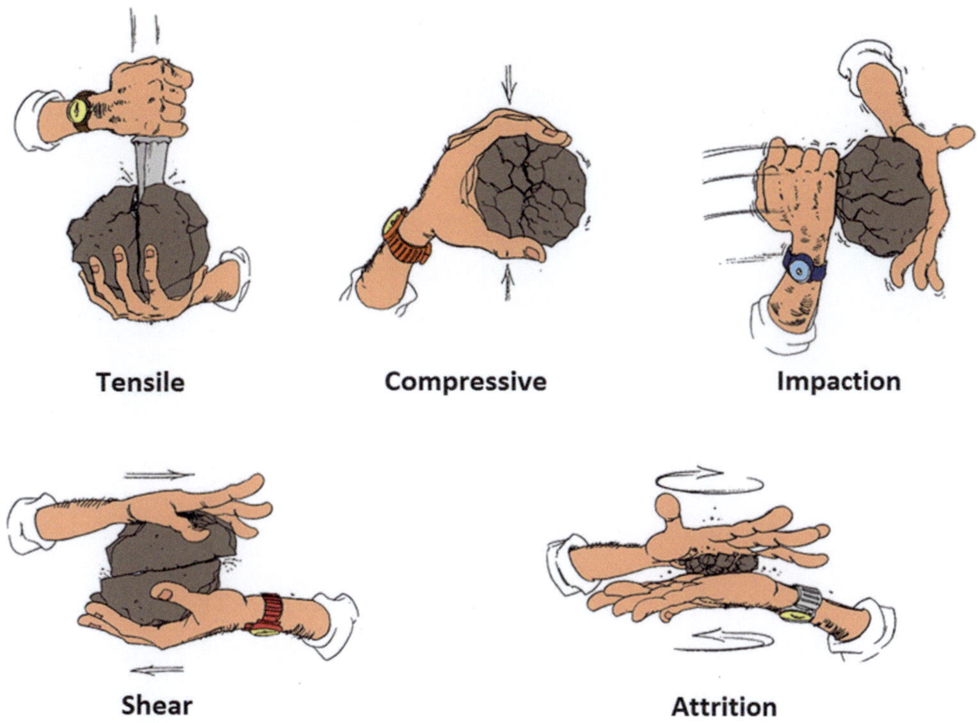

Fig. 6.59 Types of fracture and breakage (image courtesy of Metso)

influenced by the discharge opening" (NSSGA 2013). Gyratory crushers can process larger rock pieces and more tonnage per hour than jaw crushers. Jaw crushers produce a reduction ratio of 4:1–10:1 whereas the gyratory crushers produce a somewhat larger range of 3:1–11:1. As a rule, size and type of the primary crusher should be coordinated with the rock type, drilling and blasting patterns as well as the size of the loading machine.

In secondary and tertiary stages, the rock must be reduced and refined for proper size and shape. Most secondary crushers are cone crushers or horizontal-shaft impact crushers. The former works at reduction ratios ranging from 4:1 to 9:1. Very high reduction ratios are only obtained utilizing impact crushers. Secondary crushers are much lighter and smaller than the heavy-duty and large rugged primary machines. Since they take the primary rock as feed, the maximum feed size will normally be less than 15 cm in diameter. Secondary crushers are much easier to handle because most of the harmful constituents in the rock such as tramp metal (Fig. 6.60), wood, clays, and slimes have already been removed.

Tertiary crushers are generally cone crushers but some applications may require vertical-shaft impact crushers. Jaw, gyratory, and cone crusher consume the most energy while impact crusher is intermediate consumer. At the end of each crushing stage, aggregate is collected in stockpiles that have openings or draw points at the bottom.

- Jaw Crusher

As commented previously, the jaw crusher is principally utilized in primary crushing where the quantity of fines generated is typically less than 5% (▶ Box 6.5: Jaw Crusher) and operates classically in open circuit and dry conditions. It can work with a range of rock hardnesses from soft rock such as limestone to hard rock such as granite or basalt. The reduction ratio of the jaw crusher is between 6:1 and 8:1, varying from up to 10:1 for limestone and as low as 5:1 for granite.

Fig. 6.60 Metal detection (image courtesy of Luis Fueyo)

Box 6.5—Jaw Crusher

*The jaw crusher was patented by E.W. Blake in 1858. The patent states that the stone breaker "consists of a pair of jaws, one fixed and the other movable, between which the stones are to be broken". The movement of the swinging jaw is different based on whether the equipment is a single or double toggle. The jaw crusher (*Fig. 6.61*) is designed to impart an impact on a rock piece placed between the two jaws. It produces the coarsest material because of the preferential breakage of rock along inherent lines of weakness.*

*A jaw crusher is formed by two plates inclined toward each other that form a V-shaped chamber (*Fig. 6.62*). One of them is stationary whereas the second plate oscillates between two extreme positions. The gap between the jaws becomes narrower farther down into the crusher. Both plates are bolted onto a heavy block. Moving plate applies the impact forces on the material held against stationary plate. This process is repeated until the particle sizes are smaller than the smallest opening between the crusher plates at the bottom of the crusher. Fragments of breakage remain in the crushing chamber until they are sufficiently small to fall through the gap between the two surfaces. Consequently, the return motion of the moving jaw permits the feed to enter the crushing cavity and enough broken material to be discharged by gravity.*

Jaw crushers can be classified according to the method of pivoting the swing jaw. In the most used jaw crusher for aggregate comminution, the so-called Blake crusher, the jaw is pivoted at the top. This method generates a fixed receiving area and a variable discharge opening. Another option in the market is the Universal type, in which the jaw is pivoted in an intermediate position, producing a variable delivery and receiving area.

The angle between the plates, termed the nip angle, is usually 18°–23°. It is kept within a close range to en-

Fig. 6.61 Primary jaw crusher (image courtesy of Benito Arnó e Hijos, S.A.U.)

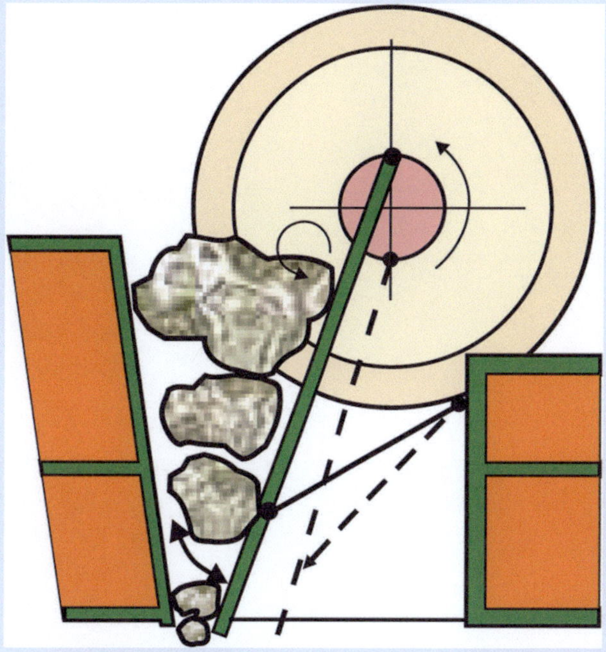

Fig. 6.62 Schematic representation of a jaw crusher (illustration courtesy of Metso)

sure that the material does not slip with the aim of achieving that maximum size reduction. The faces of the plates (Fig. 6.63) are made of hardened steel and their surfaces can be plain or corrugated. The speed of jaw crushers ranges from 100 to 400 rpm. The principal approach to determine the optimum velocity is that a particle must have sufficient time to move down the crusher throat into a new position prior to be nipped again.

Material is fed into the jaw crusher via *a hopper. The opening width is in the range of 0.8–2 m and the gap between the crusher plates is from 0.25 mm to 1.2 m. It is very common to prescreen the feed to prevent undersizing from entering the crushing chamber. With the cited method, the crusher capacity is increased and the risk of product contamination is reduced. The mechanism of movement of rocks down the crusher chamber defines the jaw crusher capacity, being a measurement of the volume or mass of crushed rock generated in the unit time of operation.*

It is very common that jaw crushers are not fed to their maximum design capacity since further crushing equipment in the processing plant does not have sufficient capacity to handle the rock amount generated if the jaw crusher was working to capacity.

Fig. 6.63 Plates of a jaw crusher

- Gyratory Crusher

Gyratory crusher is an excellent choice for a primary crusher (Fig. 6.64). It was developed by Charles Brown in 1877 and further by Gates around 1881; for this reason, it was referred to as a Gates crusher. They are sometimes used at the secondary stage due to their capability for accepting larger lump sizes.

Gyratory crushers break material in a very similar way to jaw crushers. They have larger capacities (up to 9000 tph) compared to jaw crushers (up to 2000 tph). They can operate with velocities ranging from 80 to 160 rpm. The feed mechanisms are not necessary in gyratory crushers because they are fed directly from the back of a dump truck. For a similar gape, the available feed opening area of a gyratory crusher is around three times greater than that of a jaw crusher. The principal advantage of the gyratory crusher is "the much greater unit capacity for any given gape or maximum lump size assuming that this capacity can be fully utilized to justify the substantially greater capital cost" (Smith and Collis 2001). Gyratory crusher is a machine suitable for all hard, abrasive rocks although not for soft, porous material.

In gyratory crushers, the crushing surfaces are curved and conical and the crushing bowl provides the fixed surface. Located inside this bowl is the conical crushing head, which rotates about a pivot point. Figure 6.65 is a typical sketch of a large gyratory crusher used as a primary crusher. As in the jaw crusher, the maximum movement of the head occurs near the discharge. The crushing process comprizes reduction by compression between two confining faces and a subsequent freeing movement during which the material settles by gravity until it is caught and subjected to further compression and again released. The work is continuous and balanced and, as a consequence, crushing gap between the stationary concaves in the crusher housing and gyrating crushing mantle can change constantly. The gradation of the material can be handled by setting the spacing between the mantle and the concaves at their closest point.

Fig. 6.64 Primary gyratory crusher

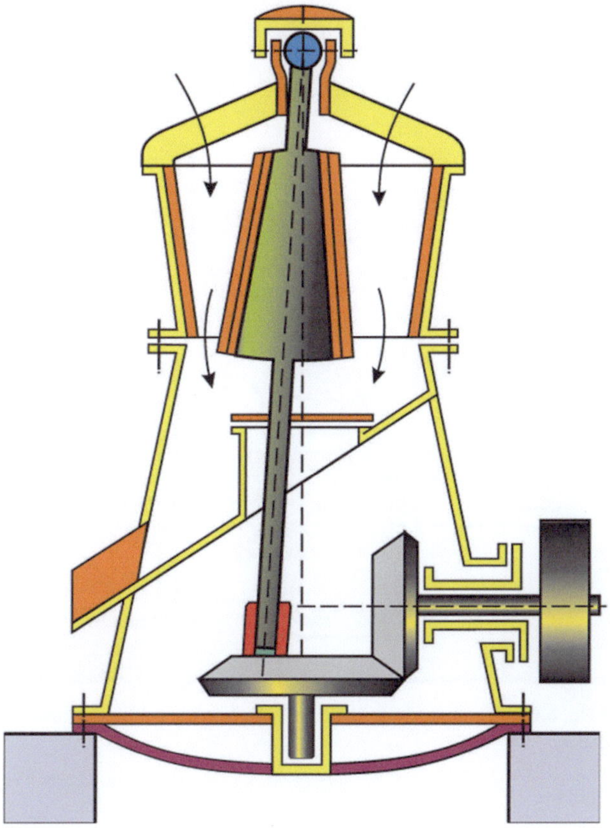

Fig. 6.65 Sketch of a gyratory crusher (illustration courtesy of Metso)

The feed material that enters the crushing chamber from the top is continuously crushed between the crushing tools as it falls by gravitation until it exits the crushing chamber at the bottom. The particles are subjected to maximum breaking forces where they are on the side with the minimum gape. This equipment handles feed material as large as 1.5 m or more to produce discharged fragments as small as 125 mm. Gyratory crushers tolerate a variety of shapes of feed particles. One additional benefit is that there is no need for prescreening of fines in the feed. The crusher can handle the all-in feed. As in a jaw crusher, the performance of a gyratory crusher can be affected by "(1) fines content (fine should preferably be less than 10%), (2) inherent and total moisture content in the rock, (3) feed distribution in the crushing area and the bulk density of the feed, (4) hardness of the rock and (5) recirculating load in the case of closed-circuit crushing" (Gupta and Yan 2006).

To decide whether a jaw or a gyratory crusher should be employed in primary crushing, the principal feature is the maximum size of the rock that the crusher will be needed for handling as well as the throughput needed. As a rule, gyratory crushers are utilized when high capacity is needed whereas jaw crushers are usually employed when the crusher gape is

Fig. 6.66 Cone crushers (image courtesy of Benito Arnó e Hijos S.A.U.)

more important than the capacity. For example, where crushing material of a certain maximum diameter is needed, then a gyratory crusher a needed gape would have a capacity approximately three times that of a jaw crusher of a similar gap. However, if a large gape is required but not a certain capacity, the jaw crusher will possible be more economical because is a smaller equipment. In summary, the general rule of thumb is to define the desired production rate. If the production rates needed are more than 800 t/h, gyratory crushers are always the preferred option. Secondary considerations in the selection process include capital and maintenance costs and the type of material being crushed. For the latter, "jaw crushers perform better than gyratories on clay or plastic materials while gyratory crushers have been found to be particularly suitable for hard, abrasive material; moreover, they tend to give a more cubic product than jaw crushers if the feed is laminated" (Wills and Finch 2016).

- Cone Crusher

The cone crusher (Fig. 6.66) was originally designed and developed by Symons approximately 1920 and is termed Symons cone crusher. It is a development of the gyratory crusher. As the mechanism of crushing in this crusher is similar to a gyratory crusher, their designs are similar although for a cone crusher the spindle is supported at the bottom of the gyrating cone whereas it is generally suspended as in larger gyratory crushers and the chamber is not as steep. As a consequence, location of optimal crushing is typically at the base of the chamber.

The material passing through the cone crusher is subjected to a series of hammer-like blows rather than being gradually compressed by the slowly moving head of the gyratory crusher. The material must be fed into the crusher utilizing some device to maintain an even distribution of well-graded material throughout the crushing chamber (i.e., a distributor). In this sense, increasing the moisture amount of the material is utilized to prevent the packing of materials in the cavity. Cone crushers also operated at much higher speeds (700–1000 rpm) than gyratory crushers. Figure 6.67 is a schematic diagram of a cone crusher.

The principal differences between gyratory crushers and cone crushers are two: "(a) cone crushers have a much shorter spindle with a large diameter crushing surface relative to its vertical dimension; and (b) the shorter spindle is not suspended in cone crushers, as in the gyratory, but is supported in a curved, universal bearing below the gyratory head or cone" (Haldar 2013). The cone crusher has a higher capacity than the gyratory crusher since it presents a greater head angle

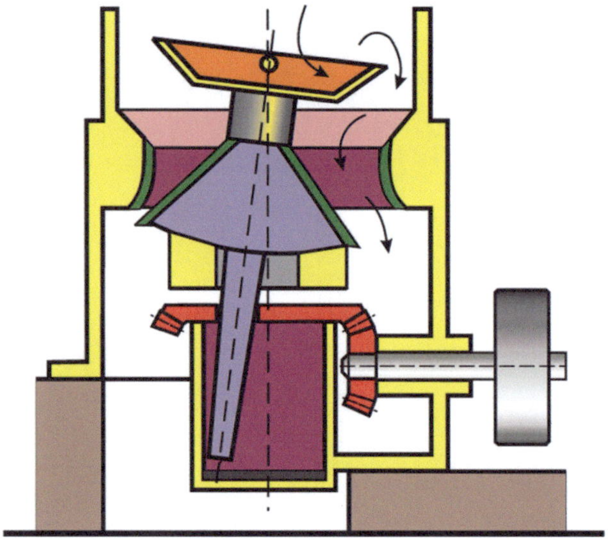

Fig. 6.67 Cone crusher schematic design (illustration courtesy of Metso)

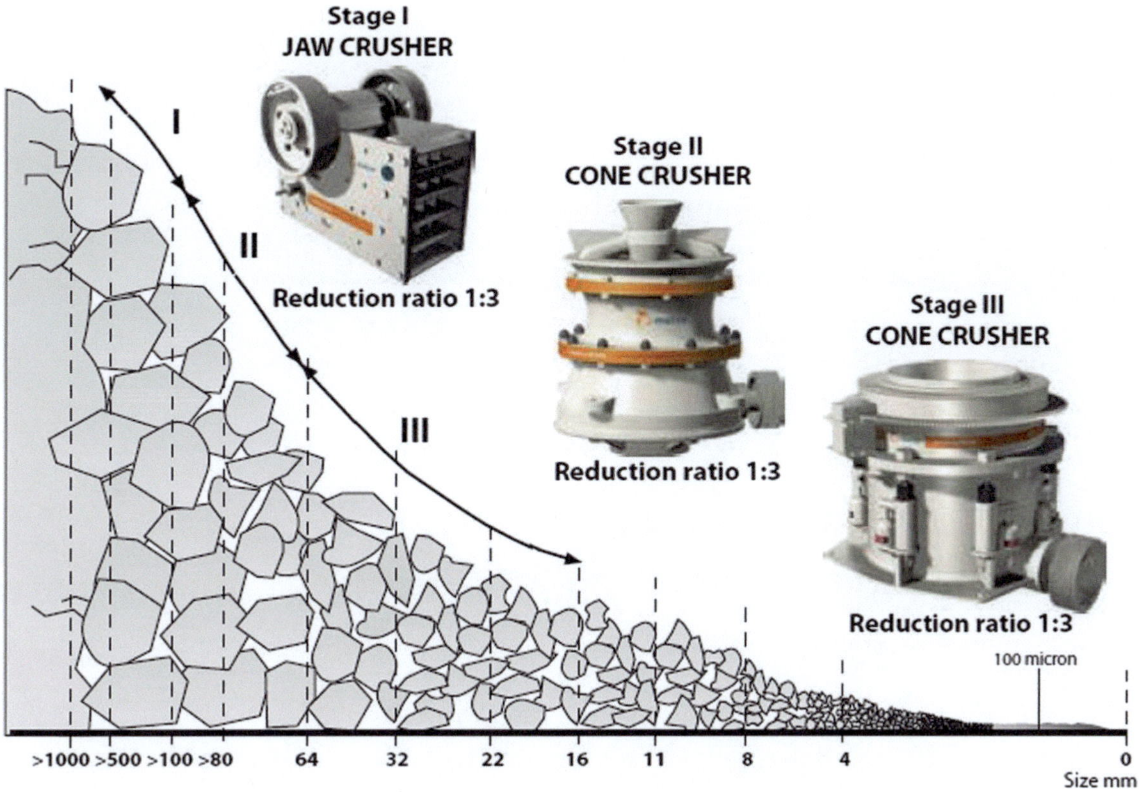

Fig. 6.68 Cone crushers in secondary and tertiary stages (illustration courtesy of Metso)

than the gyratory crusher whereas the cone crusher retains the same angle between the crushing members. A cone crusher works a continuous cycle and crushing occurs in the chamber 100% of the time.

Cone crushers are often employed in secondary and tertiary stages (Fig. 6.68). They are compression crushers providing reduction ratios of 5:1–7:1 for secondary crushing and 2:1–3:1 for tertiary crushing. In secondary roles, the standard-head cone is currently designed whereas the short-head cone is utilized in tertiary stages. A cone crusher efficiently produces materials that are high-quality and cubical in nature. On the other hand, they may crush rocks of medium to hard compressive strength, including abrasive stone. Protection of crushing surfaces takes place using high manganese steel liners.

One particular criticism of the cone crusher has been "the tendency to produce particles of a flaky shape owing to the geometry of the crushing chamber; this can be counteracted by the use of low reduction ratios and ensuring that the crushing chamber is always full of material" (Smith and Collis 2001). This situation is termed choke feed.

- Impact Crusher

An Impact crusher (Fig. 6.69) (▶ Box 6.6: Impact Crusher) is principally employed for crushing nonabrasive materials (i.e., limestone) (Fig. 6.70). This is due to the increase in the development of abrasion-resistant wear-parts. Nevertheless, the impact crusher is preferably utilized where the shape is a critical requirement and the rocks for crushing are not very abrasive.

Fig. 6.69 Horizontal impact crusher

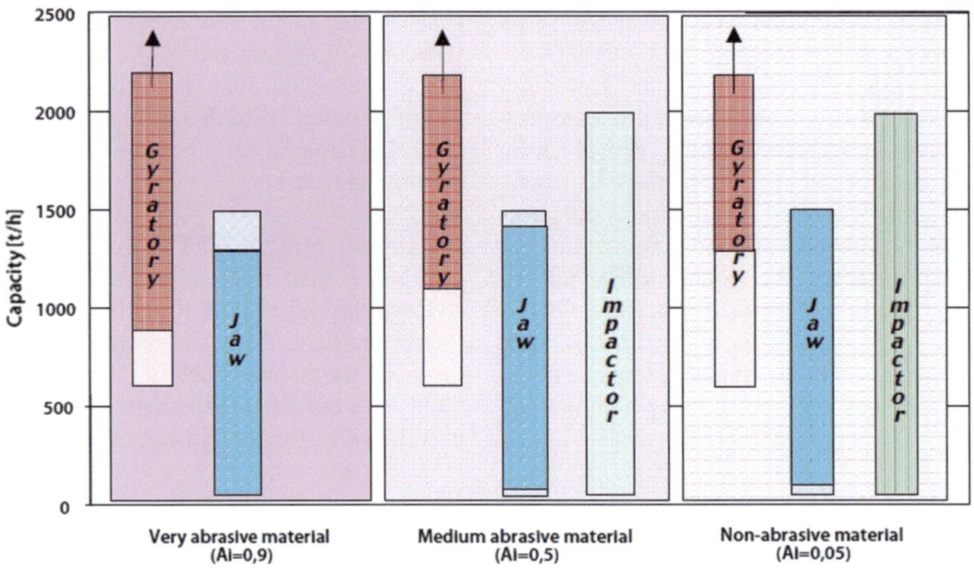

Fig. 6.70 Crushing by abrasiveness (illustration courtesy of Metso)

Box 6.6—Impact Crusher

The impact crusher applies a high-speed impact force to the feed rock. In this type of crusher, size reduction is carried out by high-speed impact rather than compression (large particles break more readily from impact than small particles). Unlike crushing by compression, impact crushed rock has no residual stress, which is important for aggregates utilized in construction. Impact crushers are also preferred in the quarry market because of their improved shape since other devices used in secondary crushing (i.e., cone crushers) tend to generate more elongated particles.

The impact crusher comprizes a lined chamber within which a rotor is revolved on a horizontal (horizontal impact crusher) (Fig. 6.71) *or vertical (vertical impact crusher) shaft. The rotor is fitted with blow bars of alloy steel that are placed by mechanical or hydraulic clamps* (Fig. 6.72). *The blow bars impart impact to the rocks, which result a good comminution that is energy efficient and a typically cubical product shape. The crushing is produced by the impact with the rotor, the chamber liners (breaker plates), and interparticulate collision. The largest degree of crushing effect is produced by the first contact, which is responsible for more than 60% of the crushing action. The remainder is produced basically by the impact against the bar as well as by interparticle collision.*

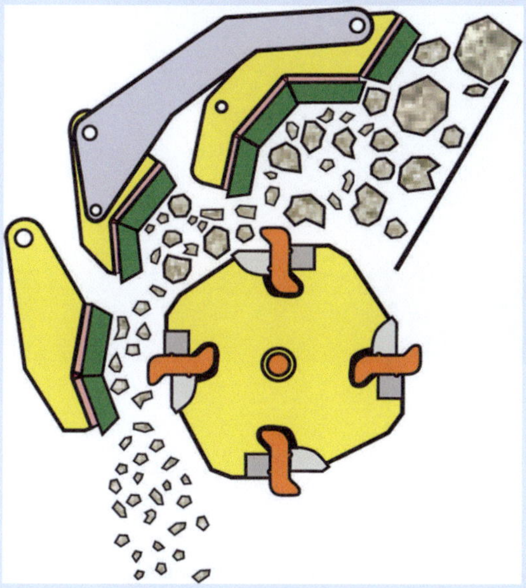

Fig. 6.71 Sketch of a horizontal impact crusher (illustration courtesy of Metso)

The material is fed to the machine tangentially into the path of the beaters, delivering a series of sharp blows to the particles. They generate an acceleration effect to the particles into the crushing chamber. Since the crushing effect produces breaks along natural cleavage planes, the final product has a good quality in terms of shape. In most cases, impact crushers tend to operate in an open circuit without control of particle sizes other than clearances between rotor and liners, rotation speed, and feed rate.

The reduction ratio can range between 20:1, in a single stage, to 40:1 for double rotor impactors. Thus, "the impactor offers several potential advantages of high reduction ratio, low capital and operating cost, ease of installation and low head room requirement and good product shape" (Smith and Collis 2001). The reduction ratio and product grading can be controlled by optimizing the rotor speed, breaker plates, rotor diameters, and crushing chamber feed. Because impact crushers depend on high velocities for crushing, wear is greater than for jaw or gyratory crushers. Major issues associated with operating impact crushers are the high wear rates of the liners or impellers, particularly when handling abrasive feeds.

A horizontal impact crusher is utilized for coarse (primary and secondary) crushing whereas a vertical impact crusher is used preferably for finer (secondary and tertiary) crushing. The former has a maximum feed size ranging from 250 to 2000 mm and capacities that range from 50 to more than 1100 tons per hour whereas the latter has a feed size ranging from 25 to 75 mm and capacities over 2000 tons per hour. The utilization of a vertical impact crusher can reduce the proportion of flaky material in aggregate by up to 50%.

Fig. 6.72 Inside an impact crusher (image courtesy of Benito Arnó e Hijos, S.A.U.)

- Other Crushers

The previously described crushers are the most used in crushed stone plants but sometimes other types of crushers such as roll crushers or hammer mills, although not widely used in the aggregate industry, can also be utilized. Roll crushers (Fig. 6.73) have limited application in the aggregate quarrying market because of their low throughput. Nevertheless, roll crushers are utilized for sedimentary, friable rock, and wet or sticky materials. Roll crushers are principally utilized in secondary or tertiary crushing stages. Their reduction ratio is limited to approximately 3:1. Roll crushers are formed by one or two cylindrical rolls placed parallel to each other and rotated in opposite directions. The rotation of the rolls draws material into the breakage zone where particles are caught, either against the breaker plate or between two rolls, and experience compressive forces. The roll surfaces play an important role in the process of catching a particle and dragging it between the rolls. The rolls can be smoothed or toothed with a pyramidal tooth. With smooth rolls the reduction ratio is approximately 3:1 while this value can be increased up to 7:1 when toothed rolls are used. The main disadvantage of roll crushers is that very large rolls are needed in relation to the size of the feed particles for reasonable reduction ratios to be achieved.

In hammer mills (Fig. 6.74), which have fallen into disuse in recent decades, a series of hammers are mounted on the rotor and operate at speeds between

Fig. 6.73 Roll crusher (image courtesy of McLanahan)

1000 and 6000 rpm. The hammers work in such a way that they may move out of the path of material entering to the crushing chamber. Because pivote hammer produce less force than it would if rigidly attached, these machines are utilized for crushing soft material (i.e., limestone). Much of the comminution in a hammer mill is the result of attrition because the particles are given high velocities. This leads to a less control of the particle sizes and a much higher amount of fines than with compressive machines. Hammer mills are typically used in secondary or tertiary crushing stages in limestone operation. A great advantage of this machine is the fact that it produces a relatively cubic product.

6.3.2 Screening, Stockpiling, and Loadout

Screening, stockpiling, and loadout in crushed stone processing plants are similar to those used in sand and gravel plants, that is, grizzlies, trommels, and mainly vibrating screens are utilized to separate the particles into the necessary sizes. After screening, trucks or conveyors move the material to stockpiles and loadout is carried out in a very similar way to in sand and gravel processing plants. ▶ Box 6.7 is an example of a crushed stone processing plant.

Fig. 6.74 Hammer mill (image courtesy of Neuman and Esser)

6

Box 6.7—La Soriana Crushed Stone Processing Plant—Text and Images Courtesy of Benito Arnó e Hijos, S.A.U.

The La Soriana crushed stone quarry and processing plant (Fig. 6.75) *is located in Huesca (Spain) and the principal objective of this installation is the production of railway ballast. In the quarry, the igneous rock (subvolcanic) is extracted by drilling and blasting and loaded with a front-end loader into a large haul truck fleet that transports the material to the processing operations. These can include crushing, screening, material handling, and storage operations. Crushing is performed in three stages: primary (first stage)* (Fig. 6.76), *secondary (second stage) and tertiary (third stage). In general, crushed rock is transported along the process line on conveyor belts.*

The run-of-mine from the quarry is dumped in a hopper and a vibrating feeder sends the material to a grizzly, which carries out the first classification of crushed elements at 100 mm size. The undersize is separated into 0/30 mm and 30/100 mm fractions, the latter being mixed with the material obtained in the jaw crusher. The oversize, including particles ranging from 100 to 900 mm, is sent to a jaw crusher (primary crusher) with a feed opening of 3645 mm × 2890 mm to produce size particles ranging from 0 to 200 mm. The output from the primary crusher is conveyed onto the primary stockpile together with the 30/100 mm fraction obtained with the undersize material. This primary stockpile feeds the secondary crusher, a cone crusher that reduces the size of the crushed stone up to 100 mm. The material is classified using two vibrating screens, each with three mesh units. Thus, four size fractions are produced: 0/6 mm, 6/25 mm, 25/63 mm, and 63/100 mm, with the third fraction (25/63 mm) used as railway ballast.

The 0/6 mm fraction is directly stockpiled whereas the 6/25 mm fraction is sent to a tertiary crusher (another cone crusher) that produces a 0/25 mm material. This material is finally also separated with a vibrating screen in three fractions: 0/6 mm, 6/12 mm, and 12/25 mm, which can be used for different applications (i.e., concrete and road).

Fig. 6.75 Aerial view of the La Soriana Crushed Stone Processing Plant

Fig. 6.76 Primary crushing stage using a jaw crusher

6.4 Questions

Short Questions

- What are the main objectives of washing processes and why?
- Explain the bucket wheel machine used in the washing process
- What is screen media?
- What is a vibrating screen for the screening process?
- What is a filter medium in the dewatering process?
- Explain the significance of "silt lagoon"
- Enumerate the main types of feeders
- Explain the jaw crusher machine used in crushing processes
- What type of crusher is most similar to a cone crusher and why?
- What is an impact crusher?

Long Questions

- Describe the hydrocyclone device used in washing processes
- Describe the vertical plate pressure filter used in dewatering processes

References

Bustillo M (2018) Mineral resources—from exploration to sustainability assessment. In: Springer textbooks in earth sciences, geography and environmental sciences. Springer International Publishing AG, Cham, 653 p

Cox C, Traczyk F (2002) Design features and types of filtration equipment. In: Mular AL, Halbe DN, Barratt DJ (eds) Mineral processing plant design, practice and control Proceedings, 1st edn. SME, Englewood, pp 1342–1357

Exbrayat JC, Mishellany A (2000) Washing of aggregates and treatment of fine aggregates. In: Primel L, Tourenq C (eds) Aggregates. A.A. Balkema, Rotterdam, pp 301–343

Gupta A, Yan DS (2006) Mineral processing design and operations. Elsevier, Amsterdam, p 718

Haldar SK (2013) Mineral exploration: principles and applications. Elsevier, Amsterdam, p 372

Metso (2014) Crushing and screening handbook, 6th edn. Metso, 298 p

NSSGA (2013) The aggregates handbook, 2nd edn. The National Stone, Sand and Gravel Association, Alexandria, Virginia, 906 p

Roberts AW (2001) Recent developments in feeder design and performance. In: Levy A, Kalman H (eds) Handbook of conveying and handling of particulate solids, vol 10. Elsevier, Amsterdam, pp 211–223

Smith MR, Collis L (eds) (2001) Aggregates: sand, gravel and crushed rock. Geological Society, London, Engineering Geology Special Publication v. 17, London, 360 p

Wills BA, Finch J (2016) Mineral processing technology—an introduction to the practical aspects of ore treatment and mineral recovery, 8th edn. Elsevier Ltd., 512 p

Environment and Sustainability

Contents

7.1 Mining, Environment and Sustainability – 226

7.2 Potential Environmental Impacts and Their Management – 229

7.2.1 Air Emission of Particulate Matter – 230

7.2.2 Increased Noise Levels – 236

7.2.3 Vibrations Associated with Blasting – 240

7.2.4 Water Condition – 244

7.2.5 Biodiversity Loss – 246

7.2.6 Visually Disturbed Landscapes and Reclamation – 251

7.3 Questions – 257

References – 257

M. Bustillo Revuelta, *The Basics of Aggregates*,
Springer Textbooks in Earth Sciences, Geography and Environment,
https://doi.org/10.1007/978-3-031-42961-3_7

Abstract

This chapter draws attention to all topics related to the environment and sustainable development in the aggregate mining world. It provides a general overview of the environmental issues associated with the design, development, operation, and closure of aggregate operations. The description starts with some comments about mining, the environment, and sustainability, including the definition of sustainable development and the importance of an environmental management system using ISO 14001 standard. The most important potential environmental impacts and their management are discussed: air emission of particulate matter, increased noise levels, vibrations associated with blasting, water affection, biodiversity loss, and visually (negative visual aesthetic) disturbed landscapes.

7

7.1 Mining, Environment and Sustainability

Many are the interactions between mines and the environment. Society needs the mining industry since products derived from this activity improve our wealth and quality of life and allow society to grow. Quarry materials (i.e., aggregates) are utilized by all sectors or the economy. However, the environmental impact of mining (◘ Fig. 7.1) is perhaps part of the price that humankind must pay for the benefits of mineral consumption because some environmental degradation due to mining is unavoidable. By its very nature, mining mineral resources such as aggregates invariably has an impact on the land upon which it operates. The impacts occur from the initial exploration and present operation of the site to its closure and rehabilitation.

In the recent past, mining was carried out with little concern for its effects on the environment, often resulting in significant environmental damage. This is because most mines need the removal of surface features during the extraction process. These activities may cause important alterations to wildlife, leading to the reduction of natural habitats. For this reason, environmental considerations are an important part of the modern mining industry. They must be included in all project planning, and "feasibility studies must account for the influence of environmental considerations on project schedules and costs" (Nelson 2011). Since economic growth cannot take place without aggregates, the rational conclusion is that the exploitation of aggregate resources is not the problem, but it must be developed in a green and modern execution. Today, aggregate mines are designed, developed, operated, and closed in an environmentally sound manner, and considerable effort is put into continually improving environmental standards.

Aggregate mining operations affect flora and fauna as well as land and water (◘ Fig. 7.2), similar to other major industrial operations. For this reason, aggregate mining companies are carrying out considerable efforts to decrease the environmental impact of mine works and diminish the footprint of their operations throughout the mining cycle, including working to reclaim ecosystems postmining. By planning the aggregate mine for a subsequent development, it is possible to improve the value of the disturbed land and help to change it to a utilization that the public will consider clearly positive. Old mine sites can be converted to wildlife habitat and refuge recreational areas (◘ Fig. 7.3), shopping malls, golf courses, airports, lakes, underground storage facilities, solid waste disposal areas, mining and power plant waste storages, museums, sites of special scientific interest and regionally important geological sites, industrial lands, pisciculture ponds, and many other economically or ecologically productive land utilizations that can benefit society. This conversion of an abandoned mine for practical commercial purposes depends upon the geological and hydrogeological conditions as well as the nature and geometry of the mining that took place.

In the last three decades, the terms sustainable development and sustainability have been in use by governments and policy makers worldwide. It is agreed that sustainable development was defined early in 1987 by the Bruntland Commission (Our Common Future, World Commission on Environment and Development, United Nations) as "a system of development that meets the basic needs of all people without compromising the ability of future generations to meet their own life-sustaining needs". In the mining world, these words were first utilized in the early 1990s at the Rio Summit (1992). Since the release of "Our Common Future", many of the major industries in the world, including aggregate mining companies, many of its governments, and the United Nations, have adopted a policy of sustainable development. Today, aggregate mining companies employ the principles of sustainable development in their environmental policies. This has resulted in positive development for the industry and has allowed aggregate mining companies to view the impacts of their operations in a more comprehensive manner. For these purposes, an Environmental Management System (EMS), according to the ISO 14001 standard (▶ Box 7.1: ISO 14001 standard), is utilized for defining the environmental policies.

Fig. 7.1 Environmental impact of mining (visual impact)

Fig. 7.2 Aggregate mining operations affect land

Fig. 7.3 Aggregate quarry reclaimed to recreational area (Las Madres, Madrid, Spain)

Box 7.1—ISO 14001 Standard

An EMS is a group set of procedures and practices that allow an organization to decrease its environmental impacts and increase its operating effectiveness, being a powerful tool for managing the unfavorable impacts of activities of an organization on environmental aspects. Applying an EMS, the company can prove to all people who they take environmental impacts actively. Nevertheless, it is important to note that environmental management systems do not by themselves define environmental objectives. Rather, they only lead the management procedure of a company to ensure that environmental programs can be efficiently developed.

The ISO 14001:2015 standard specifies the conditions for an EMS that an organization may utilize for enhancing its environmental eficiency. According to this standard, "the purpose of this International Standard is to provide organizations with a framework to protect the environment and respond to changing environmental conditions in balance with socioeconomic needs". Therefore, environmental management systems such as ISO 14001 can help companies introduce sustainable or green practices in their business and are considered a successful tool for compliance and with the upholding of environmental laws and regulations.

The ISO 14001 standard is the most important within the ISO 14000 series and it is the internationally accepted environmental management standard that certifies that an organization is committed to reducing the environmental impact of its products and operations and is constantly monitoring and seeking to identify ways of reducing that impact further. It prescribes controls for those activities that influence the environment. These include the use of natural resources, handling and treatment of waste, and energy consumption. Most mining companies are committed to managing their environmental aspects, impacts, and risks through adherence to the internationally recognized the ISO 14001 standard. In summary, ISO 14001 can be used in the implementation and operation of EMSs by aggregate producers.

The five main stages of an EMS, as defined by the ISO 14001 standard, are (◘ Fig. 7.4): "(1) commitment and policy: top management commits to environmental improvement and establishes the organization's environmental policy; (2) planning: an organization first identifies environmental aspects of its operations and then determines which aspects are significant by choosing criteria considered most important by the organization; (3) implementation: an organization follows through with the action plan using the necessary resources (human, financial, etc.); an important component is employee training and awareness for all employees; (4) evaluation: a company monitors its operations to evaluate whether targets are being met; and (5) review: top management reviews the results of the evaluation to see if the EMS is working; management determines whether the original environmental policy is consistent with the organization's values; the plan is then revised to optimize the effectiveness of the EMS".

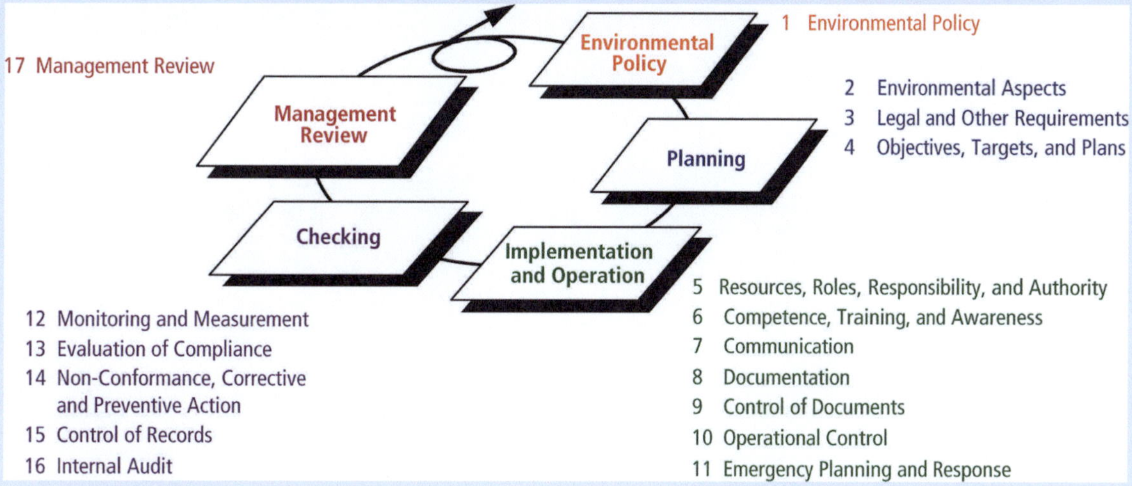

◘ **Fig. 7.4** Main stages of an EMS as defined by ISO 14001

7.2 Potential Environmental Impacts and Their Management

Mining of aggregate resources is unwelcomed near neighborhoods and communities due to the associated negative environmental impacts. Some residents near quarries object to the noise, dust, and truck traffic associated with aggregate operations (◘ Fig. 7.5). As a consequence, many citizens do not support mining and prefer that stone and sand and gravel not be mined nearby. Because aggregate mining is an extractive industry, it cannot be obtained from the landscape without causing environmental impacts, which usually receive the greatest public attention. It is important to bear in mind that, in densely populated areas, there are often land-use conflicts between aggregate quarrying and recreation, agriculture, building sites, archaeology, etc. These environmental impacts can vary significantly from site to site due to the specific characteristics of the exploitation. They are generally studied on a case-by-case basis.

Most of the environmental impacts are relatively benign (aggregate production seldom produces acid mine drainage usually generated in mining of metallic ores—◘ Fig. 7.6) but in some cases, extracting aggregates can "alter the geologic conditions, which, in turn, may alter the dynamic equilibrium of the area, resulting in cascading environmental impacts" (Langer 2001). The potential environmental impacts of aggregate production are basically determined by the geologic, climatic, and man-made characteristics of the area being mined. These impacts must be controlled, mitigated, or kept at tolerable levels during aggregate production utilizing available techniques and responsible operational practices. In general, environmental health hazards are generally rare in aggregate operations.

Environmental issues produced in aggregate mining activity can take place during all phases of the

Fig. 7.5 Truck traffic associated with aggregate operations (image courtesy of Aggregate Industries)

development proposal, from the initial exploration to its eventual closure and postclosure phases. Since there are different phases in an aggregate mining project, each phase of mining can be related to different sets of environmental impacts. The concern for environmental protection strongly influences the operation of the quarries and pits as well as the methods and type of excavation equipment.

The most obvious environmental issue of aggregate production is the conversion of land utilization to a hole in the ground (Fig. 7.7), which is linked to other environmental impacts such as loss of biodiversity, noise, dust, blasting vibrations, and visually disturbed landscape (Fig. 7.8). The most important environmental impacts associated with aggregate production are the following: (1) air emission of particulate matter, (2) increased noise levels, (3) vibrations associated with blasting, (4) water conditions, (5) biodiversity loss, and (6) visually (negative visual aesthetic) disturbed landscapes.

7.2.1 Air Emission of Particulate Matter

Suspended particulate matter, also termed dust, refers to particles of varying sizes in the air (measured in micrometers). Dust is produced by the disintegration of rocky material or soil that gives rise to fine particles that can be carried away by air currents. It is characterized by its chemical composition (i.e., siliceous or calcareous) and by its particle size. Total suspended particles (TSP) represent "particles ranging from 10 to 100 μm, while 30 μm aerodynamic diameter is commonly applied to represent TSP" (EPA 1996). Particles with sizes above 30 μm constitute a great amount of dust in quarries, and these particles are deposit within 100 m of the source. Intermediate-sized particles ranging from 0 to 30 μm are likely to travel between 200 and 500 m, whereas smaller particles, with sizes less than 10 μm, can travel much further from the source. Thus, smaller particles cause greater impacts than larger particles. Generally, the smaller the particle size is, the

Fig. 7.6 Acid mine waters (red waters) generated in mining of metallic ores

more hazardous it is to health. As a rule, dust concentrations, deposition rates, and potential impacts typically decrease quickly away from aggregate operation.

Air emission from mining activity affects the environment on a local, regional, and global scales. Dust is naturally present although its amount is usually increased due to aggregate quarry activity. The very nature of aggregate production, consisting of crushing rocks to obtain different size fractions, makes dust generation intrinsic to the aggregate activity. All dust sources in quarries, including drilling (Fig. 7.9), blasting (Fig. 7.10), crushing (Fig. 7.11), loading (Fig. 7.12), and hauling, produce coarse-grained dust that generally settles near the dust source. Numerous variables are accountable when considering dust generation. Consequently, this type of dust is a major environmental issue in many aggregate quarries when released in an uncontrolled manner. In some cases, the dust may include detrimental components (i.e., quartz) and requires special attention.

The process of crushing is suggested as the most significant source of dust. "Crushing generates more dust and more coarse particles compared to drilling" (Sairanen and Rinne 2019). Dust is easily and quickly generated during many aggregate processing operations on site. Obviously, if several crushers are in operation, the dust emissions are increased. The amount of dust generated will be dependent on "the type of material on site, weather conditions (wind and precipitation) and the controls in place at site" (OSSGA 2016a). Other factors influencing the amount of dust generated during aggregate mining are the physical size of the

Fig. 7.7 Hole in the ground of a sand and gravel pit

operation and production levels. Table 7.1 summarizes the main dust emissions that can potentially originate during quarrying operations.

The provision of an adequate air environment to promote the health, safety, and comfort of the people is a requisite for successful aggregate mining operations, especially in climatic zones with low levels of rainfall and humidity. In general, the restrictions on air dust are becoming more important and the limit values are more stringent. Dust prevention and control (Fig. 7.13), in addition to being a legal safety and environmental obligation, provides important advantages (a) reduces pollution of the environment, (b) reduces the risk of workers contracting respiratory diseases, (c) increases the useful life of facilities and work equipment, and (d) helps to obtain higher quality products. A dust management plan is a way of systematically dealing with or avoiding problem issues. It is common to require it as part of an environmental approval for operating an aggregate mine.

The location of the sources of dust generation is necessary to consider the installation of an appropriate control method. Dust control needs to be a key priority for any aggregate quarry and pit operation. A carefully prepared and implemented plan will keep dust within the needed regulatory limits. It is important to bear in mind that "education, awareness and training of staff on dust prevention, control measures, monitoring and reporting are important in reducing dust emissions at an aggregate operation" (OSSGA 2016a). The

Fig. 7.8 Changes to the visual scene

combination of measures needed varies broadly, depending on production and shipping rates, size of the site, and distance to neighboring residents. Dust emission controls must be located at (a) dust emission sources, (b) the quarry boundary to determine the environmental impact, and (c) points outside the quarry to study the possible inconvenience to third parties.

The International Finance Corporation (IFC 2007) recommended, among others, the following dust management strategies: "(a) dust suppression in roads and work areas, optimization of traffic patterns, and reduction of travel speeds, (b) exposed soils and other erodible materials should be revegetated or covered promptly, (c) new areas should be cleared and opened up only where absolutely necessary, (d) surfaces should be revegetated or otherwise rendered non-dust forming when inactive, (e) storage for dusty materials should be enclosed or operated with efficient dust suppressing measures, (f) loading, transfer, and discharge of materials should take place with a minimum height of fall and be shielded against the wind, (g) conveyor systems should be covered, (h) installation of dust extraction systems at crushers, and (i) spraying waste rock piles with sealants".

An aggregate quarry must be worked in such a way that the generation of dust is minimized. It can include the following actions: "(a) locating dust-generating activities only in areas where maximum protection can be obtained from the topography, adjacent woodland or other features, (b) considering the prevailing wind

Fig. 7.9 Dust generated by drilling

conditions and placing processes that generate dust away from sensitive areas, (c) minimizing the requirement to transport and handle materials by utilizing storage facilities close to processing areas, (d) using a water bowser to provide a short-term solution throughout the site, and (e) restricting vehicle speed to minimize dust creation" (QM 2005).

The environmental impacts from plant-generated dust can be mitigated by "either wet suppression systems (Fig. 7.14), which consist of pressurized water (or surfactant treated water) sprays located at dust-generating sites throughout the plant, or dry control systems, which include covers on conveyors (Fig. 7.15), vacuum systems, and baghouses (Fig. 7.16) that remove dust before the air stream is released to the atmosphere" (Drew et al. 2002). The control method can use wet suppression systems, dry control systems or a combination of both. The former utilizes "treated or untreated water applied to the aggregates in a spray or fog condition to wet down the airborne particles of fine dust and cause them to settle back into the process stream; dry collection systems use hoods and enclosures to contain the dust" (NSSGA 2013). The enclosure is equipped with a vacuum system to reduce the dust generated.

For its part, water trucks (Fig. 7.17) play a significant role in minimizing fugitive dust in aggregate operations because dust generated by vehicle traffic is one of the prime sources of airborne dust. To prevent this problem, the trucks periodically spray water and/

Fig. 7.10 Dust generated by blasting

Fig. 7.11 Dust generated by crushing

Fig. 7.12 Dust generated by loading processes

or chemicals on the roadways. In many cases, the utilization of chemical additives, termed surfactants, in the dust control water system is common. For drilling dust, remediation is accomplished by fitting dust collection and suppression systems to the rigs since this method is highly efficient.

7.2.2 Increased Noise Levels

Noise is an inherent health hazard in the aggregate mining industry. Most operations that take place in aggregate quarries and pits generate noise (by noise is meant an undesirable sound for the person who perceives it, being generated by mechanical vibrations that are transmitted to nearby air particles). The noise is energy transmitted through the air and it is measured as energy level or pressure, although it is generally expressed in terms of a logarithmic scale, termed the decibel (dB) scale. For instance, regulations within the European Community need that the exposure in the workplace should not exceed 85 dB in a day of work and never exceed 130 dBA.

Noise is the cheapest pollutant to produce and requires very little energy to emit. It does not generate waste and does not accumulate in the environment, but it can have permanent effects on people. Noise emissions can negatively affect wildlife and reduce the quality of the natural environment. In marine environments, the noise caused by extraction activities can affect marine mammals such as harbor porpoises and seals.

Aggregate extraction and processing (i.e., crushing and screening—Fig. 7.18) as well as truck traffic are the most noticeable sources of noise and can produce important levels of noise that affect the surrounding environment. This is because the mining process is highly mechanized, from the earliest aggregate removal to final processing, and heavy equipment is necessary at virtually every stage of operation. One of the inconveniences that accompany the mechanization of aggregate production is the noise produced by the motors, plant equipment, movement of plant equipment, etc. (most modern equipment is provided with a soundproofed cabin for the operator). Exposure to high intensities of noise is a concern for workers who drive

Table 7.1 Summary dust emission inventory for a typical quarry operation (Birch and Datson 2007)

Activity	Emission description
Drilling	Blast holes are drilled into the overburden and the mined mineral with using drill rigs. In dry conditions this drilling operation may generate significant dust. These are often treated as point sources
Blasting operations	Dust emissions are generated as the material is blasted from the bench. Loading of the broken rock from the resultant blast pile using mechanical shovels will also create dust emissions. Emissions would be generated both from the mechanical of the rock pile and from wind blown mechanisms
Handling of limestone (to primary crusher) HAUL ROADS	Once the mineral mass has been fragmented and released by blasting, it is loaded into haulage trucks. Often haulage trucks of up to 100 tonne capacity are used for the removal of the mined mineral from the open pit to the downstream processing plant. These operations will produce dust emissions during loading, unloading and during the transport of the material from the pit (from agitation of the road surface). Primarily the fugitive dust is generated, by a pick up and drop off of dust from the haulage road surface by wheels will occur. Material on the bed of the dumper will produce dust, through mechanical attrition and wind blown action
Mineral processing	The crushing and screening of the mineral into smaller size fractions produces dust emissions. A loader and gyratory crusher are often used for the primary crushing of the mineral. Here mechanical action will cause the production of a dust and fines. The crushing and screening of the ore and its subsequent transfer by belt conveyor to primary stockpiles would also be a potential source of dust generation
Overburden handling	Overburden loosened by either scrapping, bulldozers or blasting is often loaded into and transported by dumper truck. Emissions will be generated from the disturbance of soil layers and the emissions related to dumper truck movement. Repositioning and unloading from the dumper truck are also likely to generate emissions
Mound formation	Overburden material and low grade mineral deposits are often used to form mounds which are often levelled and shaped by bulldozers. Operation of dumpers, dozers and graders would generate dust emissions
Storage mounds (OB and low grade mineral)	Exposed surfaces of mineral storage/stockpiles generate dust due to wind erosion. Dust will also be generated due to mechanical impaction of dust dropped onto storage piles from conveyor belts. Dust emissions will depend on the height between the conveyor drop and top of the pile. Also truck tipping/loading and transport operations will generate mechanical movement of storage piles and thus produce further emissions
Conveyor belts	Transportation of various size fractions between downstream processing operations are commonly carried out using conveyor belt systems. The loading and discharge areas are likely to produce the greatest emissions. Uncovered conveyors will also suffer from wind blown and mechanical attrition dust emissions
Movement of other quarry vehicles	Movement of vehicles like dumper trucks, bulldozers, tankers, water bowsers, supervisors, etc. will all generate dust from the agitation of unpaved surfaces

Notes Sources used include, Chaulya et al. (2001) and Arup Environmental (1995)

Fig. 7.13 Dust control (image courtesy of Anglo American plc.)

mechanized equipment and those who operate or work near stationary equipment such as conveyor belts or crushing equipment. This exposition can even lead to labor and social conflicts.

Because aggregate operations are classically located near to their market, it is a challenge for producers to mitigate this noise. They are responsible to ensure that the noise emitted from the mine site does not exceed levels set by regulations (Fig. 7.19). Nevertheless, the proper nature of aggregate production, consisting of rock crushing and screening, makes it very difficult to significantly reduce noise levels.

Different types and intensities of noise can take place at aggregate mines. It is important to bear in mind that this noise generation can be constant noise (i.e., produced by crushing, screening and truck traffic) and intermittent but violent noise (i.e., generated from blasting). For instance, intermittent noise can be generated in a jaw crusher during the time of breaking the rock as well as during loading and unloading operations. Continuous noise is produced in a vibratory screen that operates at higher frequency. Regarding the noise generated in blasting, the sources of this noise are the release of gases confined in the borehole when the stemming is projected, the escape of gases through the cracks created in the rock, the detonation of the initiating cord, the displacement of the face of the bench, and the collision between the projected fragments.

The impact of noise depends on factors such as the sound source, characteristics of the surrounding site, climatic conditions, and sensitivity of the

Fig. 7.14 Dust sprayers for dust suppression (image courtesy of Martin Engineering)

receiver persons. Regarding latter factor, the accustomed background noise level is a significant characteristic to determine the tolerance of a person to a new noise. Exposure to noise levels above regulatory or recommended limits can result in hearing loss. It can be temporary or permanent and is gradual because of prolonged exposure to noise. It is important to emphasize that most hearing loss is preventable. Prevention can be achieved by "eliminating noise sources, substituting quieter equipment, installing appropriate engineering controls, implementing administrative controls, using personal protective equipment, and adopting effective hearing-conservation programs" (Walter 2011). Regarding hearing damage, both the noise level and duration of exposure must be taken into consideration. Good practices in the management of noise sources must be defined based on the prevailing land utilization and the vicinity of noise receptors such as communities. Acoustic maps can be generated to chart the sources of noise. It determines corrective action in a manner that the noise impacts age mitigated through various engineering techniques.

The noise generated in an aggregate processing plant by heavy equipment operations and through the movement of material can be minimized using the following techniques: "(a) low profile process systems can

Fig. 7.15 Covers on conveyors (image courtesy of APECO)

be used where the mechanical equipment is closer to the ground and noise travel can be buffered by berms or trees, (b) handling materials with rock-on-rock interfaces for transfers can reduce noise appreciably compared to rock on steel interfaces; stone boxes at transfer points reduce noise as well as maintenance costs, (c) rubber or polyurethane products for screen surfaces (Fig. 7.20), chute liners, stockpile truck liners, etc., minimize noise in materials handling, (d) building enclosures minimize machinery noise but must normally be insulated to provide effective noise reduction, and (e) non-audible backup alarms on selective pieces of equipment reduce the problem of disturbances caused by the normal audible backup alarms" (NSSGA 2013).

Other techniques of noise mitigation include (a) location of the equipment in such a way that vegetated areas, quarry walls, or topographic barriers can absorb the noise, (b) designing and laying out the workplace to create quiet workstations, (c) the truck noise can be reduced through the proper location of access roads, the utilization of acceleration and deceleration lanes, and a careful routing of trucks, (d) regular maintenance of machinery and equipment that takes account of noise, (e) haul roads can be kept away from property lines and the utilization of conveyor belts can minimize traffic within the mine, and (f) limiting the time people spend in noisy areas. Many of the successful noise mitigation measures used in quarries are a result of the existence of legislation that covers hearing protection and environmental effects and the progress by machinery suppliers.

7.2.3 Vibrations Associated with Blasting

The primary objective of blasting is the fragmentation of the rock mass and displacement of the broken rocks. Blasting is the only economic means of aggregate extraction in quarries. It is restricted to quarry operations

Fig. 7.16 Baghouse system (image courtesy of U.S. Air Filtration Inc.)

Fig. 7.17 Water truck spraying water on the roadway of the quarry (image courtesy of OSSGA)

Fig. 7.18 Crushing and screening processes

Fig. 7.19 Noise control (Bustillo 2018)

Fig. 7.20 Polyurethane screen (image courtesy of EXSTO)

because it is not utilized in sand and gravel mining. In a blasting process, most of the energy released is used to fragment the rock, but a significant percentage is wasted, producing different environmental impacts. Dangers of blasting procedures can arise from "generation of harmful gases, throw of rocks in the air, pollution by dust, vibrations generated in the ground mass (▶ Box 7.2: Blasting and Vibrations), and propagation of air shock waves" (Bustillo 2018). By far, blasting is the most studied source of ground vibrations. Meteorological conditions (i.e., wind velocity and direction, and air temperature) must be considered when blasting since they can greatly influence the environmental impacts.

Box 7.2—Blasting and Vibrations

Where an explosive is detonated in a blasthole, a pressure wave is originated in the surrounding rock mass, and as the pressure wave travels from the borehole, it generates seismic waves by displacing particles. These seismic waves traveling through the ground cause ground vibrations, which last for only a second or two in most cases. The blasting operation can have unacceptable vibration impact if not conducted correctly. "Elevated ground vibrations suggest excess confinement as well as inefficient utilization of explosive energy" (Rostami and Hambley 2011).

Blasting vibration impact is one of the most important issues in the field of engineering blasting. This is because excessive levels of structural vibrations from blasting can destroy the structural integrity of the surrounding structure in the mine area and can affect other structures far away. Blast-induced ground vibrations influence vegetation development and can generate deforestation in the future as well as affect the ecology of surrounding living communities (fauna and flora). On the other hand, ground vibrations lead to ground/slope instability.

The prediction of blast-induced ground vibrations is one of the most important challenges in mine management. For this reason, all blasting procedures must be carried out adequately by a competent person. The aggregate industry applied sequential blasting and modern technology to reduce vibrations. The main method is the utilization of delay detonators for separating explosions in each blast. Seismic equipment must be utilized to monitor ground vibrations and is the most accurate measurement technique to assess the intensity of this environmental impact.

*The intensity of ground vibration depends on controllable and uncontrollable parameters. Controllable parameters are basically related to explosive features (i.e., explosive and charge energy, initiation system and sequence, and method of loading) and blast-hole design parameters (*Fig. 7.21*), that is, hole depth, diameter, and inclination, collar height, stimming procedures, blast pattern, and many others. For its part, uncontrollable parameters are related to the geological structure and properties of the rock such as density of the rock, existence of bedding planes, joints, cracks, cavities, zones of weak rock, etc. The rock properties define the velocity at which the vibrations travel.*

Peak particle velocity (PPV) is the key predictor for ground vibration with units in mm/s. As a rule, the magnitudes of PPV causing structural damage to buildings are above 15 mm/s. As a critical parameter to measure the blasting intensity, the peak particle velocity should be accurately predicted because it may afford a helpful guideline to design blasting parameters and evaluate the damage.

7

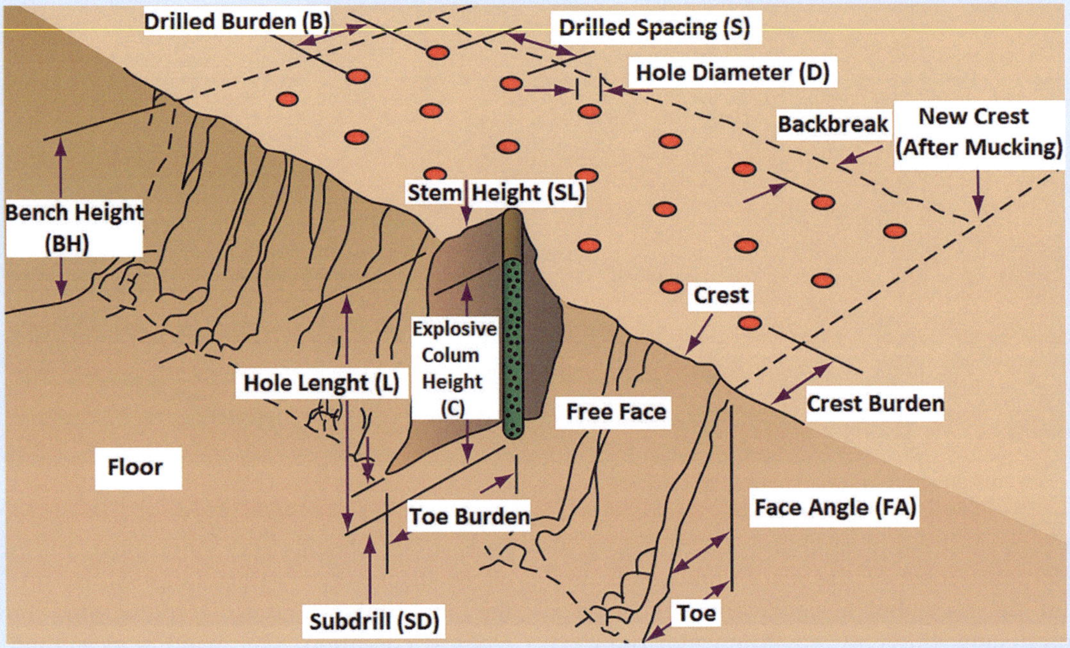

Fig. 7.21 Main blast pattern parameters in surface bench blasting (Bustillo 2018)

7.2.4 Water Condition

Water is utilized in a wide range of activities such as mineral processing, dust suppression, and employee requirements. In many cases, gravel pits need dewatering to expose the material to be extracted. Water serves many different purposes in an aggregate quarry and is utilized in high volumes. Water is essential for the aggregate industry as a significant water user. Nevertheless, aggregate extraction can have negative effects on water quality and quantity. This is because the water cycle of a mine site is interconnected with the general hydrologic water cycle of a watershed (Fig. 7.22). The main sources of water on an aggregate mine site are from precipitation, dewatering of gravel pits, and the pumping and removal of groundwater particularly around gravel pits. It is necessary to note that the use of water in mining affects not only its availability for other users but also the risk of pollution from the disposal of used water. Well over 90% of the water handled is returned to the local watershed. Cumulative effects can take place if several extraction sites are present or as a result of the combined impacts of extraction. "A cumulative impact, which may result from multiple operations extracting in the same area at the same time, is defined as the additive effect of multiple quarry dewatering operations on groundwater, surface water and ecological features" (OSSGA 2016b).

Mining aggregates naturally interfere with the environment, generating qualitative and quantitative impacts on water resources. For this reason, the protection of the water is a major challenge for aggregate operators. The impacts on water resources and the methods to avoid them may change from site to site. As a rule, these impacts can

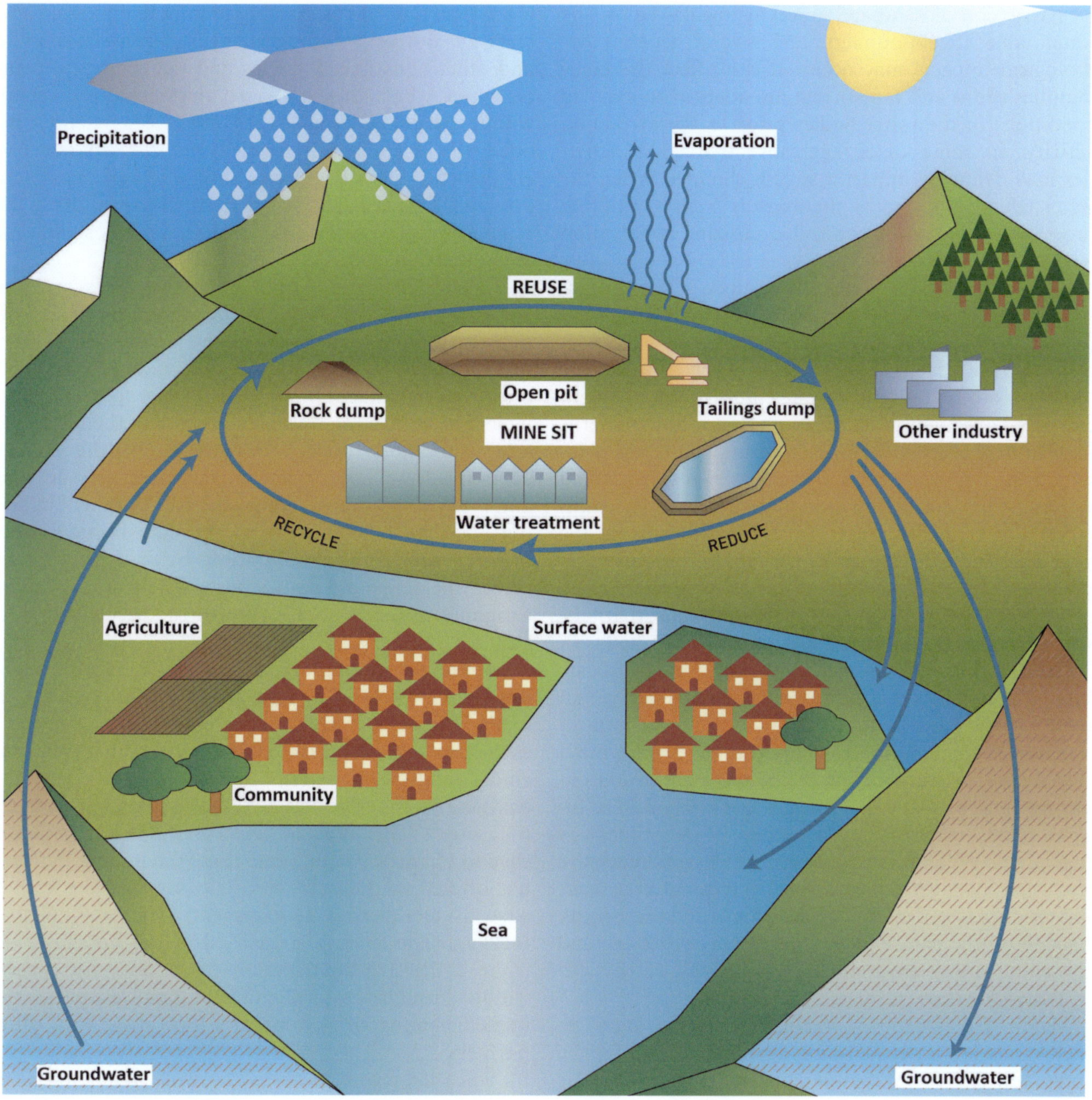

Fig. 7.22 Flows of water to and from a mine site (ICMM 2012)

affect both underground and surface waters. The different fields to be studied include (a) hydrodynamic (i.e., modifications in the water flows), (b) hydromechanical (i.e., transport of suspended matter), (c) physicochemical (i.e., variations in the quality of the waters), (d) hydrothermic (i.e., temperature changes in groundwaters or rivers), and (e) hydrobiological (i.e., modification of the biological balances in rivers, in particular eutrophy).

Potential effects that should be analyzed include "changes in the groundwater level (lowering or raising in different zones), hydrodynamic changes of the aquifer around the water body, modifications of the relationship between superficial groundwater and watercourses (i.e., reduced recharge from the aquifer which may reduce water level) and reduction of watercourse movement" (EU 2011). Regarding changes in water quality, some aggregate mineral extraction and production processes "can cause pollution as well as other changes in water quality, which could either directly affect the aquatic habitats and species present in the site and/or indirectly affect other sites that are vulnerable to this type of change" (EU 2011).

Extraction of sand and gravel aggregates below the water table affects aquifers, and several negative hydrological impacts may occur. It can affect the communities (flora and fauna) that are adapted to existing conditions in the water bodies prior to the extraction activity. In some cases, aggressive extraction activities may definitely impact a stream channel. When the water-table is affected by the impacts from dewatering processes, they can be controlled utilizing observation wells.

To avoid impacts on water resources, country regulations can prohibit penetration of the aquifer or can permit the penetration of the aquifer only if operations will not adversely affect water supplies. Moreover, "leveling out of the groundwater table in the pit pond can also cause alteration of groundwater flow patterns in surrounding areas whereas lowering of the water table during aggregate extraction may drain surrounding wetlands that are fed by groundwater discharge" (Binstock and Carter-Whitney 2011). For this reason, the aggregate industry currently principally produces aggregates from crushed rock instead of sand and gravel, avoiding the negative impacts on groundwater that take place when sand and gravel resources are extracted.

Responsible management of water by aggregate mining companies is a key ingredient in ensuring that their contribution to sustainable development is positive over the long term. It is important to bear in mind that managing water is one of the most important environmental activities at aggregate operating mines. Inappropriate management of water is underlined by inadequate water balance assessment. A comprehensive site water management plan (WMP) is essential to leading practice water management. Its size and complexity depend on the nature of the operation, hydrology, and the cultural and environmental sensitivity of the surrounding area.

Some recommended practices for water management include "(a) establishing a water balance for the mine and related process plant circuit; (b) developing a sustainable water supply management plan to minimize impact to natural systems by managing water use, avoiding depletion of aquifers, and minimizing impacts to water users; (c) consider reuse, recycling, and treatment of process water where feasible and (d) consider the potential impact to the water balance prior to commencing any dewatering activities" (IFC 2007). Different combinations of strategies can be applied, and the selection of strategies is site-specific.

7.2.5 Biodiversity Loss

The protection and conservation of biodiversity is crucial to sustainable development. The United Nations Convention on Biological Diversity defines biodiversity as "the variability among living organisms from all sources including inter alia terrestrial, marine and other aquatic ecosystems and the ecological complexes of which they are part; this includes diversity within species, between species and of ecosystems". Biodiversity is commonly defined at three separate levels: genetic diversity, species diversity, and ecosystem diversity. It is crucial that all partners constituting the mining industry admit that biodiversity has significant environmental, social, and cultural value.

In the European Community, the Birds and the Habitats Directives are the cornerstones of the EU biodiversity policy. These directives include the following objectives: (a) "protect rare and endangered habitat types and habitats of the species of European interest, including wild birds, in order to ensure their survival and restoration through designation of the Natura 2000 sites and appropriate management of the sites and (b) "protect species across the EU, both within and outside Natura 2000 network, through species protection provisions". The prohibitions by the Directives include "the deliberate destruction of, damage to, having a significant negative effect on these species and their habitats".

Aggregate mining can affect biodiversity throughout the life cycle of a project, both directly and indirectly. Direct or early adverse effects from aggregate extraction can be produced from any activity that includes land clearance (i.e., access road) or direct discharging to water bodies (i.e., clay disposal) or the air (i.e., dust). This type of adverse effect is commonly easy to identify. At the same time, the mining industry has offered considerable effort to improve the knowledge of biodiversity management. It is essential that the aggregate mining industry has the possibility to make a decisive contribution to biodiversity conservation through the production of knowledge and the implementation of actions in cooperation with other partners.

Since aggregate exploitation will often have unavoidable negative impacts on biodiversity, it is possible to offset impacts by creating benefits elsewhere to produce an overall conservation outcome that maintains the biodiversity assets of a region. In other words, an extraction site can deliver an overall net benefit for biodiversity, for example, creating new habitats for wildlife. Such offsets (i.e., habitat creation or restoration) can be direct through acquiring comparable land and managing it for biodiversity conservation. This process is sometimes referred to as biobanking (CSIRO 2014). Another form of direct offset is through funding the implementation of regional conservation plans. Biodiversity offsets can also be indirect such as by conducting relevant research for improved conservation management or through education and training that increases regional capacity for biodiversity management.

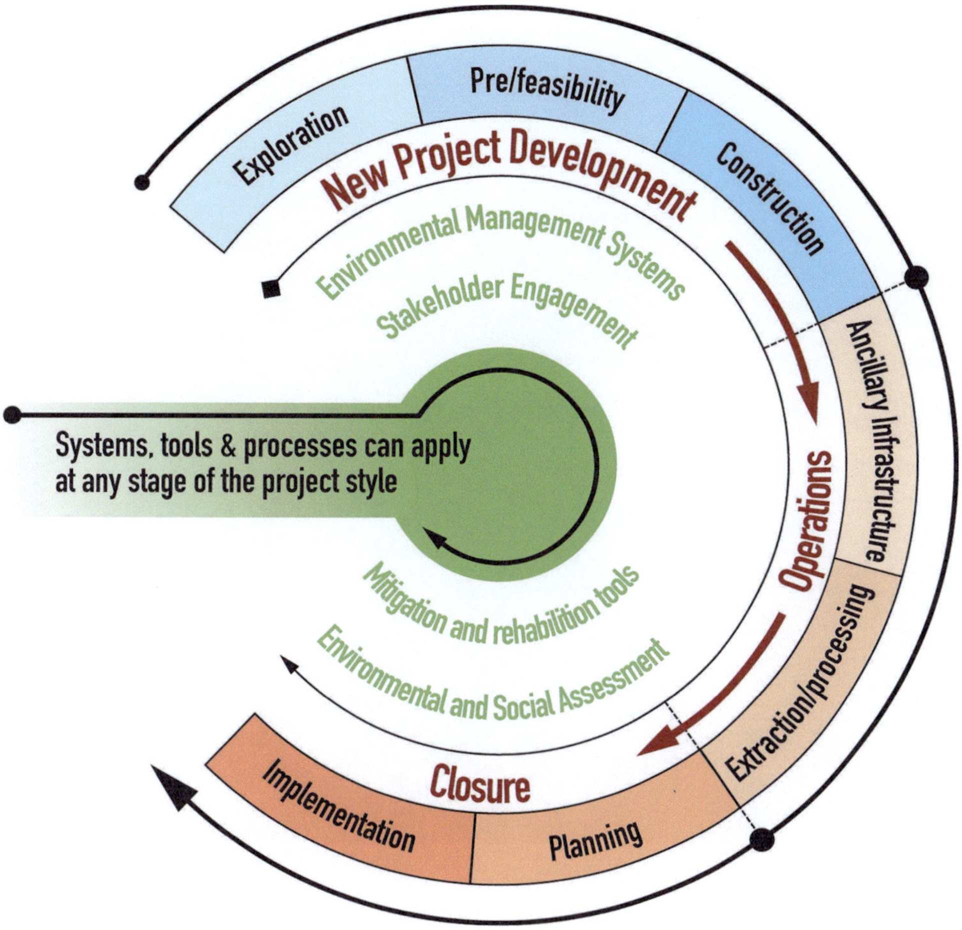

■ **Fig. 7.23** Integrating biodiversity into the mining project cycle (ICMM 2006)

The conceptual approach adopted for a good practice guidance is illustrated in ■ Fig. 7.23, showing how biodiversity is integrated into the mining project cycle.

Terrestrial habitat alterations must be diminished as much as possible and be consistent with the request to preserve critical habitats. Some controlling strategies include siting access roads in places that prevent adverse effects to critical terrestrial habitat, diminishing disruption to vegetation and soils, and implementation of mitigation techniques adequate for the type of habitat. Other strategies are "preventing the generation of barriers to wildlife movement and offering alternative migration routes if the generation of barriers cannot be avoided and manage vegetation growth along access roads and at continued above-ground facilities" (IFC 2007).

Aquatic habitats are affected by variations in surface water and groundwater flows, which generate incremental pressures on fish and wildlife communities. The sites where gravel pit lakes are located, close to wetlands and watercourses, many times can be characterized by the presence of several faunal species of naturalistic interest (i.e., migratory birds, amphibians, small mammals). In these cases, the existence of noise, dust, traffic and human presence can originate significant impacts and the abandonment of the interested areas. Water table variations around the gravel pit lake can produce an alteration of the composition in vegetation species and, in some situations, affect wetlands and species that rely on these fragile habitats.

In marine environments, aquatic habitats can also be affected by several methods to extract aggregate resources (i.e., dredge mining—■ Fig. 7.24) or other activities such as offshore loading activities or port development. It is obvious that marine aggregate extraction has become increasingly common in recent decades. Aggregate extraction methods can remove benthos, increase the turbidity of the water, and change the sediment composition as well as the hydrodynamics of sediment transport. In particular, dredging processes cause lowering of seabed levels and the removal of seabed sediments, which in turn cause the biota destruction.

Assessment and control of "adverse effects for marine environments must be in accordance with suitable

Fig. 7.24 Dredge mining (image courtesy of BMAPA)

host-country obligations to international decisions such as the United Nations Convention on the Law of the Sea" (IFC 2007). To adequately manage potential effects in aquatic habitats, it is necessary to maintain water body catchment zones equal to or comparable to predevelopment features, preventing stream channel stability by restricting in-stream and bank disturbance, and constructing, maintaining, and reclaiming watercourse crossings that are stable, safe for the intended utilization.

Therefore, habitat alteration is one of the most significant potential threats to biodiversity associated with aggregate operations. Although this alteration can take place at any stage of the mine cycle, there is no doubt that the greatest potential for temporary or permanent alteration of terrestrial and aquatic habitats occurs during construction and operational activities. Nevertheless, aggregate production sites and their activities can, if properly planned, actively contribute to biodiversity conservation. Especially "when the extraction site is in an already impoverished environment, the extraction industry can help to create new habitats for wildlife (i.e., new wetland areas for various amphibian species; cliffs as nesting opportunities for birds; habitat for insects and reptiles or disused shafts as shelter for bat populations). Some of these new habitats are in areas of low nature conservation value and form important ecological corridors between core protected areas" (Tost and Ammerer 2022). In summary, aggregate extraction and biodiversity can coexist and the aggregate industry can make a significant contribution to restoring biodiversity, which ensures its sustainable management (▶ Box 7.3: Enhancing biodiversity in Soto Pajares Gravel Pit (Madrid, Spain)). In this sense, restoration should always be the preferred way of halting the loss of biodiversity. It is achieved both through actions of mitigation and progressive restoration during operations and final restoration. From the environmental impact assessment to the rehabilitation, the aggregate industry follows the mitigation hierarchy, that is, avoid, reduce, restore.

Box 7.3—Enhancing Biodiversity in Soto Pajares Gravel Pit (Madrid, Spain) (Text and Images Courtesy of CEMEX)

*Soto Pajares is a CEMEX gravel pit that delivers aggregates to the market in Madrid, Spain (*Fig. 7.25*). The entire area near Madrid has a long tradition of gravel extraction. Several gravel quarries are already exhausted and restored, providing an excellent refuge for wildlife. The Soto Pajares gravel quarry has been in use since the early 1970s. It is located in a valley bottom and produces gravel and sand (with a production capacity of approximately one million tons per year). The extractive activity is still ongoing in the permitted areas and restoration activities are carried out in the areas where the activity is completed.*

conservation at this site. Actions ranged from measures to control invasive species, steps to increase interlake connectivity of an on-site reed bed, and educational activities for the promotion of wildlife to local communities. The main aim of the Biodiversity Action Plan is at minimum to achieve no net loss of biodiversity and, at best, to achieve a lasting and overall positive impact on biodiversity compared to the state prior to when operations began.

The plan can be summarized in two main types of actions: species protection actions and actions to enhance habitats. The former includes the following objectives: (1) surveillance and safeguard of bird colonies and other species in the extractive areas and (2) protection of nesting areas and prevention of disturbance during the breeding season.

Fig. 7.25 Soto Pajares gravel pit

The restoration model applied in the area aims at creating a mosaic of different habitats that increases the ecosystem's diversity such as the creation of shallow water habitats, riparian forests, management of landscape elements in agricultural habitats, etc. In 2012, a biodiversity baseline was developed, identifying and prioritizing habitats and appropriate measures to protect or enhance priority species and habitats. The Biodiversity Action Plan was conceived as a dynamic process designed to improve biodiversity

Regarding the first objective, at the beginning of each breeding season, the extractive areas are surveyed to locate possible breeding sites and to define operational changes to be eventually applied. In many cases it is enough to delimit and signal the breeding areas and ensure that they are not destroyed during the breeding period. It is also possible to prospect and detect in advance the areas that can be potentially used by nesting birds (e.g., sand martins) and assess which ones could be affected by the planned extractive activities. If appropriate, slopes are created for the species in safe areas of the site as an alternative to those that might be destroyed before the arrival of the birds, thereby preventing the colonies from settling in unsafe areas.

The nests of some species breeding in Soto Pajares (i.e., Ciconia ciconia—▪ Fig. 7.26, *Milvus migrans, Ardea cinerea) are usually found in the wooded or marshy areas of the site, where actions should be avoided when they involve human presence for relatively long periods. The zones that may host nests of sensitive species are delimited, and it is recommended that any activity in the vicinity of these areas is carried out outside the breeding season of these birds. Where the extraction would affect such an area (e.g., a lake with marsh vegetation), activity must be undertaken once alternative areas have been created in the site and these have been occupied by the relevant species affected by the resumption of the extractive activity. Moreover, once the extraction has ceased, the area could be restored to a lake with marsh vegetation and open waters suitable for bird species.*

▪ **Fig. 7.26** White stork (Ciconia ciconia)

The actions to enhance habitats include the recovery of riparian forests and natural vegetation, the creation of hedges and boundaries in agricultural plots, the creation of shallow water lagoons, the expansion and modification of existing lakes and the management of agricultural areas. The recovery of the riparian forest in areas surrounding the Jarama River is a priority measure since these forests have practically disappeared in the area and have declined significantly in the entire region. This action will also enhance biodiversity by increasing the availability of food and shelter for numerous species and will recover the role of these forests as ecological corridors. A significant part of the restored area is dedicated to dry farming (cereals), where the creation of hedgerows was implemented by planting woody shrubs and trees around agricultural plots. The objective of this action is to improve the ecosystem diversity and ecological connectivity, as well as crop productivity through wind attenuation, and reduction in soil erosion, and water and nutrient input retention.

The main objective of the creation of shallow water lagoons is to create a new type of lagoon ecosystem with characteristics very different from those existing on the site and its surroundings. It favors the arrival of new species and contributes to their conservation at local and regional levels. This action must be undertaken once the extractive activity is finished. Regarding the expansion and modification of existing lakes, this action will increase the biodiversity and ecological value of existing water bodies and encourage the settlement or expansion of aquatic bird populations with stricter ecological requirements. It is important to note that the extractive activity of the aggregate deposits generates water bodies that are extremely attractive to many bird species, although the lakes that currently exist have poor geomorphological characteristics due to the steep profile of their shores and their excessive depth.

Since most of the already restored quarry area is managed as farmland, a suitable management regime of agricultural areas is promoted based on alternate plowing of contiguous plots so that plots that are plowed one year, are left unplowed the following year. This favors structural complexity and provides diverse ecological niches and shelter areas with vegetation. Finally, monitoring activities are necessary to follow the evolution of natural recovery and restoration. Thus, a long-term monitoring plan is proposed to follow vegetation recovery and natural colonization, fauna, landscape evolution, soils, etc. Detection and control of invasive species and pests is also planned to allow the implementation of actions needed to combat these potential problems.

7.2.6 Visually Disturbed Landscapes and Reclamation

Aggregate activities can generate negative visual adverse effects on resources linked to other landscape utilizations such as recreation or tourism. Potential contributors to visual impacts are roads, erosion, abandoned mining installations, and deforestation. Regarding color changes, in areas where the color of the rock matches the natural color of the terrain, visual impacts will be less than with sharp color contrasts.

Landscape is more than just the visual perception of landform, vegetation, and land utilization. It can be described by way of landform, land cover, and land use elements. As commented previously, different people can evaluate the same landscape in many ways because it is a mixture of objective and subjective analysis. The latter is obviously of prime importance. In this sense, "aggregate quarries and pits are faced with NIMBY (not-in-my-backyard) reactions since some negative externalities are inevitable" (Pelekasi et al. 2012).

The impact on landscape by aggregate extraction depends on various factors. Location, size, extracted volume, and mining methods can influence the impact of aggregate activities on the visual appearance of the land. Landscape alteration can generate an adverse opinion among potential observers and compromise the possible development of the surrounding territory. In fact, the evaluation of landscape and visual impact is many times based more on the subjective perception of the people and less on the real features of the visible alteration. However, several aspects of landscape modification must be objectively assessed to estimate the magnitude of change and offer an objective evaluation of the adverse effects originating by aggregate extraction.

To prevent visual impact, a visual resource management (VRM) should be carried out. It was originally created by the Bureau of Land Management (U.S. Department of Interior), and the main goal is to manage public land in a manner that protects the scenic values of the lands. VRM includes inventorying scenic values and determining management aims for those values through the management planning procedure and assessing suggested activities to establish whether they conform to the management purposes. The VRM system is split into two parts. The first step is the identification of visual values to determine the appropriate level of management. This step, called the VRM inventory, has three components: scenic quality evaluation, sensitivity analysis, and distance zone measures. The second part of the VRM system is the analysis stage. It includes establishing whether the potential visual adverse effects from suggested surface-disturbing activities will meet the management goals defined for the area or whether design adjustments will be requested.

Aggregates extraction close to the markets where they will be utilized is very limited, and thus, "there is often a conflict between the preservation of environmentally significant landscapes and maintaining a steady and affordable supply of aggregates for construction materials" (Binstock and Carter-Whitney 2011). The locations of the aggregate mining operation and processing plant as well as the associated installations (i.e., stockpile areas, haul roads and customer loading areas) are generally established to minimize any impact on the landscape. Nevertheless, as a rule, aggregate industry generates a disturbed landscape. However, this impact is foreseeable and controllable since aggregate operations are designed facilities, and a good design is very important. Changing a landscape from undeveloped or agricultural land to a quarry is clearly an impact of aggregate mining; it is difficult to discuss the associated visual impacts in absolute terms. The nature of this modification depends on the surface forms, slopes, natural ground cover, and facility type. The reclamation of the disturbed land and its ecosystem must be a part of the aggregate extraction plan.

The aggregate industry must take into account the following considerations to be successful in gaining approval for the utilization of land during aggregate extraction: "(1) ensuring that the planned use for the property is compatible with both current and any planned uses of land surrounding the operation, (2) determining whether the land includes or is near sensitive environmental receptors or ecosystems such as listed threatened or endangered species, wetlands or jurisdictional waters, cultural or historic resources or others so the operation can be designed at the front-end to either avoid impacting the areas or obtain permits or authorization needed to address anticipated impacts, (3) determining who the stakeholders are in the permitting process and engaging in open dialog regarding their concerns and expectations in advance of the project evaluation and decision making process and (4) developing a comprehensive mining plan and reclamation plan for the facility that incorporates the issues and concerns addressed in the permit application" (NSSGA 2013).

From a landscape viewpoint, minimizing unsightly changes can be achieved: (a) through proper landscape analysis, design, and operations, (b) by extracting aggregates from the most suitable deposits, and (c)

Fig. 7.27 Super quarry Glensanda (Scotland, UK) (image courtesy of Aggregate Industries)

through the creation of superquarries (Fig. 7.27) and the use of underground mines (Fig. 7.28), depending on local conditions. Designing a single large operation or superquarry at an environmentally acceptable site is clearly preferable to many smaller operations at scattered locations. Screening of workings can be achieved through the construction of berms, tree plantings of sufficient density to provide a visual barrier (Fig. 7.29), fencing, or other landscaping techniques. The orientation of the working faces can reduce the visual impact from sensitive, long-distance views.

An aggregate operation is a temporary land utilization and the site is likely to be converted into another beneficial use when mining is completed. The after-mining utilization of the land is necessary to reduce the environmental impacts of aggregate exploitation. When the aggregate resource is depleted, physically disturbed land can be reclaimed for other residential, commercial, and natural uses. It is not possible to return ecosystems exactly to their original structure. Instead of returning an area to its original condition, a more realistic approach is to approximate the new habitat as closely as possible to its original function and to recapture the landscape character. Thus, it is now recognized that restored mineral workings can make a major contribution to both biodiversity and geodiversity.

Reclamation of land from which aggregates have been extracted is an essential part of the viability of aggregate mining operations. The most common form of reclamation and restoration is the filling of the excavation using imported waste such as domestic, commercial, and industrial wastes. The degraded land is restored using the topsoil removed and stored during the initial quarry exploitation. In many cases, it is common to use topsoil from other quarries. The process of restoration starts from the beginning of an operation, and the process must be in accordance with the final restoration plan. This ensures that the process is easier and more effective.

Fig. 7.28 Underground Kanpanzar quarry (image courtesy of Calcinor)

Reclamation can also focus on human needs including residential, business, and recreational uses providing significant social value to the community associated with the continued utilization of the quarry. Restored aggregate quarries also afford a group of recreation facilities that would otherwise be unavailable. Once backfilled to the original topographic level, the aggregate quarry or pit can be the site for many different purposes such as leisure activities, urban development as well as agriculture. In the present expanding suburban areas, mined-out aggregate quarries are converted into secondary uses such as wildlife habitats, golf courses (Fig. 7.30), recreational areas, shopping malls, agricultural areas, sites of special scientific interest and regionally important geological sites, parks, botanical gardens, school grounds, museums, underground storage facilities, solid waste disposal areas, mining and power plant waste storages, high-quality lakefront housing sites, water-storage reservoirs, piscicultures, and a myriad of other economically or ecologically productive land utilizations that can benefit society. Aggregate quarries can even provide valuable nesting sites for birds on rock faces (Fig. 7.31) or in sand faces and a range of habitats and associated flora and fauna on silt and clean water ponds.

For example, Thorpe Park (London, UK) is an amusement park located on the site of a sand and gravel exploitation. In Steilacoom (Washington, USA), an ancient glacier sand and gravel site was converted after reclamation in the Chambers Bay Golf Course. Oakwood Lake (California, USA) is a commercial waterslide park in a reclaimed aggregate operation. In San Antonio (Texas, USA), the Beckman quarry was reclaimed as Six Flags Amusement Park (Fig. 7.32) and the La Cantera shopping center (Fig. 7.33). In British Columbia (Canada), Butchart Gardens (Fig. 7.34) were reclaimed from 50 acres of an exhausted aggregate quarry.

Fig. 7.29 Trees providing a visual barrier

Fig. 7.30 Ancient quarry converted into a golf course (image courtesy of Cemex)

Fig. 7.31 Nesting site on a rock face (image courtesy of Birds and Blooms)

Fig. 7.32 Six Flags amusement park (image courtesy of the Park)

Fig. 7.33 La Cantera shopping center (image courtesy of the shopping center)

Fig. 7.34 Butchart gardens (image courtesy of Canada Tours)

7.3 Questions

Short Questions

- Define sustainable development.
- List the most important environmental impacts associated with aggregate production.
- Define total suspended particles and the influence of the size in this concept.
- Explain the sequence of dust control techniques.
- List the most important sources of noise in aggregate extraction and processing.
- Definition of peak particle velocity in ground vibration related to blasting.
- What is biodiversity? Explain the relationship between mining and biodiversity.
- List the potential contributors to visual impacts in mining.
- What does not-in-my-backyard reaction mean?

Long Questions

- Explain the significance of an environmental management system and the ISO 14001 standard.
- Comment the potential effects related to the impacts on water resources as well as the recommended practices for water management.

References

Arup Environmental (1995) The environmental effects of dust from surface mineral workings. Arup Environmental, PECD 7/1/468, Report on behalf of the Department of the Environment (DOE), HMSO, London

Birch W, Datson H (2007) Reducing the environmental effect of aggregate quarrying: dust, noise & vibration. Sustainable Aggregates, MIRO, UK, 182 p

Binstock M, Carter-Whitney M (2011) Aggregate extraction in Ontario: a strategy for the future. Canadian Institute for Environmental Law and Policy, 95 p

Bustillo M (2018) Mineral resources—from exploration to sustainability assessment. In: Springer textbooks in earth sciences, geography and environmental sciences. Springer International Publishing AG, Cham, 653 p

Chaulya SK, Chakraborty MK, Singh RS (2001) Air pollution modelling for a proposed limestone quarry. Water, Air and Soil Pollution, 126:171–191

CSIRO (2014) Morton S, Sheppard A, Lonsdale M (eds) Biodiversity: science and solutions for Australia. CSIRO Publishing, Australia, 226 p

Drew LJ, Langer WH, Sachs JS (2002) Environmentalism and natural aggregate mining. Nat Resour Res 11:19–28

EPA (1996) Air quality criteria for particulate matter. Final report. U.S. Environmental Protection Agency, Washington, D.C.

EU (2011) Non-energy mineral extraction and Natura 2000—guidance document. European Union, European Commission, Luxembourg, 144 p

ICMM (2006) Good practice guidance for mining and biodiversity. International Council on Mining and Metals, London, UK, 144 p

ICMM (2012) Water management in mining: a selection of case studies. International Council on Mining and Metals, London, UK, 32 p

IFC (2007) Environmental, health and safety guidelines for mining. International Finance Corporation, World Bank Group, 33 p

Langer WH (2001) Environmental impacts of mining natural aggregate. In: Bon RL, Riordan RF, Tripp BT, Krukowski ST (eds) Proceedings of the 35th forum on the geology of industrial minerals. Utah Department of Natural Resources, Miscellaneous Publication 01-2, pp 127–138

Nelson MG (2011) Site environmental considerations. In: Darling P (ed) SME mining engineering handbook, 3rd ed. Society for Mining, Metallurgy, and Exploration, Inc., pp 1643–1664

NSSGA (2013) The aggregates handbook, 2nd ed. The National Stone, Sand and Gravel Association, Alexandria, Virginia, USA, 906 p

OSSGA (2016a) Dust management at pits & quarries. Ontario Stone, Sand & Gravel Association, Mississauga, Ontario, 6 p

OSSGA (2016b) Water management in quarries. Ontario Stone, Sand & Gravel Association, Mississauga, Ontario, 6 p

Pelekasi T, Menegaki M, Damigos D (2012) Externalities, NIMBY syndrome and marble quarry activity. J Environ Plann Manage 55:1192–1205

QM (2005) Controlling dust and erosion in quarries. In: Quarry management, October 2005, pp 41–43

Rostami J, Hambley D (2011) Blasthole drilling. In: Mining engineering handbook, 3rd ed. Society for Mining, Metallurgy, and Exploration, Inc. (SME), pp 435–441

Sairanen M, Rinne M (2019) Dust emission from crushing of hard rock aggregates. Atmos Poll Res 10:656–664

Tost M, Ammerer BA (2022) Sustainable supply of aggregates in Europe. Montanuniversität Leoben, Chair of Mining Engineering and Mineral Economics, Leoben, Austria, 123 p

Walter GE (2011) Noise hazards and controls. In: Darling P (ed) SME mining engineering handbook, 3rd ed. Society for Mining, Metallurgy, and Exploration, Inc., pp 1633–1640

Standard

ISO 14001: 2015. Environmental management systems—requirements with guidance for use

Applications

Contents

8.1 Introduction – 260

8.2 Concrete and Mortar – 260
8.2.1 Concrete – 260
8.2.2 Mortar – 272

8.3 Roads – 277
8.3.1 Unbound Materials – 278
8.3.2 Bituminous Mixtures – 281

8.4 Railway Ballast – 286

8.5 Armourstone – 291

8.6 Gabions and Matresses – 295

8.7 Filter Media – 298

8.8 Decoration – 300

8.9 Questions – 303

References – 303

M. Bustillo Revuelta, *The Basics of Aggregates*,
Springer Textbooks in Earth Sciences, Geography and Environment,
https://doi.org/10.1007/978-3-031-42961-3_8

Abstract

The aim of this chapter is to review the main applications of aggregates in construction. Aggregates can be used as a construction material in two large types of applications: unbound applications (the aggregate is not bound) and bound applications (mixes containing binding agents such as cement or bitumen). The largest proportion of aggregates is used to manufacture concrete, which is the most widely used building material in the construction industry. Aggregates constitute, in percentages greater than 90%, most of the layers of road surfaces. They are the principal material in pavement construction. One of the most demanding applications for crushed aggregate is railway ballast. Armourstone consists of coarse aggregates utilized in hydraulic structures and other engineering works. Gabion is a robust structure comprizing a double-twist wire-mesh basket filled with hard, durable stone. It performs a great variety of functions within coastal, estuarial, and fluvial environments. Drainage and filter aggregates are sand, gravel, or crushed stone or mixtures thereof. They play essential roles in many engineering projects. Finally, decorative/ornamental aggregates can be used in a wide variety of spaces such as commercial and residential spaces, domestic pathways, driveways, car parks, architectural landscaping, gardening (patios and rockeries), sports grounds, etc.

8.1 Introduction

Aggregates are used in many applications, some of which are essential for modern society. The life of human beings today would be unthinkable without the existence of concrete or roads. For this reason, aggregates are the material most used by humans after water. Aggregates can be used as a construction material in two large types of applications: unbound applications (the aggregate is not bound) and bound applications (mixes containing binding agents such as cement or bitumen). They are an essential part of public and private buildings, bridges, roads and highways, airports, waste disposal and treatment plants, etc. In order of importance, the main applications of aggregates are (a) concrete (and mortar), (b) unbound and bound (bituminous) pavements, and (c) railway ballast. Other important uses are for armourstone and gabions and as filter media. Finally, decorative/ornamental uses can be an example of the infinite applications of aggregates.

8.2 Concrete and Mortar

8.2.1 Concrete

The largest proportion of aggregates is used to manufacture concrete. It can be defined as "a mixture of aggregate, cement, and water, which hardens" (ISO 6707-1 standard), "a composite material that consists essentially of a binding medium within which are embedded particles or fragments of aggregate" (ASTM C125 standard), and a "material formed by mixing cement, coarse and fine aggregate and water, with or without the incorporation of admixtures and additions, which develops its properties by hydration of the cement (EN 206-1 standard)". These are the definitions of concrete based on the most important regulatory standards worldwide. More specifically, concrete is a mixture of cement, water, fine aggregate (sand), and coarse aggregate (gravel or crushed rocks) in which cement and water have hardened by a chemical reaction (hydration) to bind the nonreactive aggregate. Other materials that can often be added to the previous mixture are additions, which are fine powders that can substitute some of the cement, and small quantities of chemicals, known as admixtures, which mainly can improve some properties. Concrete is the most widely used building material in the construction industry, being used in all types of buildings (from residential to office blocks) (◘ Fig. 8.1) and in infrastructure projects (i.e., roads, bridges, and dams—◘ Fig. 8.2).

The major factors responsible for the wide utilization of concrete are moldability, early hardening, high early compressive strength, development of required properties with admixtures to be used in complex and adverse situations, and durability. The concrete properties can be controlled within a broad range by using appropriate components and by special mechanical, physical, and chemical processing techniques. Generally, a mixture to make concrete is approximately 10–15% cement, 60–80% aggregates and 15–20% water. Cement is an essential component of concrete (although it occupies the smallest proportion by volume). Cement, when hydrated, binds the aggregates together. Water is also an essential component in concrete because it intensively takes part in the chemical reaction with cement.

Aggregates (sand and gravel) constitute approximately 80% of the concrete weight and 70% of the concrete volume. The aim is to utilize as much aggregate as possible to bind the elements together with the

Fig. 8.1 Building concrete structure

Fig. 8.2 Hoover dam (USA) (image courtesy of Luis Fueyo)

8

hardened cement paste. Thus, aggregates form the skeleton of the concrete. Porosity within this skeleton must be occupied with cement as completely as possible. Concrete aggregates should have a strong bond with the hardened cement paste, should not interfere with cement hardening, and should not have a negative effect on concrete durability. Aggregate in concrete acts as a filler, reduces cost, provides strength and durability, and reduces the change in volume caused by the hydration and/or drying processes. The aggregate used in concrete production can be natural, manufactured, or recycled. In Europe, aggregates for concrete are defined in the EN 12620 standard. Two main size groups are defined: fine aggregate (sand) (◘ Fig. 8.3) and coarse aggregate (gravel) (◘ Fig. 8.4). Fine aggregate is the designation for size fractions lower than 4 mm, that is, the particles pass a 4 mm sieve, and coarse aggregate is the designation for size fractions greater than 4 mm, that is, particles are retained in a 4 mm sieve.

Concrete producers fill most of the concrete volume with cheaper ingredients, the aggregates, so they first carefully separate the different grain sizes (sand and gravel) and then mix them in desirable proportions. In the resulting particle distribution, the successively smaller particles fill the voids between the larger parts, being referred to as good gradation. The ASTM C33/33 M standard for concrete aggregates "defines the requirements for grading and quality of fine and coarse aggregate for their utilization in concrete". ◘ Figure 8.5 shows the limits for fine aggregate and one specific size of coarse aggregate in the standard cited.

It is important to remember that aggregates, according to their density, are also classified as lightweight, normal weight, and heavyweight aggregates. The majority of aggregates utilized to make concrete fall into normal weight and lightweight types. ASTM C330/330 M, C331/331 M, and C332 standards cover "the requirements of lightweight aggregates intended for use in different types of concrete applications in which prime consideration is reduced density while maintaining the compressive strength of the concrete". These specifications cover "two main types of lightweight aggregates: (a) aggregates manufactured by expanding, pelletizing or sintering products such as blast-furnace slag, clay (▶ Box 8.1: Expanded Clay), diatomite, fly ash, shale, or slate, and (b) aggregates prepared by processing natural materials such as pumice (◘ Fig. 8.6), scoria or tuff". The ASTM C637 standard specification for aggregates for radiation-shielding concrete covers special aggregates utilized in radiation-shielding concretes. In these situations, the composition of the aggregates and their high specific gravity, or both, are of prime consideration. Finally, the ISO 19595 standard points out the "properties and requirements of aggregates obtained by processing natural materials and mixtures of these aggregates for their utilization in concrete".

◘ **Fig. 8.3** Fine aggregate (sand) used in concrete

Fig. 8.4 Coarse aggregate (gravel) used in concrete

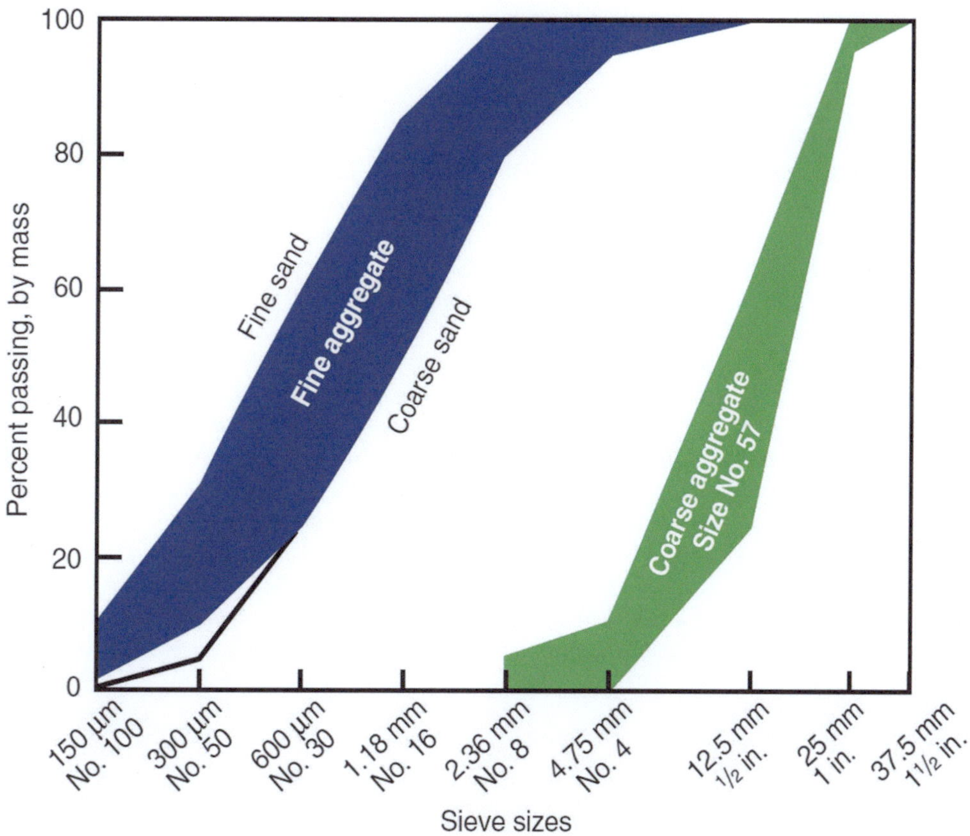

Fig. 8.5 Curves indicate the limits specified in the ASTM C33 standard for fine aggregate and one used size number (grading size) of coarse aggregate (Kosmatka and Wilson 2011)

Fig. 8.6 Pumice aggregates (image courtesy of Hess Pumice)

8

Box 8.1 – Expanded Clay

Lightweight aggregates are used in many industries due to their numerous advantages in comparison to many other materials. In this sense, light weight expanded clay aggregate (LECA) "is one of the materials with the greatest compressive strength among lightweight aggregates" (Ozguven and Gunduz 2012). *In the manufacturing process, after heating at 1150 °C in a rotary kiln, the expansion of the clay reaches a value of around four to five times its original size, taking the shape of pellets. As a consequence, 1 m^3 of fresh clay produces 5 m^3 of durable, expanded clay. Expanded clay is a well-proven, high-quality, efficient, and durable lightweight aggregate that has been utilized for more than 50 years, and is suitable for a wide range of applications in the construction sector.*

The expanded clay aggregates are spherical and brown with an internal black core. This is attributed to the carbon content and the oxidation state of iron. "The microstructure is smooth but, on the microscale, it is relatively rough with open pores, being the structure of the pores approximately 10–200 μm in size; the specific gravity of expanded clay range from 0.66 to 1.65 g/cm^3, being this specific gravity around 20–45% less than that of the normal aggregates" (Vijayalakshmi and Ramanagopal 2018).

*According to LECA (a Saint-Gobain Brand), the eight steps in expanded clay manufacturing process are the following: "(a) clay extraction: clay is extracted from quarries located close to the plants, keeping haulage costs and carbon emissions to a minimum; the clay is brought into the clay barn (*Fig. 8.7) *after separating larger stones from the clay, (b) crushing: the smaller stones are crushed and the clay is pressed through small holes, (c) kneading: the clay is kneaded with some water to make processing easier (d) drying: clay runs through the drying kiln, where the water is evaporated and the clay is heated to 400 °C; this is where the expanded clay pellets get their specific size dimensions (e.g. 0–4 mm, 4–10 mm, 10–20 mm) as the clay cracks into smaller particles, (e) burning (*Fig. 8.8)*: in the next part of the kiln (the burning kiln), the clay is heated to 1150 °C initiating the expansion process and creating light round lightweight clay pellets, which have a hard ceramic shell and a porous core; after passing through the kiln, the expanded clay is cooled: a correct cooling process is essential to ensure a high-quality product, which also significantly improves the energy performance (as the hot clay cools, cold air is warmed and this heated air is used to dry, heat and expand the clay in the kiln), (f) sifting: when the lightweight aggregate clay pellets are taken out of the kiln, they are placed in a sieve system that divides them into different fractions based on their grain size, and (g) packing: in the last process, the lightweight aggregate clay pellets (*Fig. 8.9) *are packed into bags or filled directly onto the trucks and ships".*

*In the construction industry, LECA is utilized extensively in the production of lightweight concrete blocks (*Fig. 8.10)*. It can be also used as both a sound and thermal insulation material. If LECA is used in structural backfill against foundations, retaining walls, and bridge abutments, the reducing earth pressure can reach values of 75% compared with conventional materials and increase stability while reducing settlement and land deformation.*

Fig. 8.7 Clay barn (image courtesy of Leca Danmark AS)

Fig. 8.8 Inside the burning zone (image courtesy of Leca Danmark AS)

Fig. 8.9 Expanded clay after burning

Fig. 8.10 Sample of lightweight concrete using expanded clay

The physical, chemical, and mineralogical properties of aggregates must be known before mixing concrete to obtain a desirable mixture. This is because the characteristics of the aggregates selected cannot adversely influence the performance of the manufactured concrete, not only in the fresh state but also in the hardened state. It could be said that "mineral characteristics of aggregates affect the physical properties of the concrete in its hardened state while the geometric characteristics mostly influence concrete's behavior in its fresh state" (Vieira 2015). The properties of aggregates commented on in the following paragraphs are of prime concern. These properties include size gradation, shape and texture, particle density, water absorption, impurities, and durability.

The size distribution or grading of aggregates is a significant feature since it defines the paste requirement for a workable concrete as well as the water demand of the concrete. The coarser the grading, the lower the cement content needed for a determined workability and water/cement ratio. For equal absolute volume, the void content decreases when different sizes are combined. Consequently, the necessary amount of paste also decreases (◘ Fig. 8.11). This means that the largest possible aggregate size should be used and a continuous range of particle sizes from fine sand up to coarse stones is desirable (◘ Fig. 8.12). ◘ Figure 8.13 illustrates a theoretical ideal of uniform gradation in concrete. The optimum utilization of the aggregate size improves the concrete quality. In general, the maximum particle size of coarse aggregate can be 10, 20, or 40 mm.

The shape and texture of the aggregate principally influence the characteristics of fresh concrete. Smooth and rounded aggregates (◘ Fig. 8.14) give a more workable mix while rough and angular aggregates (◘ Fig. 8.15) make concrete harder to place, work, and compact. Moreover, angular particles need more cement to maintain the same water-cementing material ratio. In turn, particle surface texture mainly affects the bond between the aggregate and the cement paste. Because crushed stone produces angular and elongated aggregate particles, the improvement of aggregate particle shape can result in an enhancement of the concrete durability, although it increases production costs.

The particle density of an aggregate is a property of particular value in concrete mix design for establishing weight-volume relationships. However, it is not often utilized as a measure of aggregate quality. The majority of natural aggregates have particle densities between 2.4 and 2.9. Changes in the relative density of aggregate produce the concrete density to vary, which is undesirable if a minimum concrete density is required (i.e., in heavyweight concrete used in nuclear radiation shielding). Aggregate water absorption is also a significant property because it may influence the amount of water required in a concrete mix to achieve the given water/cement ratio. This parameter can be small (i.e., dense fine-grained rocks) or high (i.e., porous materials).

Aggregate particles must be free of absorbed chemicals or clay coatings as well as other fine materials that could generate the deterioration of concrete. Fine materials increase the water demand to

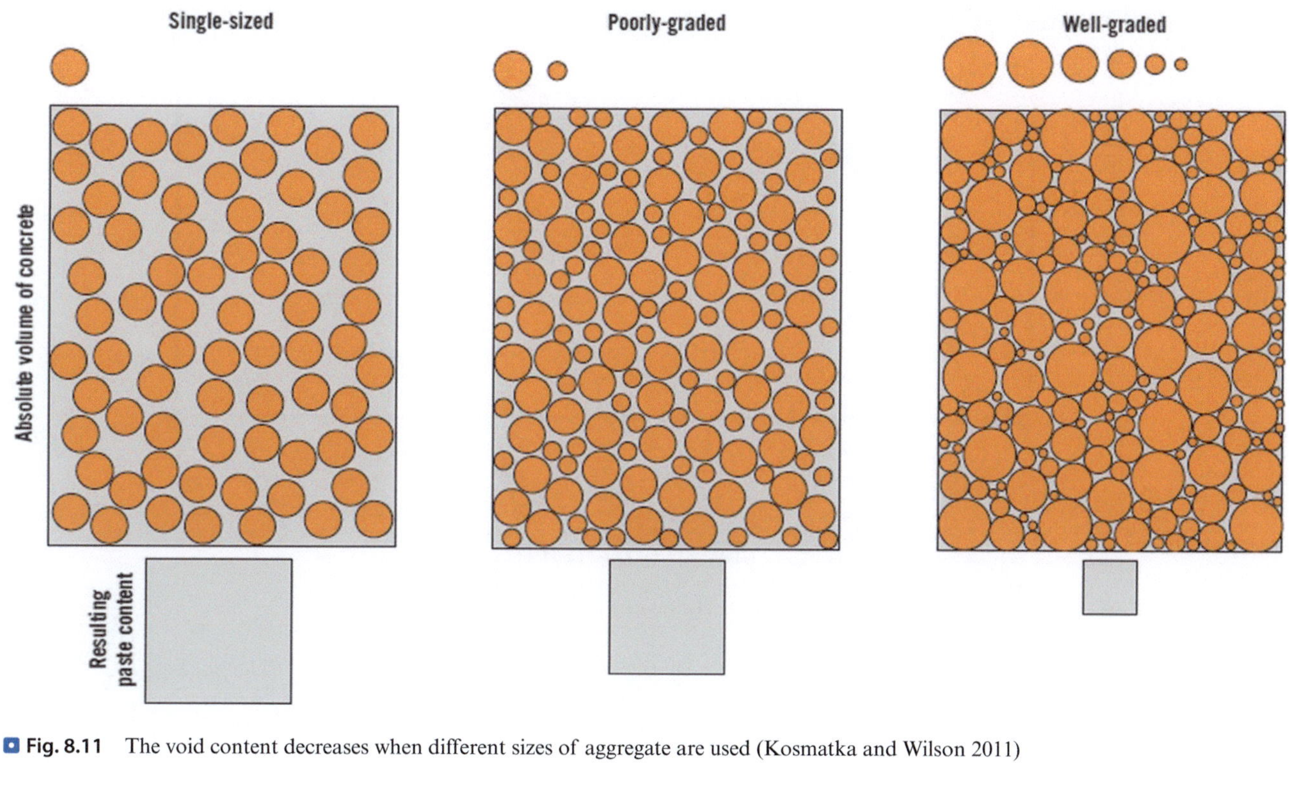

◘ **Fig. 8.11** The void content decreases when different sizes of aggregate are used (Kosmatka and Wilson 2011)

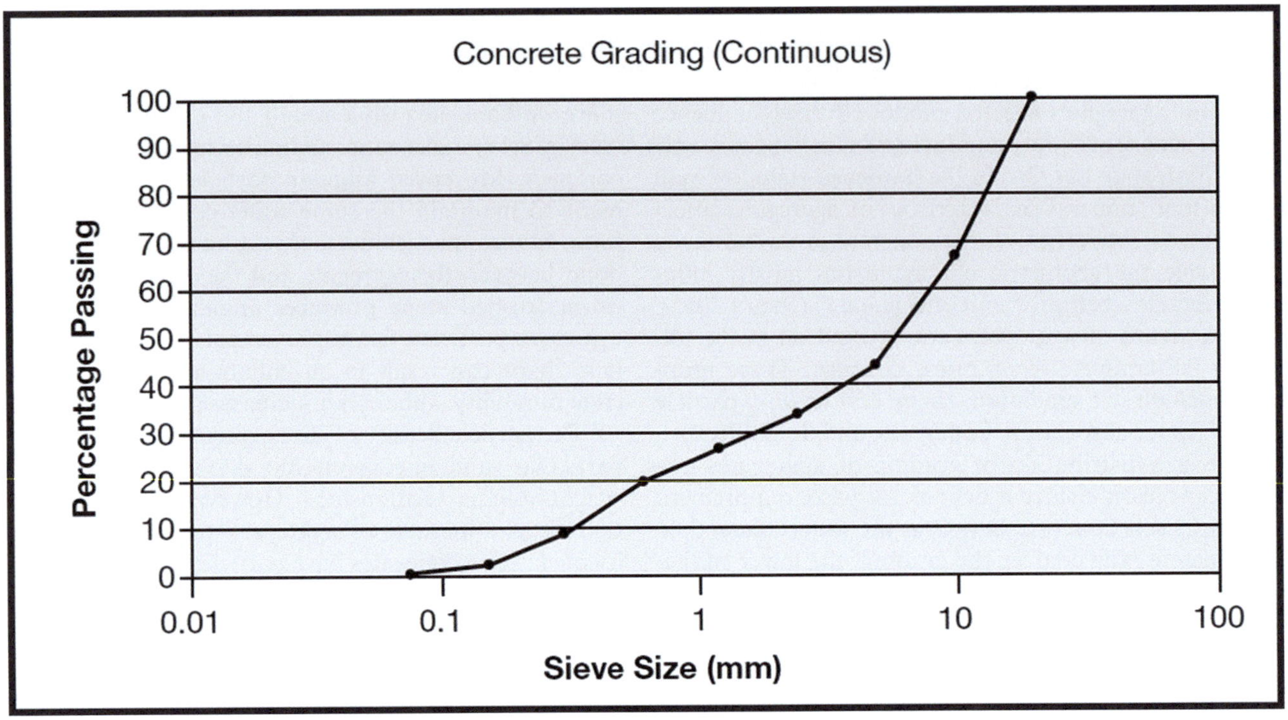

Fig. 8.12 Concrete grading (continuous) curve

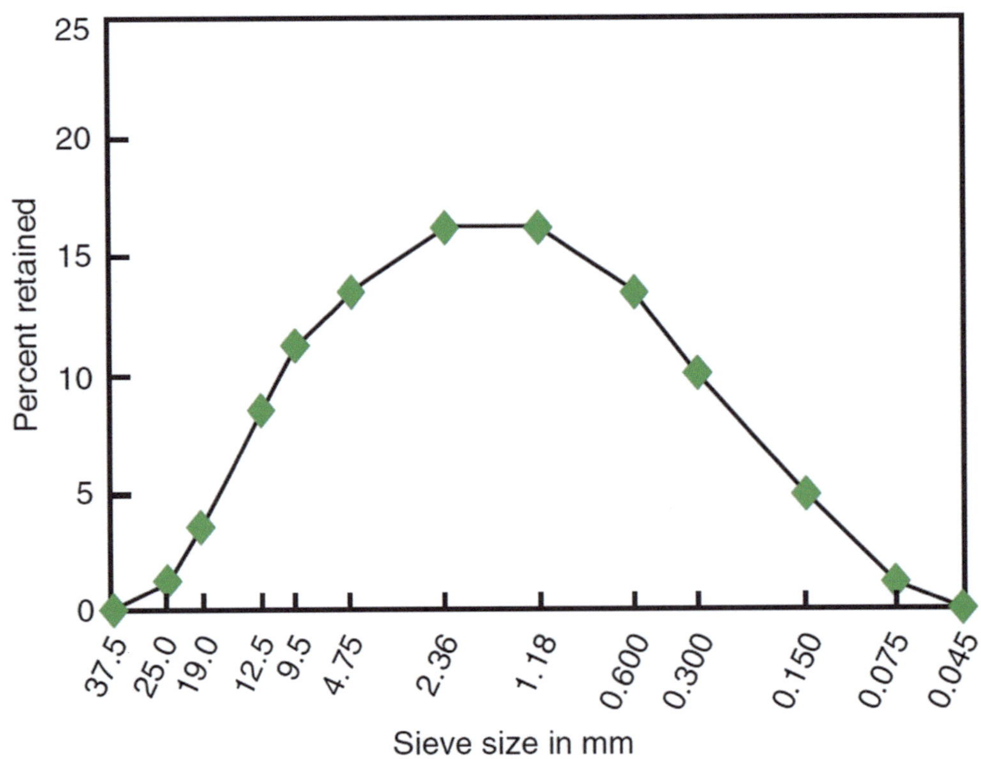

Fig. 8.13 An ideal combined aggregate grading for concrete (Kosmatka and Wilson 2011)

8

Fig. 8.14 Rounded aggregates

achieve a given workability and their presence as coatings impairs the aggregate-matrix bond. Other harmful substances that can be present in aggregates are organic components, iron oxide, coal lignite, mica, chlorides, and sulfates. For example, organic impurities can "delay setting and hardening of concrete, interfere with the interaction of chemical admixtures, reduce strength gain, and in unusual cases, cause deterioration" (Kostmatka and Wilson 2011). Table 8.1 shows the main harmful materials and the corresponding effects on concrete.

Regarding aggregate chemical resistance and durability, reactive aggregates can result in serious damage to the material. Most types of aggregate utilized for concrete manufacturing are inert, that is, they do not react chemically with the cement and/or hydration products. However, several aggregate types react chemically with the alkali in the cement, affecting the integrity of the concrete. Aggregates in concrete, among other components, can incorporate silica, silicates, and carbonates. In certain mineral forms, these components may react with the alkaline hydroxides in the pore water derived from the sodium and potassium oxides in the cement. The general term for this process is the alkali-aggregate reaction (AAR), which has two forms: alkali-silica reaction (ASR) (▶ Box 8.2: Alkali-Silica Reaction) and alkali-carbonate reaction (ACR). The former is the most common and important and involves active silica. These reactions are another cause of concrete degradation through expansive internal processes. The deterioration of numerous major concrete structures has been caused by these reactions, specifically the alkali-silica reaction, many times after years of apparently adequate service.

Fig. 8.15 Angular aggregates

Table 8.1 Harmful materials in aggregates and the corresponding effects on concrete (Kostmatka and Wilson 2011)

Substances	Effect on concrete
Organic impurities	Affects setting and hardening, may cause deterioration
Materials finer than the 75-μm (No. 200 sieve)	Affects bond, increases water requirement
Coal, lignite, or other lightweight materials	Affects durability, may cause stains and popouts
Soft particles	Affects durability
Clay lumps and friable particles	Affects work-ability and durability, may cause popouts
Chert of less than 2.40 relative density	Affects durability, may cause popouts
Alkali-reactive aggregates	Causes abnormal expansion, map cracking, and popouts

Box 8.2—Alkali-Silica Reaction

*The alkali–silica reaction (ASR) is a material degradation mechanism that takes place over time in concrete between the alkaline cement paste and the reactive amorphous silica present in aggregates. It is of more concern because aggregates with reactive silica components are very common. In this reaction, silica occurs in its reactive form as the minerals opal (*Fig. 8.16*), chalcedony, crystobalite, tridymite, and as volcanic glasses. They are commonly amorphous materials (lack of long-range order at the atomic level). The product of ASR is in the form of a gel and, in absorbing water, these gels can swell and induce enough expansive pressure to damage concrete, which is one of the major causes of cracking of concrete (*Fig. 8.17*). The gel is also amorphous and contains silicon, alkalis, calcium, and water.*

Fig. 8.16 Scanning electron microscopy (SEM) micrograph of opal CT (image courtesy of María Ángeles Bustillo)

The cracks formed through this process can range from 0.1 mm to as much as 10 mm in width. Cracking occurs in zones with a frequent supply of moisture. For this reason, alkali-silica reactivity can be significantly diminished by keeping the concrete as dry as possible. Because the alkali-silica reaction is a localized process, the degree of reaction and the extent of further deterioration can change from location to location in the same concrete structure. The cracks resulting from ASR may provide direct paths to penetrate aggressive components. In extreme cases, the expansion caused by the alkali-silica reaction could lead to the yielding of the steel reinforcement.

The whole process is slow and cracking can take years to develop in structural concrete. The principal factors influencing the quantity and rate of reaction in ASR are the following: (a) the amount of alkali and active silica available, (b) the aggregate particle size, (c) the availability of moisture and (d) the ambient temperature. Consequently, the best way to avoid ASR includes (a) avoiding the use of reactive aggregates, (b) testing aggregates for their potential reactivity, (c) controlling the amount of alkali in the cement, (d) combining the Portland cement with an admixture (silica fume, fly ash, and/or ground-granulated blast furnace slag), and (e) limiting the total alkali content of the concrete. Petrography, when utilized for studying a reacted concrete, may confirm the existence as a result of alkali-silica reaction. Polished sections of concrete may be examined using scanning electron microscopy. The great advantage of this technique is that the gel can be analyzed utilizing X-ray microanalysis.

Fig. 8.17 Cracking in a bridge pylon of concrete due to ASR (image courtesy of Sika)

Regarding ACR, reactions with certain dolomitic rocks are associated with this type of reactivity. Dedolomitization (transformation of former dolomite to calcite) is linked to expansive alkali-carbonate reactivity because this reaction produces crystallization of brucite, which can generate considerable expansion. The damage caused by ACR is quite similar to that originating from the alkali-silica reaction. However, this reaction is rare since aggregates susceptible to ACR are less common.

8.2.2 Mortar

This heading devoted to aggregates used in mortars necessarily introduces some degree of overlap with the heading concerned with aggregates for concrete. Nevertheless, there are important differences between mortars and concretes, principally based on the manner of their utilization. The earliest documented use of mortar (a mix of sand, binder, and water) was around 4000 B.C.; the use of that primitive mortar is mentioned several times in the Bible (i.e., Deuteronomy 27:2—*set up large stones and coat them with plaster*). Thus, mortars have a long tradition of thousands of years. Mortars were used up to the 1950s exclusively as job-site-mixed mortars (Fig. 8.18). The search for better performance, workability, and durability for mortars has led to an absolute substitution of work site mixed mortars by factory-made ones in the Western countries since the 1960s. The introduction of dry-mix mortar technology and the use of silo transport and machine application of mortars made it possible that from 1960 to the present the amount of mortar applications incremented exponentially. Currently, factory-made mortars are manufactured in modern automated factories, providing hundreds of mortar types for all types of applications with a warranty for continued quality and performance.

Since mortar has numerous applications in building and construction, there are also many different definitions of this term, although often with similar meanings. For instance, ASTM standards include more than 10 definitions of terms with mortar words based on the committee involved. The simplest definition is probably that of the ASTM A1096/A1096 M standard, which defines mortar as "a mixture of cement, fine aggregate, and water". The ISO 6707-1 standard considers mortars as "a mixture of binder, fine aggregate, and water, which hardens and which is normally used as a jointing product". Finally, the EN 998-1 and -2 standard definitions of mortar are also quite simple: "mixture

Fig. 8.18 Job-site mixed mortar

formed by one or more inorganic binders, aggregates, water, and sometimes admixtures and additives". Classification of mortars can be carried out according to different concepts such as composition, properties, requirements, production method, supply method (Fig. 8.19), and applications (Fig. 8.20).

As in the case of concrete, mortar is made of mixing aggregates, cementitious materials, water, and admixtures and additives if needed. The main differences between concrete and mortar components are the types of cementitious materials and the aggregate size. Cementitious materials in mortars can include several types of binders while concrete can only incorporate cement as a binder. Regarding their size, mortar aggregates use only fine aggregate (sand) while concrete is produced by using fine and coarse aggregate (gravel).

Aggregates for mortars consist of natural or manufactured sand, which is the main constituent of mortars because it accounts for at least 75% of the mortar volume. Sand acts as an inert filler, offering economy (it is a cheap material), workability, and reduced shrinkage as well as influencing the compressive strength of the mortar. Some screeds incorporate coarse aggregates (usually 10 mm maximum size), but sand is typically the major constituent. Sand particles are bonded together as the paste hardens to afford the needed structural properties. The quality of the sand is controlled by several factors such as the mean particle size, grading (Fig. 8.21), presence of impurities, shape, and texture. In some cases, lightweight aggregates such as perlite, vermiculite, expanded clay, and pumice are utilized as additional aggregates with the aim of decreasing the density of the mortar and increasing its insulation properties. Due to their low density (ranging from 90 to 500 kg/m^3), just a few percent by weight of these lightweight aggregate types is added to the mixture.

In plastering and rendering mortars, cleaning of the sand is essential. The sand aggregate must be free of clay materials, organic matter, and compounds (e.g. iron sulfides) that will decrease the durability of the mortar, and components that can affect the finish of the mortar and its mechanical strength and hardness. In this type of mortar, sand should have a maximum size appropriate to the thickness and finish of each coat. It is important to note that an excess of fine particles increases the requirement for water and,

Fig. 8.19 Bags of dry mortar (image courtesy of Grupo Cementos Portland Valderrivas)

Fig. 8.20 Floor screed (self-levelling) mortar (image courtesy of Miguel Martín)

◘ **Fig. 8.21** Sieving aggregates for mortar manufacturing (image courtesy of Grupo Cementos Portland Valderrivas)

consequently, the mortar shrinkage increases in the drying process. The excess of fines also reduces the adherence of the mortar, so a risk of microcracking exists.

The European standard EN 13139 specifies "the properties of aggregates and filler aggregates obtained by processing natural, manufactured, or recycled materials and mixtures of the aggregates for use in masonry mortars (◘ Fig. 8.22), floor/screed mortars, surfacing of internal walls (plastering mortar), rendering of external walls (◘ Fig. 8.23), special bedding materials, repair mortar, and grouts for buildings, roads, and civil engineering works". This standard enumerates the prevailing aggregate sizes for mortar (0/1 mm, 0/2 mm, 0/4 mm, 2/4 mm, and 2/8 mm), including helpful annexes about the effects of different chemical components of aggregates on the mortar in which they are incorporated.

Regarding the utilization of recycled materials, it is stated that mortars containing this type of fine aggregate require more mixing water to reach the same level of workability in comparison to conventional mortars. Moreover, the density of the mortar using recycled aggregates is also affected because their particle density is lower than that of conventional aggregate. However, "the mortar with incorporated fine recycled concrete aggregate proved to be a viable solution with great potential for use in the construction industry" (Neno et al. 2014).

The ASTM C144 standard only deals with aggregates used in masonry mortars ("natural or manufactured sand, with the latter being the product obtained by crushing stone, gravel, or air-cooled iron blast-furnace slag specially processed to ensure suitable gradation"). To adjust the grain size distribution, different

Fig. 8.22 Masonry mortar

Fig. 8.23 Blue rendering of external walls

granulometric fractions of aggregates are required. This standard specifies limits of fine aggregate gradation to be used in masonry mortar (◘ Table 8.2 and ◘ Fig. 8.24). It is important to note that well-graded aggregates decrease separation of the components in fresh mortar, reducing bleeding and improving workability.

As mentioned above, the main factors affecting sand quality in mortars are (a) average and range of particle size, (b) shape of the particles, and (c) impurities such as organic material or clay content. The shape and grading of the sand particles display a clear influence on the workability, or handling properties, of a mortar mix. The average and range of particle size can be rapidly defined using sieve analyses. Particle shape, while easily ascertained for coarse aggregates (i.e., in concrete), is more difficult to establish for particles of a finer size such as those present in mortars. For this reason, microscopic studies are essential to determine particle shape (i.e., rounded and angular). In this sense, the geology of the aggregate deposit in which the sands are found indicates fine aggregate petrology. Excessive quantities of flaky or elongated particles can be problematic since the generated mortars can be difficult to work on the trowel and the joints can prove more difficult to finish. This tendency can also be applied to angular crushed aggregates in comparison to rounded granular aggregates, which are easier to work.

The aggregates of mortar can be necessarily tested if most harmful organic compounds are detected. Organic matter may affect the cement hydration processes and the strength and durability of mortars. Consequently, their presence must be avoided if possible. Excessive clay content can generate a higher water demand in the mortar for a given level of consistency and produce a clear adverse effect on strength. Other problematic components in aggregates such as chlorides, sulfates, pyrite, lignite, coal, and mica must also be tested.

◘ **Table 8.2** Fine aggregate gradation to be used in masonry mortars (ASTM C144 standard)

Aggregate size (mm)	% Pass
4.75	100
2.36	95–100
1.18	70–100
0.600	40–75
0.300	10–35
0.150	2–15
0.075	0–5

8.3 Roads

Aggregates constitute, in percentages greater than 90%, most of the layers of road surfaces. They are the principal material in pavement construction. This implies a large consumption of these materials in road construction, not only if the road is new but also in its rehabilitation. In fact, the construction of roads is the second most important application of aggregates after concrete production. As with concrete, the characteristics of the aggregates will define the future behavior of the road. Recycled aggregates from construction and demolition waste (CDW) can also be used in the construction of unbound and bound layers of road pavements (see ▶ Chap. 9).

The pavement of a road is defined, according to the current regulations, as the set of layers made with selected and treated materials, which constitutes the superstructure of the platform, resists traffic loads, and allows circulation to take place safely and comfortably. Pavements are formed by the following layers (◘ Figs. 8.25 and 8.26): asphalt layer, base, subbase, and subgrade. The base lies immediately below the asphalt layer and provides support for it, providing the main support of the pavement structure; it is the essence of a successful pavement. The subbase (◘ Fig. 8.27) is defined as the layer of granular material located between the base and the subgrade and its function is similar to that of the base. The main function of these layers is to spread the stresses (i.e., decrease the stresses with depth) applied to the pavement surface by traffic loading. Both layers, subbase and base, provide rigidity to the pavement, limit tensile stresses on the lower surface of the treated layers located above, support the loads and transmit them to the lower layers, and allow the drainage of the water that infiltrates (the water, if allowed to remain within the base, considerably shortens the life of the pavement). The subgrade is the foundation layer and is formed by the existing soil compacted to a specified density before road construction starts.

For its part, the surface (asphalt) layer (◘ Fig. 8.28) is the upper layer of a bituminous mix pavement, being directly in contact with vehicles and atmospheric agents. It contains the highest quality materials and its functions are diverse (a) provide good grip and resistance to tire slippage, (b) waterproof to prevent excessive water penetration into the underlying layers, (c) give regularity to the surface, (d) generate adequate performance concerning noise, visibility and rain, and (e) establish adequate durability in terms of resistance to plastic deformation, fatigue breakage or the action of water and ice. The layers described are different in composition and are placed in separate construction operations.

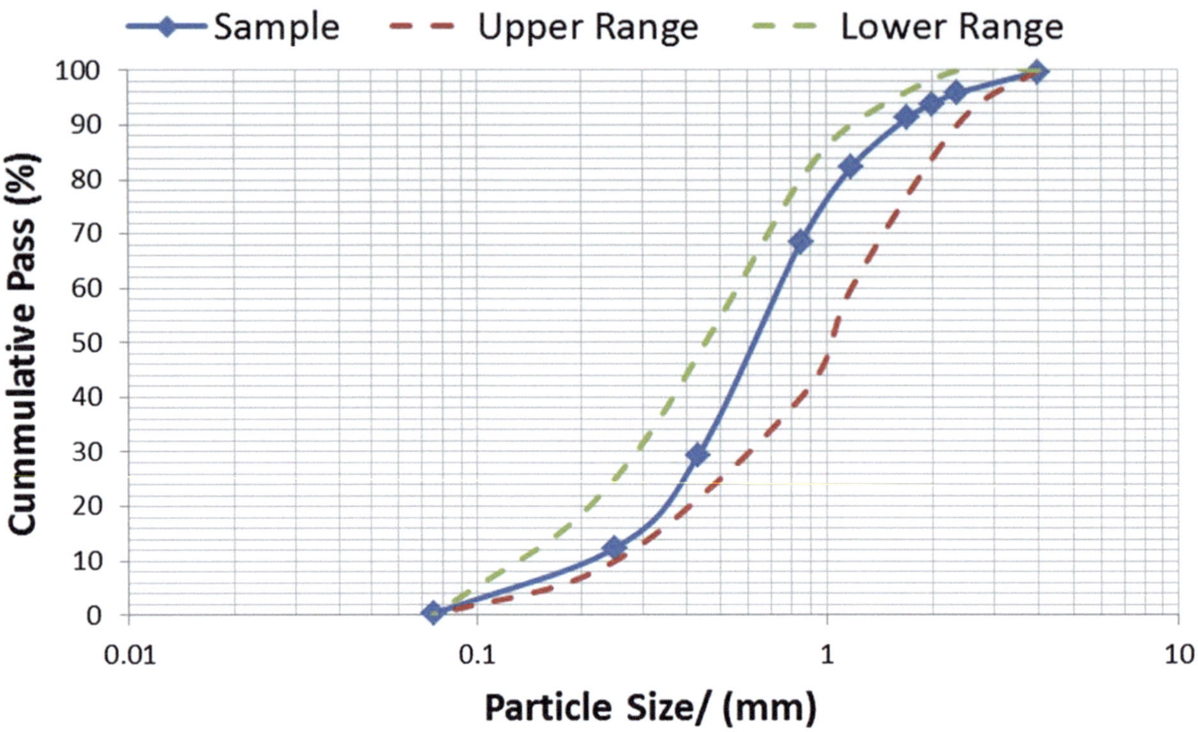

Fig. 8.24 Lower and upper limits in fine aggregate gradation according to the ASTM C144 standard (including a sand example with adequate particle sizes)

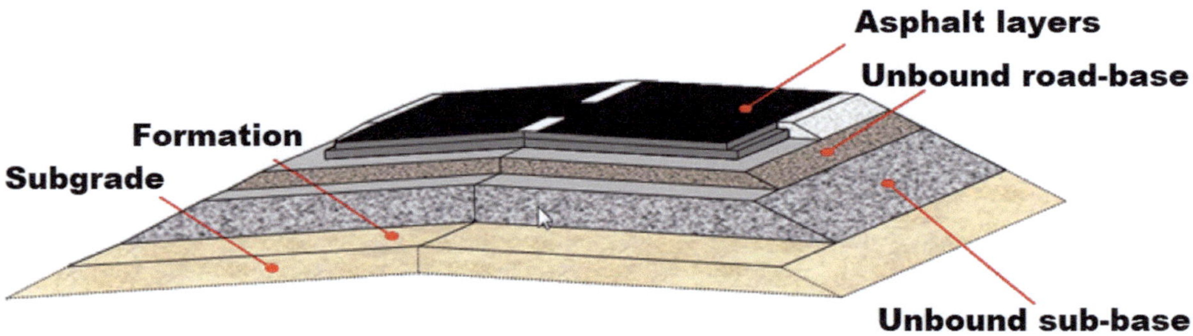

Fig. 8.25 Layers of pavement (illustration courtesy of EAPA)

One of the most important factors that controls the characteristics of the pavement layers, materials used, and thicknesses of the different layers is the type of traffic. For this reason, worldwide legislation has established different categories of traffic. Table 8.3 is an example of this type of classification.

8.3.1 Unbound Materials

Unbound aggregates have formed the basis of all pavements since engineering began. They are utilized in road pavements as base and subbase layers and play an equally important role in the pavement structure as those employed for the asphalt layer. Although unbound materials are intermediary materials of the pavement structure, the adequate behavior of these layers is essential. ASTM D2940/D2940 M and EN 13242 standards specify the characteristics of aggregates used in base and subbase layers.

The unbound material can be a natural aggregate such as gravel, sand, or crushed rock or it can be a by-product of an industrial process (i.e., crushed slag or construction and demolition waste). Strength and durability are the main features needed for the long-term performance of unbound pavement materials. Strength is needed for resisting repeated loading during construction and under traffic. Adequate durability is

Fig. 8.26 Pavement with different layers

required for resisting particle breakdown or degradation caused by changing environmental conditions. As a rule, the degradation of aggregates is the most common cause of unbound granular pavement failure.

The performance of aggregates in an unbound granular layer depends on (a) its intrinsic properties, (b) the particle shape and grading, and (c) the method of placing. The principal characteristics of a suitable material used as an unbound aggregate are the following: "(a) hardness (i.e., resistance to abrasion and attrition during mixing, handling and transporting), (b) toughness (i.e., resistance to crushing and impact loads during compaction and under traffic), (c) soundness (i.e., resistance to breakdown from weathering by chemical and physical processes) and (d) pore size distribution which, if unfavorable can increase frost susceptibility" (Smith and Collis 2001). The basic aggregate characteristics to be analyzed focus on the mineralogical composition, the granulometry and content and quality of the fines, the size and shape of the particles, the presence of harmful substances (organic matter, clays, sulfur, etc.), and resistance to fragmentation.

From the viewpoint of granulometry, it is recommended to use granulometries that are as continuous as possible, to obtain maximum compactness, which will positively influence the stability of the layers. The optimal gradations for road pavements include distributions of particle sizes that produce the maximum density when compacted (the aggregates of a base or subbase must be compacted to their maximum density to minimize constant deformations from applied loads). Regarding the maximum size of aggregate particles, "large granular materials provide improved strength and skid resistance. However, larger aggregates also exhibit reduced workability, including increased segregation potential and an irregular surface finish" (Griffin 2015).

The content of fines and, specifically, their quality, must also be a pursuable objective as well as the study of the shape and fracture faces of the particles. Flat and elongated aggregate particles break during the compaction of the bases or subbases or later during traffic. It is recommended to utilize, whenever possible, cubic shapes. It should not be forgotten that this factor can be improved during the crushing process in crushed aggregates. Regarding the fracture faces or angularity of the aggregates, these properties influence the resistance and stability of the structural layers of the pavement. This is because they contribute decisively to internal friction, that is, the greater the number of fracture faces of the

Fig. 8.27 Sub-base formed by granular material

aggregates, the greater the internal friction between the particles. Once again, this factor can be improved both from the crushing process of the aggregates and with stockpiling. Angular particles with sharp edges are preferred for promoting aggregate interlocking.

From the compositional viewpoint, the main types of rocks used for the construction of road pavements include igneous, metamorphic, and sedimentary rocks. “Care should be exercised in the use of igneous rocks in pavement applications due to potential alkali-silica reactivity, low abrasion resistance and relative solubility” (Griffin 2015). In general, aggregates of calcareous origin are used in the subbases, while aggregates of siliceous origin tend to be used preferably for the wearing course, although they have many times a lower adhesiveness with hydrocarbon binders than those from calcareous sources (i.e., limestone).

Regarding the mechanical characteristics of the aggregates, they must have values of the Los Angeles coefficient lower than 40–50 because mechanical resistance is one of the factors that most influences the future behavior of the bases and subbases during the useful life of the road, affecting the aggregates through the mechanical resistance of the mineral skeleton. In contrast, aggregates with coefficients lower than 20 are excellent materials, with sufficient wear resistance for any application, particularly for bituminous wearing courses that must withstand heavy traffic.

The base layer must generally contain strong and durable aggregates that are not be damaged by moisture or frost action. The aggregate layer shall be thick enough for protecting the subgrade layer from excessive stress. Particle roughness and microtexture have an important influence on the mechanical behavior of base aggregates. Particles with a rough surface texture are preferred. This is because they influence the interparticle friction and the resulting shear strength. Packing also has a major influence, and is controlled by grading and density. In this sense, “the voids in the aggregates should not, however, be overfilled with fines to completely block drainage, resulting in reduced strength, and create other problems” (NSSGA 2013). The content of fines must be adequately calculated since permeability decreases with increasing fines, but an increase in the amount of fines increases stability until an optimum level. In addition, the presence of impurities in the structural layers of the pavement, such as clay, dust, or organic matter, directly affects their resistant behavior.

Fig. 8.28 Asphalt layer of a pavement

Table 8.3 Classification of roads based on traffic volume

Type of road	Vehicles per day
High traffic	More than 1000
Medium traffic	400–1000
Light traffic	Below 400

As unbound roadbed materials, recycled aggregates can be used for roads with medium and low average daily traffic volumes; this utilization produces a low value-added product but consumes large amounts of recycled materials with low embodied energy.

8.3.2 Bituminous Mixtures

Asphalt is a mixture of approximately 90–95% well-graded aggregates together with filler and, in some cases, additives; the proportion of coarse aggregates is prominent. Bitumen makes up the remaining less than 5% of the mixture, where it constitutes the black, adhesive component that binds the aggregate together to form a cohesive mixture (Fig. 8.29). In an asphalt mixing plant, aggregates of different sizes are dried, heated, and further mixed with bitumen to produce a saleable product.

An asphalt mixing plant can be defined as a "set of equipment for asphalt mix production" (ISO 15642 standard). Its capacity has increased from a few tons per hour to 750 tons per hour of asphalt. On the other hand, production plants can be portable (Fig. 8.30) and stationary (permanent in one location and with greater production capacity). Selection of the type of asphalt plant is conditioned by several factors: "location, market requirements, plant capacity, competition, budget, operational costs, site restrictions, and customer preference" (Moore 2015). The complete plant operation is monitored from the control house of the plant.

In the asphalt plant, after the transportation and weighing of aggregates, the next step occurs in the dryer, where the aggregates are blended, heated, and dried. In sequence, the hot aggregates are screened, weighed, and stored in hot bins, in units mounted at the top of the central mixing plant. As a final step,

Fig. 8.29 Aggregates and bitumen (image courtesy of Nynas)

mixing of the aggregates with bitumen (Fig. 8.31) is performed in the mixing unit. The EN 13108 European multipart standard classifies the different mixtures according to the grading of the aggregate (mixture type), their upper sieve size (maximum nominal aggregate size), the intended use of the material, and the binder used in the mixture. The selection of aggregates for utilization in asphalt pavement depends on many factors such as the availability, cost, quality of the material and the type of construction. Grading, particle shape and angularity, absorption, cleanliness, fines content and quality, composition, and durability are the main properties that define the suitability of aggregates for this application.

The aggregates utilized in a bituminous mixture are formed by either coarse or fine aggregates or, more commonly, a mixture of both sizes. Coarse and fine aggregates must consist of sound, tough, durable, angular-shaped particles. Grading plays a significant role in the durability of asphalt pavement because it determines the most important properties such as permeability, stiffness, workability, stability, permeability, frictional resistance, fatigue resistance, and resistance to moisture damage. Changes in gradation outside the

Fig. 8.30 Portable asphalt production plant

Fig. 8.31 Different types of bitumen in a portable asphalt batch plant

allowable tolerances may influence other properties such as the voids content within the asphalt, which can generate defects in the asphalt (i.e., rutting). Grading can be achieved using different methods. As an example, Table 8.4 shows the aggregate gradation requirements specified by the Federal Aviation Administration in the United States.

Regarding the shape of the particles, it controls the frictional properties of the aggregate structure. Flat and elongated particles tend to break during mixing and compaction as well as under traffic. This characteristic is essential in the mix design of asphalt pavement for avoiding premature pavement failure. Values of less than 35% flaky particles are common in many standards worldwide. For its part, broken faces obtained by crushing can provide better skid resistance than rounded particles. Thus, angularity increases mixture rutting resistance. The porosity of an aggregate "permits the aggregate to absorb asphalt and form a bond between the particle and the asphalt; a degree of porosity is desired, but aggregates that are highly absorbent are generally not used" (MAPA 2014).

Foreign or deleterious substances make some materials unsuitable for paving mixtures. The presence of certain impurities (i.e., coatings of clays or organic matter) impairs the adhesion between aggregate particles and bituminous binder. As in the case of aggregates for bases and subbases, the content of fines and their quality is of special relevance. In bituminous mixtures, a size fraction less than 0.063 mm is especially important since some relevant characteristics of the mixture depend on the mastic formed by the union of the mineral powder and the binder. Aggregates coated with excessive amount of dust can also cause poor bonding, which results in stripping (the separation of the asphalt film from the aggregate).

Igneous and metamorphic aggregates are certainly suitable for use as coarse aggregates in wearing courses since their qualities of resistance to polishing make them ideal for guaranteeing the necessary surface texture over a long period, even with heavy traffic. Regarding aggregates with a calcareous composition, their use is widespread in all the layers of the pavement, except as coarse aggregate in wearing courses since it is an easily polished rock (with intense traffic they cause slippery surfaces in a very short time).

Finally, the aggregate utilized in the bituminous mixture must be durable enough (resistance to

8

Table 8.4 Aggregate gradation requirements specified by the Federal Aviation Administration in the United States (NSSGA 2013)

Aggregates bituminous pavements				
Sieve size	**Percentage by weight passing sieves**			
	1½″ max	**1″ max**	**3/4″ max**	**1/2″ max**
1½ in. (37.5 mm)	100	–	–	–
1 in. (25.0 mm)	86–98	100	–	–
3/4 in. (19.0 mm)	68–93	76–98	100	–
1/2 in. (12.5 mm)	57–81	66–86	79–99	100
3/8 in. (9.5 mm)	49–69	57–77	68–88	79–99
No. 4 (4.75 mm)	34–54	40–60	48–68	58–78
No. 8 (2.36 mm)	22–42	26–46	33–53	39–59
No. 16 (1.18 mm)	13–33	17–37	20–40	26–46
No. 30 (0.600 mm)	8–24	11–27	14–30	19–35
No. 50 (0.300 mm)	6–18	7–19	9–21	12–24
No. 100 (0.150 mm)	4–12	6–16	6–16	7–17
No. 200 (0.075 mm)	3–6	3–6	3–6	3–6

abrasion, polishing, stripping, and weathering) to accept long-term loading whereas maintaining its physical, mechanical, and chemical properties even with changing climatic conditions, including wetting and drying cycles. It is very important, from the point of view of durability, to study the possible attack by salts and the corresponding deterioration of pavement; very often salts are used during the winter to prevent the formation of ice (Fig. 8.32).

The utilization of recycled aggregates is an attractive approach for constructing new pavements sustainably since their incorporation generates many advantages (i.e., preservation of natural resources, lower energy consumption, lower harmful gas emissions and cost benefits). However, recycled aggregates with lower strength and higher porosity have negative effects on the different properties of asphalt mixes. To enhance the characteristics and/or performance of recycled aggregates incorporated in asphalt mixes using treatment methods is one of the most broadly employed improvement techniques. Asphalt used in roads can also incorporate recycled pavements, which include aggregates, because asphalt pavements are recycled (▶ Box 8.3: Recycled Asphalt Pavement).

Fig. 8.32 Using salt to prevent ice formation (image courtesy of NTB Scanpix)

Box 8.3—Recycled Asphalt Pavement

Recycled (or reclaimed) asphalt pavement (RAP) can be defined as removed and/or reprocessed pavement materials that incorporate asphalt and aggregates. The ASTM D8 standard defines reclaimed asphalt pavement as "asphalt pavement or paving mixture removed from its original location for use in recycled asphalt paving mixture". These materials are generated when asphalt pavements are removed for reconstruction or resurfacing (Fig. 8.33). *If adequately crushed and screened, RAP is composed of high-quality, well-graded aggregates coated by asphalt cement. Most of the RAP produced is recycled and later utilized. Recycled asphalt pavement is almost always returned to the roadway structure in some form, and is incorporated into asphalt paving employing hot or cold recycling. However, RAP can also be used as an aggregate in base or subbase construction. As an example of the importance of asphalt recycling, as early as 1993, the US Environmental Protection Agency and Federal Highway Administration identified asphalt pavement as America's No. 1 recycled product* (Bustillo 2021).

Beginning in the 1970s, the recycling of asphalt pavement into new asphalt mixtures was a common practice. This is because the utilization of recycled materials in pavement gained a wider acceptance for more sustainable construction and rehabilitation of transport infrastructures. The asphalt industry is one of the pioneers of recycling and reuse, and commercial products manufactured with bitumen are one of the most recycled materials in the world. In many countries, bituminous pavements that are removed are not currently landfilled. As commented previously, the process of recycling contributes to sustainability since it decreases waste disposal and conserves new resources (aggregate and bitumen).

Asphalt is 100% recyclable, which means that all RAP is either reused or recycled. The asphalt pavement is removed and crushed at specific dimensions so that some percentage of the asphalt pavement will then be incorporated into new paving material. Pavements are recycled in conventional asphalt pavement plants with minimal modifications. Increasing economic and environmental requirements will bring many new technical refinements to recycling old asphalt pavements.

Road recovered materials are carried to an asphalt plant (Fig. 8.34) *for sorting and batching to ensure that the mix incorporates an adequate ratio of bitumen to aggregate and an appropriate proportion of aggregate size and air voids. The percentage of RAP incorporated in an asphalt mix depends on several factors. The asphalt mixture specification in each country indicates the quantity of recycled asphalt pavement enabled and the methods to calculate the contribution of the old (hardened) bitumen to the new mixture in the pavement application. Recycling percentages often range from 5% to approximately 30% for highway pavements, but they can reach 60% for certain applications. Quantities recycled directly back into road surfaces vary from country to country but can be as high as 85%. There are many alternatives for recycling and the selection to be carried out depends on technical, environmental, and economic requirements established for each case. Approaches to recycling are different in various countries based on national requirements, technologies, resources, knowledge, etc.*

Fig. 8.33 Asphalt pavement for recycling

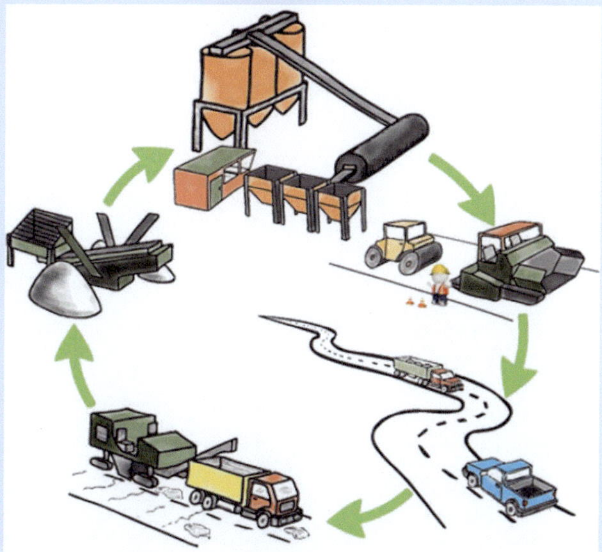

Fig. 8.34 Illustration of recycling asphalt pavement (Zaumanis and Frank 2014)

8.4 Railway Ballast

One of the most demanding applications for crushed aggregate is railway ballast (◘ Fig. 8.35), also known as railroad ballast. Rail tracks are built on compacted granular structures termed ballast, which are laid on natural materials. The term ballast was originally derived from a nautical term, in particular the ballast used on ships. It was the stone utilized to stabilize the British coal ships on their return voyage. According to EN 13450 European Standard, railway ballast is "aggregate where 100% of the surface of the particles can be described as totally crushed used in the construction of the upper layer of railway track, with sizes e.g. 22/40 mm, 31.5/50 mm or 31.5/63 mm". Railway ballast is a selected crushed and graded aggregate placed upon the railroad roadbed (◘ Fig. 8.36) to provide drainage, track stability, flexibility, and uniform support for the rail and ties. This type of aggregate also provides distribution of the track loadings to the subgrade (provides the required cover for the underlying subgrade) as well as facilitating maintenance. In addition, it deters the growth of vegetation that might interfere with the track structure.

The quality of the rock utilized is an important feature in railway track construction. Two types of ballast are recognized: top ballast and bottom ballast. Top ballast is the top section of the ballast structure that is exposed to tamping whereas bottom ballast is the bottom and lower part of the structure that supports the overall structure. Ballast rests, in turn, on the bottom ballast, which extends over the leveling of the track. The term subballast (or blanket) is also used and it is a layer of specified coarse-grained material interposed between ballast and formation. The main objectives of the subballast are to distribute the load on the formation to eliminate mud pumping and for containing the seasonal moisture content variations in the subgrade. In summary, the subballast provides solid support for the top ballast.

Ballast specifications include requirements for the size, shape, and quality of the rock particles. The aim is to obtain a thick and angular material so that its fragments fit together to form a layer with good

◘ **Fig. 8.35** Crushed aggregate used as railway ballast

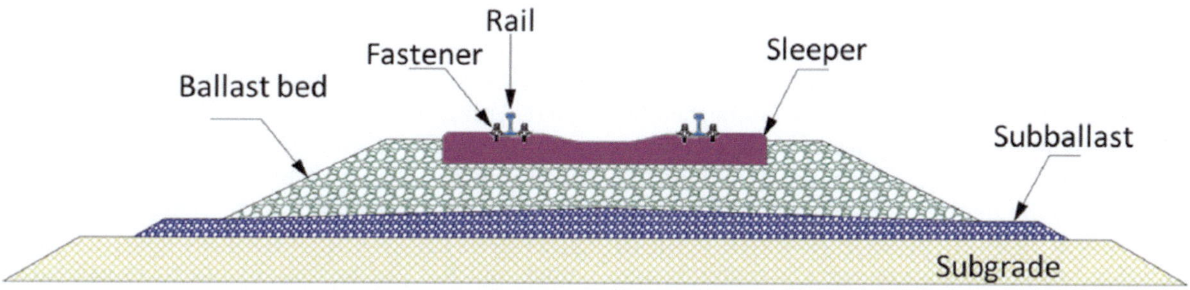

Fig. 8.36 Common railway ballasted track (Guo et al. 2022)

dimensional stability. As a consequence, ballast layer shall be thick enough to hold the track in position. The thickness of the ballast layer depends on the size and spacing of the ties and the type of train traffic but it usually ranges from 150 mm in lightly trafficked lines to 500 mm on lines with heavy traffic or high-speed trains. The individual fragments must be strong enough to resist wear and tear from the hammering effect of rail traffic. On the other hand, it is necessary to exclude fine materials because their presence reduces the stability of the assembly and the slip resistance of the ballast. Good ballast must be strong, hard-wearing, stable, easy to drain, easy for maintaining clear, workable, and resistant to deformation under load. The selection of an adequate ballast material may slow the deterioration of the ballast layer performance.

According to Selig and Waters (1994) and Mundrey (2009), the most important functions of ballast structure are the following: "(a) support the actions coming from the sleepers (the rectangular support pieces kept perpendicular to the tracks—Fig. 8.37) to the substructure and retain the track in its correct position; those actions can be grouped as uplift, lateral and longitudinal forces, (b) pressure reducing ballast structure to the allowable stress for the underlying structure just below the sleeper, (c) act as resiliency and energy absorbent for the track structure, namely, in what concerns to noise and vibrations, especially in high speed lines, (d) the voids between the aggregates provide space for movement and accumulation of crushed aggregates due to fouling, (e) provide quick rainwater drainage system down to the structure and contribute to the elasticity of the railway, and (f) facilitate lining operations and correction of geometry defects by the possibility of rearrange ballast particles with tamping; this eases and speeds up the maintenance operations". If ballast becomes contaminated with other materials, it may lose its capacity to fulfill the functions mentioned. For this reason, it is a common practice to clean the ballast over time, ensuring extended life of aggregates and minimizing load concentrations on rail, sleepers, and track structures.

Fig. 8.37 Railway sleepers

Ballast is required to perform these tasks without exceeding permissible limits of degradation produced by the load and the environment to which it is exposed. It must also resist the entrance of fine particles because the mechanical degradation of ballast particles generates fine particles of varying sizes. Ballast degradation is referred to as a change in the particle size distribution of the ballast layer as the result of particle breakage and abrasion, and migration of fine-grained soil from the subgrade. It is important to note that increased levels of degradation can lead to ballast differential settlement, thus altering the track geometry.

For aggregates for railway ballast, there are many types of national specifications worldwide. Examples of national specifications are the European standard (EN 13450), the American Railway Engineering Maintenance-of-Way Association in USA (AREMA), the Australian Standard for Railway Ballast Specifications or the Canadian National Railways Specification of Crushed Ballast. These specifications cover the types, characteristics, property requirements, and manufacture of mineral aggregates for processed ballast as well as testing requirements. The description of requirements indicates that the selection of suitable aggregates for railway ballast involves similar considerations all over the world.

The properties of railway ballast aggregate are essential to the effective load-carrying capacity of the rail structure and the subsequent in-service life of the rail track structure. Strict standards must be established for providing a high-quality track structure. It is important to bear in mind that the importance of ballast has grown exponentially through the increase in axle loads and train speeds. Different types of materials are used according to the applied requirements, availability, and cost of making ballast. In this sense, one particular material can possess most of the desirable features for good ballast while a deposit of apparently similar material located in the same geographical region will not meet the applicable specification requirements.

Substantial amounts of coarse aggregate are used as railway track ballast since it is typically made of crushed rock with sizes ranging between approximately 20 and 60 mm (▪ Fig. 8.38). The particle size distribution has a significant influence on the mechanical behavior in track conditions. ▪ Tables 8.5 and 8.6 show the different categories for ballast grading in the European EN 13450 standard and AREMA, respectively. Differences in particle size distribution are related to different latitudes of railway lines. It is important to consider that ballast particle size can vary after extraction processes mainly due to transport and handling, thus changing the particle size distribution.

Track ballast is required to be strong, clean, and angular with a high resistance to abrasion. If the aggregates used are rounded granular materials, the pebbles can roll over or slide against one another when a train passes. Thus, aggregates with angular edges must be utilized. They ensure the particles interlock with each other and the ties to fully secure them against movement. Another important characteristic of ballast is its weathering resistance because ballast must resist the degradation caused by environmental agents such as sun, rain, or ice. This weathering resistance is closely linked to other aggregate properties such as absorption and permeability.

Regarding the nature of rock used as ballast, different types of rocks (i.e., limestone, basalt, granite, gneiss, rhyolite, dolomite, and quartzite) are utilized worldwide. However, most railway ballast is sourced from siliceous (i.e., igneous) rocks and many countries prohibit the use of sedimentary rocks (i.e., limestones or dolostones). Some siliceous rocks such as shales and slates, which produce flake or elongated particles, cannot be used since the particles do not interlock adequately with each other and aggregates deteriorate

▪ **Fig. 8.38** Ballast andesite aggregate (sizes ranging approximately between 20 and 60 mm)

Table 8.5 Different categories for ballast grading (EN 13450)

Sieve size (mm)	Railway ballast size 31.5–50 mm	Railway ballast size 31.5–63 mm			Railway ballast size 22–40 mm
	Percentage passing by mass				
	Grading category				
	G_c RB A	G_c RB B	G_c RB C	G_c RB D	G_c RB E
80	100	100	100	100	–
63	100	95–100	95–100	93–100	–
50	70–99	65–99	55–99	45–70	100
40	30–65	30–65	25–75	15–40	90–100
31.5	1–25	1–25	1–25	0–7	60–98
22.4	0–3	0–3	0–3	0–7	15–60
16	–	–	–	–	0–15
8	–	–	–	–	0–2
31.5–50	≥ 50	–	–	–	–
31.5–63	–	≥ 50	≥ 50	≥ 85	–

Note The requirement for passing the 22.4 mm sieve applies to railway ballast sampled at the place of production

Table 8.6 Different categories for ballast grading (AREMA)

Sieve size (mm)	Sieving rate (mass percentage, %)			
	Grade A: 63.5	Grade B: 63.5	Grade C: 50.8	Grade D: 50.8
76.2	100	100	—	—
63.5	90–100	80–100	100	100
50.8	–	60–85	95–100	90–100
38.1	25–60	50–70	35–70	60–90
25.4		25–50	0–15	10–35
19.1	0–10	–	–	0–10
12.7	0–5	5–20	0–5	–
9.5	–	0–10	–	0–3
4.75	–	0–3	–	–
2.38	–	–	–	–

rapidly when subject to vibration. Petrographic study of the materials is especially relevant (i.e., the presence of microfractures) since it allows to advance possible later problems of very diverse origins. Similar rocks can show different test results due to differences in their respective micro and macro petrographic features. Moreover, many rock properties such as abrasion resistance or crushing rate are largely determined by their mineral composition, which can be clearly established with a petrographic study using microscopy.

To measure the properties of ballast, different tests have been developed. For instance, the corresponding European EN standard considers geometrical (grading, content of fine particles, fines content, particle shape, and particle length), physical (resistance to fragmentation and impact, resistance to wear and particle density, and water absorption), chemical and durability (magnesium sulfate soundness, freeze–thaw resistance, and electrical conductivity) tests. European standardization also considers the presence of the so-called *Sunburn* or *Sonnenbrand* alteration, which is a type of defect often found in basalt rock that has been affected by atmospheric conditions. It begins with the appearance of gray/white spots, typically capillary cracks that extend outward from the spots and connect them with one another. This phenomenon weakens the rock's mineral structure and the rock breaks down into smaller particles. ◘ Table 8.7 shows the test methods commonly used to determine each property of ballast materials. These tests are concerned with establishing a quantitative estimate of the aggregate ballast resistance.

Due to the importance of ballast features, quarries supplying these materials must notify the railway company when new open faces in the quarry can change the rock quality and send new samples for testing. For instance, rocks at different depths in the same quarry can show different qualities due to weathering. Finally, the disadvantage of ballasted tracks on high-speed railways is the eventual existence of ballast flight (▶ Box 8.4: Ballast Flight).

◘ **Table 8.7** Ballast material selection and testing methods (Guo et al. 2022)

Property classification	Ballast properties	Testing methods	Evaluation terms
Mechanical properties	Hardness, strength	Los Angeles Abrasion test	Abrasion, wear and crushing
	Hardness	Deval abrasion test: dry and wet grinding	Surface wear
	Hardness	MDA test	Surface wear
	Strength	Single particle crush test	Crushing resistance
	Hardness	Scratch hardness	Surface wear-resistance into fines
	Strength	Point load strength test	Fragmentation into small pieces
	Impact toughness	Drop weight test	Resistance to impact loading and crushing
	Hardness	Dorry abrasion test	Surface wear-resistance into fines
	Hardness	Mill abrasion test	Surface wear-resistance into fines
Physical properties	Grading	Gradation measurement	Drainage, compactness
	Grading	Small particle/dust measurement	Drainage properties
	Physical stability	Particle density measurement	Displacement resistance, indirect strength
	Physical stability	Bulk density measurement	Porosity, drainage, track stability (lateral longitudinal, vertical)
	Shape characteristics	Particle profile measurement	Breakage into multiple small pieces, drainage
Chemical properties	Chemical stability	Damp-dry resistance test	Durability, weathering resistance
	Chemical stability	Freeze–thaw test	Durability, weathering resistance
	Chemical stability	Particle porous structure test	Durability, weathering resistance
	Chemical stability	Particle water absorption	Saturation, weathering resistance
	Chemical stability	Sodium and magnesium sulphate solution test	Durability, weathering resistance
	Chemical stability	Clay clods and triable particles	Durability, initial abrasion level
Ballast layer properties	Ballast layer profile	Bed thickness shoulder and slope measurements	Rail stability, elasticity, damping
	Layering	Geometric position testing	Rail stability, comfort

Box 8.4—Ballast Flight

With the development of high-speed lines on ballasted tracks, a significant safety concern termed ballast flight (◘ Fig. 8.39) *has been taken into account and considered a problem since the 1980s. The problem of flying ballast is well-documented in high-speed rail operations. It is the phenomenon of a ballast particle being displaced from the track bed and causing damage to the rolling stock as well as the track infrastructure (other adjacent structures close to the right of way of the railroad can also be affected). This is because the train above a certain speed produces powerful airflows that cause the ballast flying occurrence to increase. In general, when the train speed is over 280–300 km/h, the probability of ballast flight will increment dramatically. It is a very complex problem because ballast flight is related to both railway infrastructure and train characteristics, involving not only mechanical effects but also aerodynamic effects. Ballast flight significantly affects the safe operation of high-speed railways, resulting in major maintenance costs and safety concerns. It is important to bear in mind that this problem can also be noted at low speeds produced by the presence of ice or snow in cold seasons, or other materials on the line.*

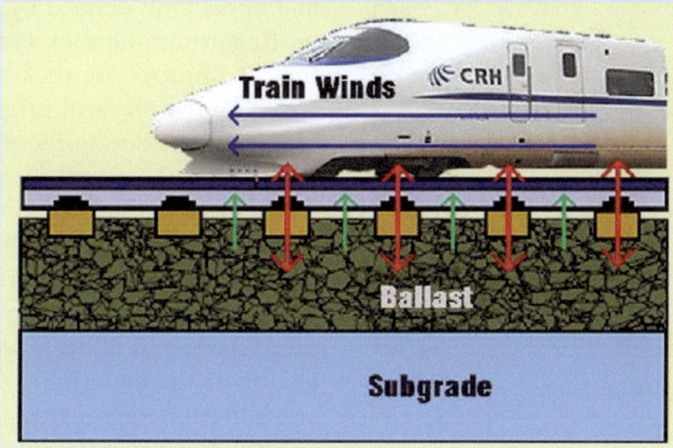

◘ **Fig. 8.39** Ballast aerodynamic and dynamic effects (Jing et al. 2012)

8.5 Armourstone

Armourstone, according to EN 13383-1 European Standard that specifies the properties of aggregates for use as armourstone, consists of "coarse aggregates used in hydraulic structures and other civil engineering works" (rip rap is a similar term that is commonly used in North America). This standard considers three types of armourstone (a) natural armourstone: "armourstone from mineral sources that have been subjected to nothing more than mechanical processing", (b) manufactured armourstone: "armourstone of mineral origin resulting from an industrial process involving thermal or other modification excluding concrete armour units" (◘ Fig. 8.40), and (c) recycled armourstone: "armourstone resulting from the processing of inorganic or mineral material previously used in construction". According to data from the European Aggregates Association, armourstone represents appoximately 3% of aggregate production in Europe.

In the context of coaltal engineering, armouring means permanent protection to a structure in water, providing safety for both the environment and people. Armourstone is principally selected in breakwaters, which are permanent structures built in coastal areas (harbors and shores) for protecting against tides, currents, waves, storms, etc. (◘ Fig. 8.41). Breakwaters can be constructed in many ways but the most common type is the rubble mound breakwater. It is formed by three layers, the outer one built of armourstone. Armourstone is also utilized in river embankments, dam walls, retaining walls, decorative uses, slope stabilization, and bridge abutments.

Natural armourstone is the most used type worldwide. It is a natural quarried stone (◘ Fig. 8.42) obtained by drilling and blasting and selected for its durability and robustness against degradation and wear. It is many times a byproduct of quarries, being the main product aggregates for concrete, roads, or ballast. However, the required sizes can be difficult and slow to obtain for a quarry operator as the production

Fig. 8.40 Concrete armour units (image courtesy of María Ángeles Morán)

of large blocks in a quarry is very low, diminishing the production of other products. Several blasting methods such as an increase in blast hole spacing with permanent burdens or using single rows of blast holes for avoiding the utilization of delay blasting can be employed to increase the efficiency of the armourstone production. Rock blocks can be degraded during transport from the quarry to the production site. This issue can be controlled by reducing the travel distance of machines transporting quarried rocks from the faces.

There is a wide range of design factors to take into account when considering the supply of armourstone to a project and the following are the most important: "(a) stability of the embankment, (b) wave action, wave height, the wave extreme and the wave pattern or wave train, (c) transition zones, (d) armourstone weight and (e) permeability" (CCAA 2008). The most important items for a correct field evaluation of armourstone in the quarry are the following: "(a) geological setting, (b) rock type, (c) regional in situ stress regime, (d) weathering grade, (e) groundwater conditions, (f) discontinuities, (g) production methods, (h) armourstone integrity as seen in stockpiles, (i) sampling, and (j) block integrity testing" (CIRIA 2007).

Armourstone is classically formed by large equant blocks of rocks with masses generally greater than 0.25 tons. They are substantially coarser than traditional aggregates for other applications (i.e., for concrete or roads). The rocks are laid to maximize the interlocking among the blocks, therefore minimizing the spaces. Armourstone is available in standardized stone classes that are formed by a lower and an upper value of the stone mass. For instance, the EN 13383-1 European Standard establishes the acceptable size and mass ranges for consignments of rock materials. Light grading is the designation of grading "with a nominal upper limit defined by a mass between and including 25 kg and 500 kg" and heavy grading is the designation of grading "with a nominal upper limit defined by a mass of more than 500 kg".

Regarding the rock types and mineralogy, the selection of appropriate rock materials must be carried out based on the physical properties of the rock and the geological considerations of the quarry. The best source to produce large armourstone size materials can come from areas formed by igneous or high-grade thermal metamorphic rocks. Sedimentary rocks are not adequately used in armourstone projects although certain types of limestone can be an exception. The presence of cleavage and jointing weaknesses can produce the rejection of the material.

Although aggregates can have very variable properties due to their different geological origins, the properties of armourstone tested include intrinsic properties such as particle density and water absorption, abrasion resistance, resistance to freezing and thawing and Sonnenbrand effects, manufacturing properties such as particle size and mass distribution and construction-induced properties such as aesthetics or armourstone integrity. For example, in the abrasion resistance test using the Los Angeles coefficient, values lower than 25 are recommended for high-risk projects. In some structures, wave attack can originate movement of the rock blocks; consequently, high-durability materials must be required (the durability of armourstone can be quantified by the rate of loss of material performance attributable to changes in armourstone properties). In manufacturing properties, the shape is particularly important for armouring material that is placed above water because the shape may significantly affect the armour layer stability. Finally, armourstone integrity (construction-induced properties) can be evaluated using nondestructive (sonic waves) and destructive (standard loading tests) techniques. Some of these properties can change with time as a function of loadings from the physical, chemical, and biological environment. The main properties of natural armourstone are listed and classified in Table 8.8.

Fig. 8.41 Armourstone for protecting a coastal area (image courtesy of Carolina Bustillo)

Fig. 8.42 Natural quarried stone blocks for armourstone

8

Table 8.8 Properties of armourstone (CIRIA 2007)

Property	Intrinsic	Production-induced	Construction-induced	Category of property
Aesthetics	**	–	*	Physical and environmental
Petrography	**	–	–	Physical, chemical and environmental
Rock density	**	–	–	Physical
Rock porosity	**	–	–	Physical
Rock water absorption	**	–	–	Physical
Rock resistance to attrition and wear	**	–	–	Mechanical
Rock resistance to weathering agents	**	–	–	Physical and mechanical
Rock fabric strength	**	*	–	Mechanical
Size of armourstone piece	**	**	–	Geometrical
Integrity of armourstone piece	**	**	–	Physical and mechanical
Shape of armourstone piece	**	**	–	Geometrical
Armourstone mass or size distribution	*	**	–	Geometrical
Armourstone layer (or volumetric) porosity	–	*	**	Geometrical and environmental
Armourstone layer thickness	*	*	**	Geometrical
Armourstone integrity	*	**	*	Geometrical

*Relevant; **very relevant; – not relevant

Fig. 8.43 Box gabions (image courtesy of GEOTECH)

8.6 Gabions and Matresses

Gabion is a word of Italian origin meaning “big cage”. It is a robust structure comprizing a double-twist wire-mesh basket filled with hard, durable stone. The utilization of gabions to prevent erosion is not a modern concept. Thus, “gabions were applied to the banks of the Nile River in ancient Egypt more than 2000 years ago; the Egyptians used cylindrical willow baskets filled with small stones to protect the overflooding riverbanks of the Nile” (NSSGA 2013).

The ASTM D6711 American standard defines gabion (in erosion control) as “a wire mesh container of variable size and interconnected with other similar containers; it is filled with rock to form flexible, permeable, monolithic structures”. Gabions can be classified according to their dimensions into box gabions (i.e., 2 m length × 1 m width and 1 m height) (◘ Fig. 8.43), gabion mattresses (◘ Fig. 8.44) or simply mattresses (i.e., 6 m × 2 m × 0.15–0.30 m), and sack gabions (◘ Fig. 8.45) (tubular gabions or rock rolls used principally in fluvial engineering). ASTM D6711 standard defines a mattress (in erosion control) as “a wire mesh container uniformly partitioned into internal cells with relatively small height in relation to other dimensions; mattresses have smaller mesh openings than the mesh used for gabions”.

Gabion baskets (◘ Fig. 8.46) are rectangular (box-like) wire baskets of various sizes. They are basically made using double-twisted wire mesh to enable the completed structure to deform adequately without failing and to prevent unraveling if the mesh is cut. The corrosion protection of the mesh is obtained by galvanizing the wire and with an additional polymer coating (i.e., PVC). The use of hard rock material as filler provides a secure, sound, and reliable method for retaining soil and earth. Protection is important because the service life of the basket is related to the wire durability and ability to hold the rock together. The life of these structures is also controlled by the durability of the rock used.

Gabions may perform many different functions in coastal, estuarial and fluvial environments. Gabions are utilized for constructing retaining structures, scour protection, channel linings and weirs for hydraulic structures as well as for erosion protection. They need proper design and installation to perform correctly, especially when they are used in high-energy hydraulic environments. ◘ Figure 8.47 shows the main gabion elements, their required properties, and some classical applications.

◘ **Fig. 8.44** Gabion mattress (image courtesy of Geoart)

Fig. 8.45 Gabion sacks (image courtesy of Aqua Terra Solutions)

Fig. 8.46 Gabion basket

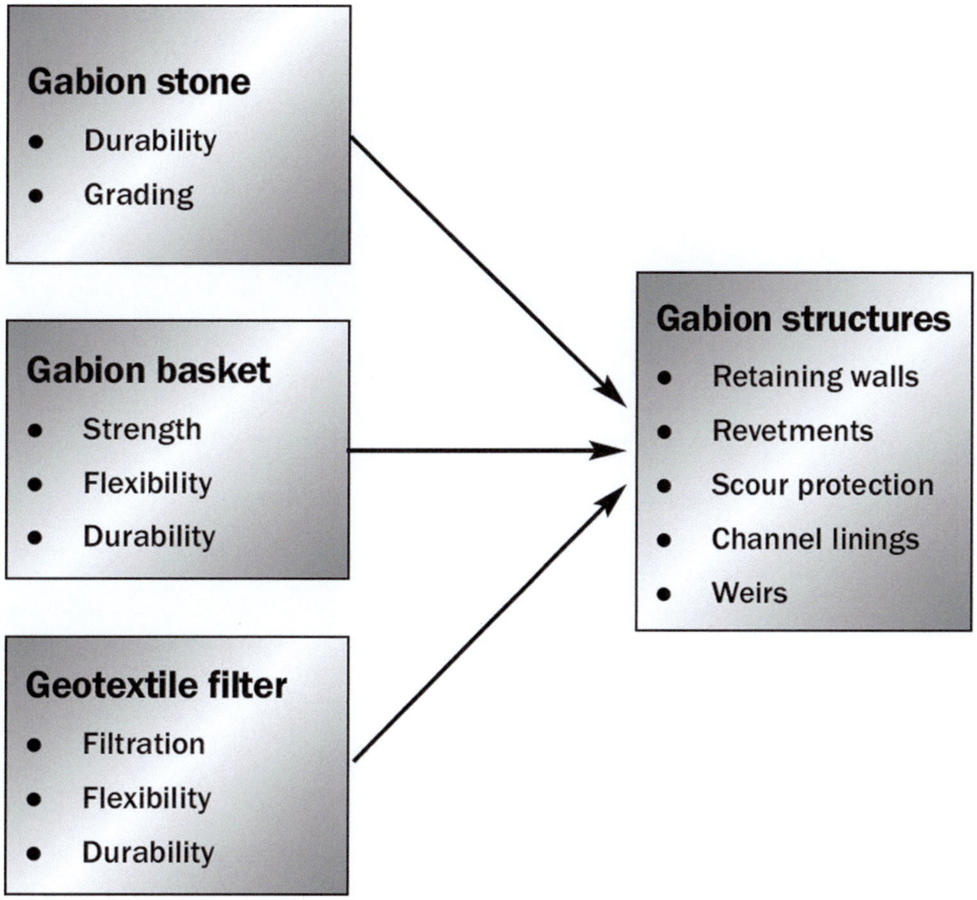

Fig. 8.47 Key features and applications of gabion structures (CIRIA 2007)

Applications for gabions include but are not restricted to "retaining walls (Fig. 8.48), culvert protection, wing walls, bridge abutment protection, landscaping (Fig. 8.49) and erosion control" (CCAA 2009). When gabions are joined together, they act as a monolithic mass for retaining-wall construction. These structures are employed where soil retention or steep slope stability is of concern. Gabion mattresses are utilized for riverbank and scour protection, channel linings for erosion control, and embankment stability.

According to the ASTM D6711 American standard, "rock to fill gabions or mattresses may be any natural deposit of the required sizes or may be crushed rock produced by any suitable method; rocks shall be hard, angular to round, durable, and of such quality that they shall not disintegrate on exposure to water or weathering during the life of the structure". The aggregate must be durable since it is exposed to a range of atmospheric and other cyclic conditions that threaten to break it down while in service.

Clean crushed concrete can be utilized as an alternative when rock is not available. Mudstones and other argillaceous weak rocks should be avoided if possible because these materials tend to degrade once placed. Aggregates should sustain enough angularity to provide adequate interlocking contact with one another; point-to-point contact of rock surfaces must be maintained. The rock must be free of adhering dust or deleterious matter and not include flat or elongated particles.

Mattresses incorporate larger-sized rock than that utilized in gabion baskets, and are utilized as an alternative to traditional armourstone for protecting against soil erosion and scour. Rocks must be delivered to the work site minimizing the reduction in sizes (breakdown) caused by the handling processes.

For gabion baskets, "the rock size shall be a minimum of 100 mm and a maximum of 250 mm. For mattresses, the rock size shall be a minimum of 75 mm and the maximum size shall be 2/3 the thickness of

Fig. 8.48 Gabions as a retaining wall (image courtesy of María Ángeles Morán)

the designed mattress thickness or 250 mm, whichever is the lesser" (CCAA 2009). EN 13383-1 European standard establishes the coarse grading 90/180 mm as the most suitable for gabion utilization. Grading is selected to prevent individual rocks from falling through the mesh of the gabions or mattresses. Obviously, these rocks can never breakdown to sizes smaller than the wire mesh opening dimensions.

The void porosity of gabions can change depending on the type of rock fill. Values "can vary from 25 to 35%; values of 25% would be appropriate when stone fill is carefully hand-placed while 35% would be typical of gabions filled for the most part by mechanical means" (CIRIA 2007). Gabions are highly permeable as a consequence of their high porosity.

8.7 Filter Media

Filtration is a process in which water is passed through the beds of granular material (i.e., sand) to remove impurities. Filters play essential roles in many engineering projects although the total amount of aggregates utilized for these applications is comparatively small. Table 8.9 shows the main applications of aggregates as filter media. The most common applications of aggregates as filter media are in water filtration, treatment of effluents and as drainage filters for earthworks, and other civil engineering structures (Fig. 8.50). Regarding the processing methods used for the production of these aggregates, the high quality of aggregates needed as filter material for water and effluent works is not always readily available from commercial manufacturing processing. For its part, the drainage layers of civil engineering works (i.e., embankment dams) are designed to take advantage of the available natural materials with minimum processing.

Drainage and filter aggregates are sand, gravel, crushed stone, or mixtures thereof. Typically, sand media is used in filtration for solid–liquid separation. Crushed concrete particles from construction and demolition waste (see ▸ Chap. 9) can also be used as filler for water treatment filters since these particles have a great number of voids, very high-water absorbance and a great number of open pores.

Fig. 8.49 Garden gabions (image courtesy of ENZAR)

The key properties of filter media aggregates are grading, shape, coatings, strength, and chemical reactivity. In general, aggregates must be formed by clean, hard, durable mineral particles. They must be free from substances that would interfere with the aggregate filtration properties. Grading is essential to the design of a filter. The aggregates shall conform to the specified grading limits after being placed. A broad range of particle sizes, from fine sand up to boulders, are utilized for the different parts of a filter structure. The shape is also fundamental in a filter aggregate. Thus, it is desirable to be basically equidimensional because it facilitates the packing of the coarser and fine constituents, improving the effectiveness of the filter. Another significant feature related to shape is the roundness because it is better for the filter aggregate to have rounded particles.

Coatings can vary in thickness and are composed of clay, silt, iron oxides, gypsum or other components. They are usually present but must be removed. Filter aggregates must be strong enough to avoid a breakdown during structure construction and when in use. It is important to remember that the global properties of a filter depend on aggregate particles due to the absence of a binder component. Finally, the aggregates must not be affected by the chemical properties of the liquid to be treated and the components must not be leached from the filter medium into the filtrate.

Although the purification techniques always utilize aggregates and are only one part of the process, all used methods of water and wastewater treatments utilize a filter in the process. Within the water filtration market, the filtration process is performed through the utilization of a slow sand filter (Fig. 8.51) or rapid filters (Fig. 8.52). The former is characterized by slow filtration rates and requires relatively infrequent cleaning while rapid filters, which are coarser, need frequent cleaning because they operate at higher filtration

Table 8.9 Main applications of aggregates as filter media (Smith and Collis 2001)

Category of works	Filter application	Principal function		
		Drainage	Pressure relief	Filtration
Land drainage	Surface blanket drainage	*		
	Ditch/trench fill	*		
	Drain/soakaway fill	*		
	Pipe filters	*		*
	Pipe envelopes	*		
Structure drainage	Foundation drainage	*	*	
	Roadbed drainage			
	Runway drainage	*		
	Retaining structure drainage	*	*	
	Embankment dam drains	*	*	*
	Slope drainage	*		
Remedial and stabilization works	Uplift pressure relief		*	
	Seepage force prevention		*	
	Saturation control	*	*	
	Piping prevention			
Water supply	Well filters			*
	Well envelopes	*		
Filtration and purification works	Fine and coarse filter beds			
	Gravel underdrain envelope	*		*
Marine works	Bedding layers			
	Transition layers			*
	Pressure relief		*	

rates. To have equal cleansing potential, "the slow sand filter must have a much larger surface area in contact with the raw water than the rapid sand filter" (NSSGA 2013). In slow sand filters, the bed is formed by 0.6–1.2 m of fine sand and several layers of graded gravel. Rapid filters comprize coarse sand and layers of fine to coarse gravel. The particle size of gravel used in slow sand filters ranges from 4.75 to 60 mm. In both applications, the sand should be hard, abrasion resistant, and free from contaminants. Specialist silica sand can be used as rapid gravity potable water filters, swimming pool filters and slow sand filters.

Water has a major influence on the stability of engineering soil structures. For this reason, draining the water out of the soil structure improves its stability. Protective filters are formed by "one or more layers of free-draining sand and/or gravel materials placed on or within a less pervious soil to remove seepage water and prevent erosion of the soil or damage to the overlying structures from uplift pressure" (Smith and Collis 2001). The filter layer that meets the seeping water first is of fine material and sands of increased coarseness conform to the subsequent layers. The last layer from which water comes out of the filter is made of gravel. Deficiencies in grading can be compensated by changing the thickness of the filter. In any case, filter drains should be designed in such a way that all the seeping water through the structure is effectively drained off.

8.8 Decoration

Decorative/ornamental aggregates are small pieces of stone known as gravel, chipping, or pebbles (▶ Box 8.5: Ornamental Aggregates).

Fig. 8.50 Placing filter material in a dam (image courtesy of Waimea Water)

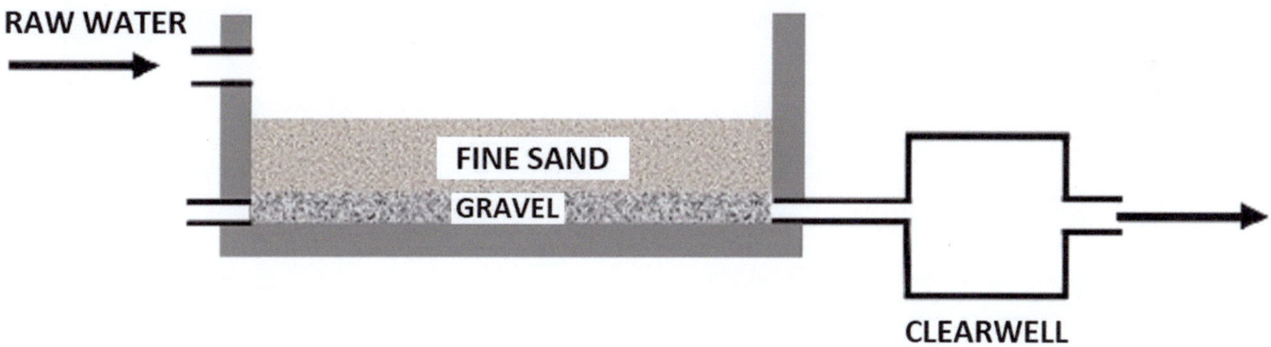

Fig. 8.51 Scheme of the slow sand filter

Fig. 8.52 Rapid sand filtration (image courtesy of Cropaia)

Box 8.5—Ornamental Aggregates

Ornamental aggregates are fundamental resources for economic, urban, and social development. These aggregates can be used in a wide variety of spaces such as commercial and residential, domestic pathways, driveways, car parks, architectural landscaping, gardening (patios and rockeries) (Fig. 8.53), *sports grounds,* etc. *Aggregates have become a staple in landscaping applications due to their durability, practicality, and beauty. Landscaping with aggregates forms a framework for movement and visual definition, establishing different areas in the landscape* (Fig. 8.54). *Decorative aggregates can be*

Fig. 8.53 Aggregates used in a garden (image courtesy of Manuel Jorge Bustillo)

used to frame bedding plants and add color or texture to the garden or for creating garden paths or driveways. In gardens, ornamental aggregates with larger sizes can stand alone for decorative purposes whereas smaller pieces may be utilized to define borders, enhance other features, or to fill empty areas where the grass is either not present or is not practical.

Because decorative aggregates are natural products, they can show some variations in size and color caused by geographical and processing variables. These products are available in a broad variety of colors and sizes. Natural sands and gravels are used for decorating gardens, terraces and flowerpots as well as for defining boundaries (i.e., around flowerbeds, walkways and passageways) while different types of marble chips are utilized for outdoor decoration in gardens, planters, and indoor decoration.

Fig. 8.54 Landscaping with decorative aggregates (image courtesy of Ornamental Stone)

8.9 Questions

Short Questions

- What is the expanded clay?
- Explain the influence of fine materials in fresh concrete
- Define the alkali-carbonate reaction
- What are the main functions of sand in mortar?
- Define the pavement of a road
- Comment the ratio of aggregate versus bitumen in an asphalt mix
- Enumerate the main characteristics of unbound aggregates in road pavements
- What is the railway ballast?
- Define armourstone. What is the size of its components?
- Define gabion and comment on its main applications

Long Questions

- Describe the alkali-silica reaction
- Explain in detail the most important functions of a ballast structure

References

Bustillo M (2021) Construction materials—geology, production and applications. Springer, Cham, 603 p

CCAA (2008) Guidelines for the specification of Gabions, technical note 72. Cement Concrete and Aggregates Australia, 7 p

CCAA (2009) Aggregate for Gabion baskets and wire mattresses, technical note 74. Cement Concrete and Aggregates Australia, 6 p

CIRIA (2007) The rock manual. In: The use of rock in hydraulic engineering, 2nd edn (C683). Construction Industry Research and Information Association, London, 1304 p

Griffin J (2015) P11 review of unbound pavement material specifications: wet/dry strength variation. Queensland Department of Transport and Main Roads, Queensland Government, 74 p

Guo Y, Xie J, Fan Z, Markine V, Connolly DP, Jing G (2022) Railway ballast material selection and evaluation: a review. Constr Build Mater 344:128218

Jing GQ, Zhou YD, Lin J, Zhang JP (2012) Ballast flying mechanism and sensitivity factors analysis. Int J Smart Sens Intell Syst 5:928–939

Kosmatka SH and Wilson ML (2011) Design and control of concrete mixtures, EB001, 15th edn. Portland Cement Association, Skokie, 460 p

MAPA (2014) Asphalt pavement design guide. Minnesota asphalt Pavement Association, New Brighton, 92 p

Moore J (2015) Asphalt production plants. In: Gerlis R (ed) The shell Bitumen handbook, 6th edn. Shell International Petroleum Company, ICE Publishing, London, pp 378–412

Mundrey JS (2009) Railway track engineering, 4th edn. Tata McGraw-Hill Publishing Company Limited, New Delhi, p 483

Neno C, Brito J, Veiga R (2014) Using fine recycled concrete aggregate for mortar production. Mater Res 17:168–177

NSSGA (2013) The aggregates handbook, 2nd edn. The National Stone, Sand and Gravel Association, Alexandria, Virginia, 906 p

Ozguven A, Gunduz L (2012) Examination of effective parameters for the production of expanded clay aggregate. Cement Concr Compos 34:781–787

Selig ET, Waters JM (1994) Track geotechnology and substructure management. Thomas Telford Publications, London, p 446

Smith MR, Collis L (eds) (2001) Aggregates: sand, gravel and crushed rock. Geological Society, London, Engineering Geology Special Publication v. 17, London, 360 p

Vieira M (2015) Concrete. In: Goncalves MC, Margarido F (eds) Materials for construction and civil engineering. Springer International Publishing, pp 185–236

Vijayalakshmi R, Ramanagopal S (2018) Structural concrete using expanded clay aggregate: a review. Ind J Sci Technol 11:12 p

Zaumanis M, Frank R (2014) 100% hot-mix asphalt recycling. Asphalt professional, technical paper, Nov 2014, pp 28–33

Standards

ASTM A1096/A1096 M–15. Standard test method for evaluating bond of individual steel wire, indented or plain, for concrete reinforcement

ASTM C33/C33 M–18. Standard specification for concrete aggregates

ASTM C125–20. Standard terminology relating to concrete and concrete aggregates

ASTM C144–18. Standard specification for aggregate for masonry mortar

ASTM C330/C330 M–17a. Standard specification for lightweight aggregates for structural concrete

ASTM C331/C331 M–17. Standard specification for lightweight aggregates for concrete masonry units

ASTM C332–17. Standard specification for lightweight aggregates for insulating concrete

ASTM C637–20. Standard specification for aggregates for radiation-shielding concrete

ASTM D8–20. Standard terminology relating to materials for roads and pavements

ASTM D2940/2940 M. Standard specification for graded aggregate material for bases or subbases for highways or airports

ASTM D6711/D6711 M–2022. Standard practice for specifying rock to fill gabions, revet matresses, and gabion matresses

EN 206-1:2013+A1:2016. Concrete. specification, performance, production and conformity

EN 998-1:2016. Specification for mortar for masonry—part 1: rendering and plastering mortar

EN 998-2:2016. Specification for mortar for masonry—part 2: masonry mortar

EN 12620:2013. Aggregates for concrete

EN 13139: 2013. Aggregates for mortar

EN 13242:2013. Aggregates for unbound and hydraulically bound materials for use in civil engineering work and road construction

EN 13383-1:2002. Armourstone—part 1: specification

EN 13450:2013. Aggregates for railway ballast

ISO 6707-1:2020. Buildings and civil engineering works—vocabulary—part 1: general terms

ISO 15642:2008. Road construction and maintenance equipment—asphalt mixing plants—terminology and commercial specifications

ISO 19595:2017. Natural aggregates for concrete

Recycled Aggregates

Contents

9.1 Introduction – 306

9.2 Construction and Demolition Waste – 306
9.2.1 CDW Composition – 311
9.2.2 Management of CDW – 317
9.2.3 Valorization Plants – 326

9.3 Recycled Aggregates – 335
9.3.1 Properties of Recycled Aggregates – 336
9.3.2 Applications – 343

9.4 Questions – 346

References – 347

M. Bustillo Revuelta, *The Basics of Aggregates*,
Springer Textbooks in Earth Sciences, Geography and Environment,
https://doi.org/10.1007/978-3-031-42961-3_9

Abstract

This chapter introduces the reader to the basic characteristics of construction and demolition waste (CDW) as well as recycled aggregate, which is one of the most important products obtained from construction and demolition waste. Composition of CDW, including its most interesting component (concrete), as well its management is discussed. Valorization plants (stationary and mobile) for the treatment of construction and demolition waste are explained. Recycled aggregates, reprocessed materials previously used in construction, are further discussed and their main properties commented. Finally, the applications of recycled aggregates are briefly commented: production of concrete and pavements as well as general fill and embankment material, and for erosion control.

9.1 Introduction

The growth of the world population and the economic conditions of developing countries have considerably augmented the development of the building and construction industry. However, while their evolution has contributed to world economic growth, they also originate multiple sources of waste. Thus, the construction sector is perceived as a major contributor to environmental degradation. Old buildings have been demolished to make new constructions (◘ Fig. 9.1), creating large amounts of debris worldwide, which generates severe environmental pollution problems as well as serious disposal issues.

The building and construction market consumes most of the total resource consumption of the world. This sector generates approximately one-third of all waste (◘ Fig. 9.2), "which represents a main contributor to CO_2 emission to the atmosphere" (Lauritzen 2019). Construction materials by weight account for more than 40% of the resources utilized in the United States, nearly 60% in Japan, and 70% in Germany. From 30 to 50% of the materials utilized in Europe are for housing; around 70% of the amount of aggregates and approximately 20% of the total metals are used in the construction sector. Based on volume, "waste from construction and demolition is the largest waste stream in the EU, representing about one-third of all waste produced; in a lifetime, an average European citizen generates not less than 160 tons of this type of waste, an amount which is increasing over time" (European Commission 2016).

Because of this intense activity, building and construction material consumption generates environmental impacts at every step of the building and construction processes: mining raw materials, processing, manufacturing, transportation, construction, and disposal at the end of the useful life of buildings. In many countries worldwide, part (or all) of this waste stream is unlawfully deposited on land (fly-tipping) (◘ Fig. 9.3). Consequently, the building and construction sectors focus their resource management on resource efficiency, reduction, and recycling of waste materials. Since the bulk of these residues is inert, they were classically landfilled in the 1970s and 1980s. Over time, they were diverted from landfills to recycling whereas currently the focus is on reuse and prevention.

9.2 Construction and Demolition Waste

Common wastes originating from the construction and renovation of houses, buildings, and civil-engineering structures are termed construction and renovation wastes. If they are demolished, the wastes produced are termed demolition wastes. Thus, according to its generation phase, these wastes can be subdivided into three categories: construction waste (CW), renovation waste (RW) and demolition waste (DW). All of them are grouped into so-called construction and demolition waste (CDW) or construction and demolition debris (CDD), which includes waste materials from construction and demolition operations. Thus, CDW is a general term that can be defined as the waste material produced in the process of construction, maintenance, renovation or demolition of buildings and civil engineering structures, representing a large proportion of the overall waste generated in industrialized countries. In general, the demolition waste of construction works accounts for 70% of CDW.

CDW also includes the materials that can all of a sudden be originated by natural disasters (i.e., floods, earthquakes, hurricanes, and tsunamis) (◘ Fig. 9.4). Modern societies generate huge amounts of construction and demolition waste annually and continues to increase over time. For instance, each year around 3 billion tons of waste are generated in the EU 27. Of this, as noted above, approximately one-third (1 billion tons) comes from construction and demolition activities. Construction can also need excavation works such as groundworks and tunneling that generate softer materials such as soils, subsoils, and clays (◘ Fig. 9.5); these materials can also be considered CDW. The suggested theoretical approach to defining CDW is "to take into account both its nature (materials used in buildings and constructions) and the activities that originate it (construction and demolition activities), regardless of who performs these activities" (BioIS 2011).

Fig. 9.1 Old building being demolished (image courtesy of Luis Fueyo)

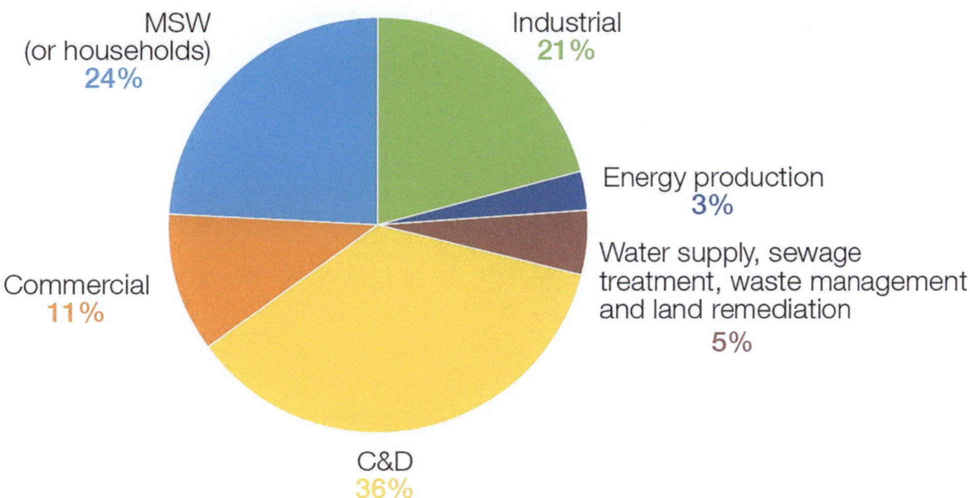

Fig. 9.2 Amounts of waste from several sources in the material and product life cycle (the data are for the OECD countries and exclude agricultural, forestry, mining, and quarrying wastes—UNEP 2015)

The adverse impacts of CDW production are very varied, including the use of large quantities of land resources for waste landfilling, harming the surroundings by hazardous pollution, and wasting natural resources. Although many building and construction materials can be recycled from the CDW stream (i.e., asphalt, brick, wood, and metals), heavy materials originating in large amounts cost more to dispose at landfill ("more than 50% of all CDW generated per year in the world is concrete"—Lauritzen 2019). This

9

Fig. 9.3 Illegal disposal (fly-tipping) of building and construction materials after demolition

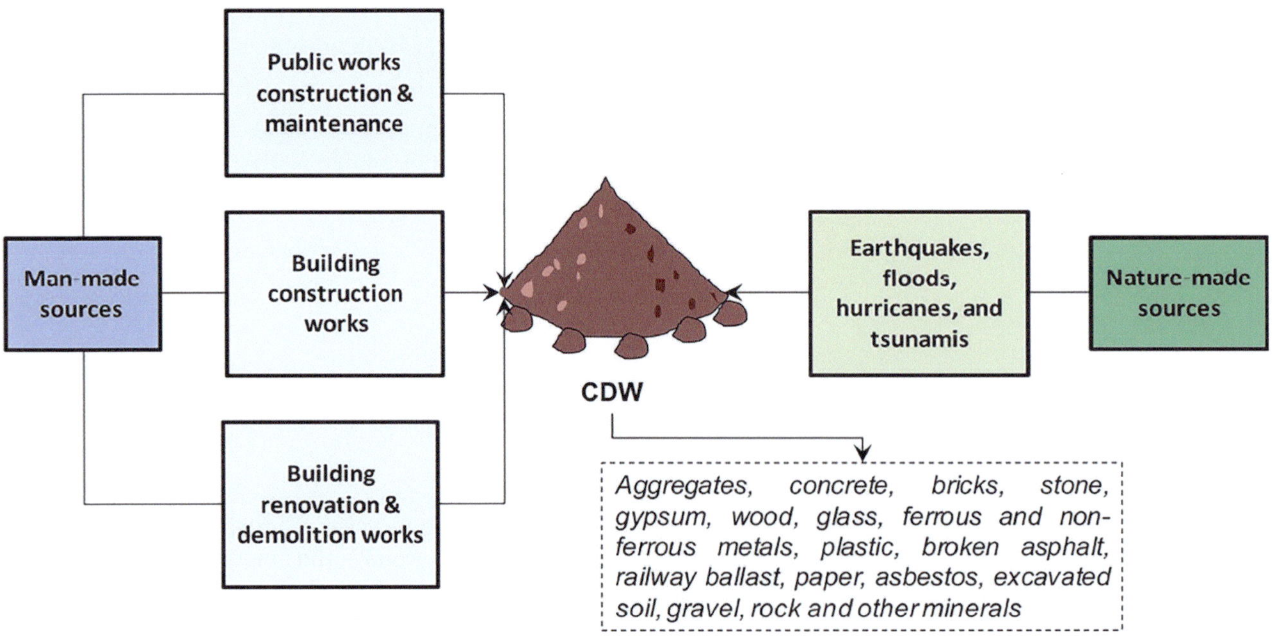

Fig. 9.4 Classification of construction and demolition waste (CDW) according to the source of origin (Simões dos Reis et al. 2021)

Fig. 9.5 Excavation works

results in these materials receiving priority attention for recovery and recycling. From a global perspective, the recycling of concrete is the most challenging development because of the large amounts of consumption of raw materials and various opportunities for recycling. In principle, all concrete waste is inert and can be recycled. Full implementation of recycling concrete waste all over the world can save billions of tons of natural resources. Access to natural aggregates (NA) in some countries is becoming scarce and alternative sources of NA are needed.

Since the manufacture of concrete uses large quantities of coarse and fine aggregates, the production of recycled aggregates is obviously the principal option for recycling concrete. The Union Européenne des Producteurs de Granulats (UEPG) defines recycled aggregates as "reprocessed materials previously used in construction". Obviously, the first benefit of using recycled aggregates is the reduction of the environmental impacts associated with the extraction of virgin resources. This means that uses for recycled aggregates other than downcycling are needed. However, the market acceptance of recycled aggregates by the concrete industry is very limited. "As presented in the 2019–2020 annual review of UEPG (the European Aggregates Association), only 10.6% of all aggregates produced in the EU + EFTA (the European Free Trade Association) were RA, most of which were concrete recycled aggregate" (Pacheco and de Brito 2021).

Construction and demolition waste includes those materials used in the construction of buildings, structures, roadways, and other landscaping materials, including soil, brick, drywall, mortar, cellulose, wood, bituminous materials, and metals. The high speed of urbanization entails the increased demand for housing and transportation and, therefore, the amount of CDW is permanently growing. CDW consists mostly of relatively inert and nonbiodegradable products that are heavy and characterized by their high density. However, if not separated at the source, CDW can contain hazardous waste, that is, waste that has hazardous properties and can prove to be harmful to human health or the environment (Fig. 9.6), the mixture of which can pose risks to the environment and can hamper recycling.

Fig. 9.6 Hazardous CDW containing asbestos (image courtesy of Luis Fueyo)

Examples of hazardous substances are lead-based paints, polycyclic aromatic hydrocarbons (PAHs), asbestos, and phenols. Table 9.1 shows some examples of building components in which harmful substances are found. At least, the inert mineral waste (i.e., bricks and concrete) is necessarily separated from the mixed construction and demolition waste. Some metals, such as reinforcement bars, are recovered in the recycling process and inert CDW can be easily recycled into recycled aggregates.

Compared to other resources and products, building and construction products have a relatively long-life and most materials are valuable and recyclable. "EU-28 generated 820 million tons of CDW in 2017, 46% of the total waste generation, representing an average of 1.7 tons per person" (Gálvez-Martos et al. 2018). The European Commission considers CDW a priority waste stream caused by the large quantities that are continuously originated. For this reason, article 11.2 of the Waste Framework Directive (2008/98/EC) notes that "Member States shall take the necessary measures designed to achieve that by 2020 a minimum of 70% (by weight) of nonhazardous construction and demolition waste, excluding naturally occurring material defined in category 17 05 04 in the List of Wastes (Table 9.2), shall be prepared for reuse, recycled, or undergo another material recovery" (including backfilling operations that use waste to replace other materials). The arrival of the economic crisis in 2007/2008 complicated the fulfillment of this objective. Currently, European countries recycle from less than 10% to over 90% of these CDW, depending on the state (EU countries use different definitions of CDW, which makes country comparisons very difficult).

According to the Environmental Protection Agency, the United States "generated 600 million tons of CDW in 2018, which accounts for an average of 1.5 tons per person (demolition represents more than 90% of total CDW generation whereas construction represents less than 10%)". To enhance the recycling amount of CDW, different countries have ratified some governmental laws and specific regulations. As a consequence,

▣ **Table 9.1** Examples of building components in which harmful substances are found (Strufe 2005)

Substance	Building components and/or materials
Asbestos	Roots and tiles Glue Sound insulation Fire-resistant sealing Wall plaster
PVC	Gutters and pipes
Lead	Roofs and tiles Electrical cables
Cadmium	Plastic (e.g., cable, pipes, and plates) Occurring with zinc Occurring with concrete
Mercury	Fluorescent tubes Switches and relays (electrical installations) Others (e.g., concrete)
Nickel	Stainless steel Surface treatment
Chromium	Stainless steel Others (e.g.. painted surfaces)
Copper	Cables and wires Permanent installations Temporary installations Roofs, pipes, etc. Screws, locks, etc. Pigments and dyes
Zinc	Gutters/pipes and galvanized products Plastic (especially gutters and pipes)
PCB	Small capacitors and electrical installations Double-glazed windows (glue) Sealant (softener) Paint (pigments) Fire-resistant additive (paint, glue/binder)
Chlorinated paraffin	Plastic (in general) Sealant (softener) Others (e.g., glue)
CFCs	Thermal insulation—PUR foam Other insulation materials

the recycling level of CDW in some countries reached around 90% of the total quantity produced. From a purely economic point of view, the recycling of construction and demolition waste is only attractive if the recycled products are competitive with natural products concerning cost and amount.

9.2.1 CDW Composition

The amount and composition of construction and demolition waste can change between countries and regions depending on a great number of factors: population growth, legislation, regional planning, the construction industry of the country or region, climate conditions, etc. Hence, it is very difficult to univocally define a composition representative of a country or even a region and some caution is required when reviewing statistics about its production around the world. The amount and quality of construction and demolition waste are controlled by internal (i.e., type, age, construction technologies, and construction materials) and external factors (i.e., demolition technologies, management capabilities, population growth). The composition of CDW varies broadly as a function of the type of work. For instance, the construction of a road generates a large quantity of excavated materials that will become waste whereas the demolition of a building will originate a large quantity of waste concrete.

The composition of CDW is "also changing during time, due to ageing of the existing buildings and to the low-quality structures, especially build between the 1960s and 70s, which are coming to the end of their lifetimes and needing demolishing" (Pellegrino and Faleschini 2016). As an example, Pacheco-Torgal et al. (2013) reported the following CDW composition per square meter: "(a) 120.0 kg of construction of new buildings, (b) 338.7 kg of rehabilitation, (c) 1129.0 kg of total demolition, and (d) 903.2 kg of partial demolition". ▣ Figure 9.7 shows the 2017 generation composition for construction and demolition waste (debris) in the United States, ▣ Fig. 9.8 illustrates the diversity of CDW in Europe and ▣ Table 9.3 provides an overview of the range of components of European CDW. These components can be subdivided into major and minor components, as shown in ▣ Table 9.4.

Concrete is the principal component of CDW (excavated materials excluded) (► Box 9.1: Concrete in CDW). Bricks (▣ Fig. 9.9) are often presented as mixed masonry with concrete and, like source-separated concrete, "that material of the CDW stream is relatively straightforward to process, with similar end markets for aggregate products" (Bustillo 2021). As a consequence, concrete and bricks make up the heaviest portion of the CDW stream.

9

Table 9.2 Chapter 17 of the European waste catalog related to the construction and demolition wastes

17 01	*Concrete, bricks, tiles and ceramics*
	– 17 01 01 concrete
	– 17 01 02 bricks
	– 17 01 03 tiles and ceramics
	– 17 01 06 mixtures of, or separate fractions of concrete, bricks, tiles and ceramics containing dangerous substances
	– 17 01 07 mixtures of concrete, bricks, tiles and ceramics other than those mentioned in 17 01 06
17 02	*Wood, glass and plastic*
	– 17 02 01 wood
	– 17 02 02 glass
	– 17 02 03 plastic
	– 17 02 04 glass, plastic and wood containing or contaminated with dangerous substances
17 03	*Bituminous mixtures, coal tar and tarred products*
	– 17 03 01 bituminous mixtures containing coal tar
	– 17 03 02 bituminous mixtures other than those mentioned in 17 03 01
	– 17 03 03 coal tar and tarred products
17 04	*Metals (including their alloys)*
	– 17 04 01 copper, bronze, brass
	– 17 04 02 aluminium
	– 17 04 03 lead
	– 17 04 04 zinc
	– 17 04 05 iron and steel
	– 17 04 06 tin
	– 17 04 07 mixed metals
	– 17 04 09 metal waste contaminated with dangerous substances
	– 17 04 10 cables containing oil, coal tar and other dangerous substances
	– 17 04 11 cables other than those mentioned in 17 04 10
17 06	*Insulation materials and asbestos-containing construction materials*
	– 17 06 01 insulation materials containing asbestos
	– 17 06 03 other insulation materials consisting of or containing dangerous substances
	– 17 06 04 insulation materials other than those mentioned in 17 06 01 and 17 06 02
	– 17 06 05 construction materials containing asbestos
17 08	*Gypsum-based construction material*
	– 17 08 01 gypsum-based construction materials contaminated with dangerous substances
	– 17 08 02 gypsum-based construction materials other than those mentioned in 17 08 01
17 09	*Other construction and demolition wastes*
	– 17 09 01 construction and demolition wastes containing mercury
	– 17 09 02 construction and demolition wastes containing PCB (for example PCB containing—sealants, PCB-containing resin-based floorings, PCB-containing sealed glazing units, PCB-containing capacitors)
	– 17 09 03 other construction and demolition wastes (including mixed wastes) containing dangerous substances
	– 17 09 04 mixed construction and demolition wastes other than those mentioned in 17 09 01, 17 09 02 and 17 09 03

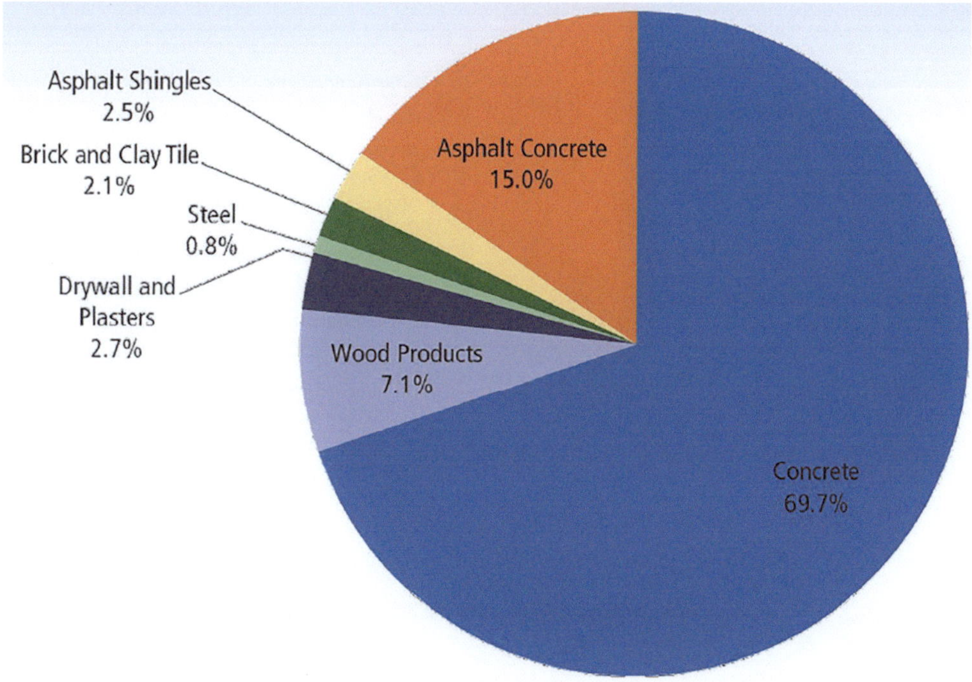

Fig. 9.7 Composition of construction and demolition debris in 2017 in the United States (EPA 2019)

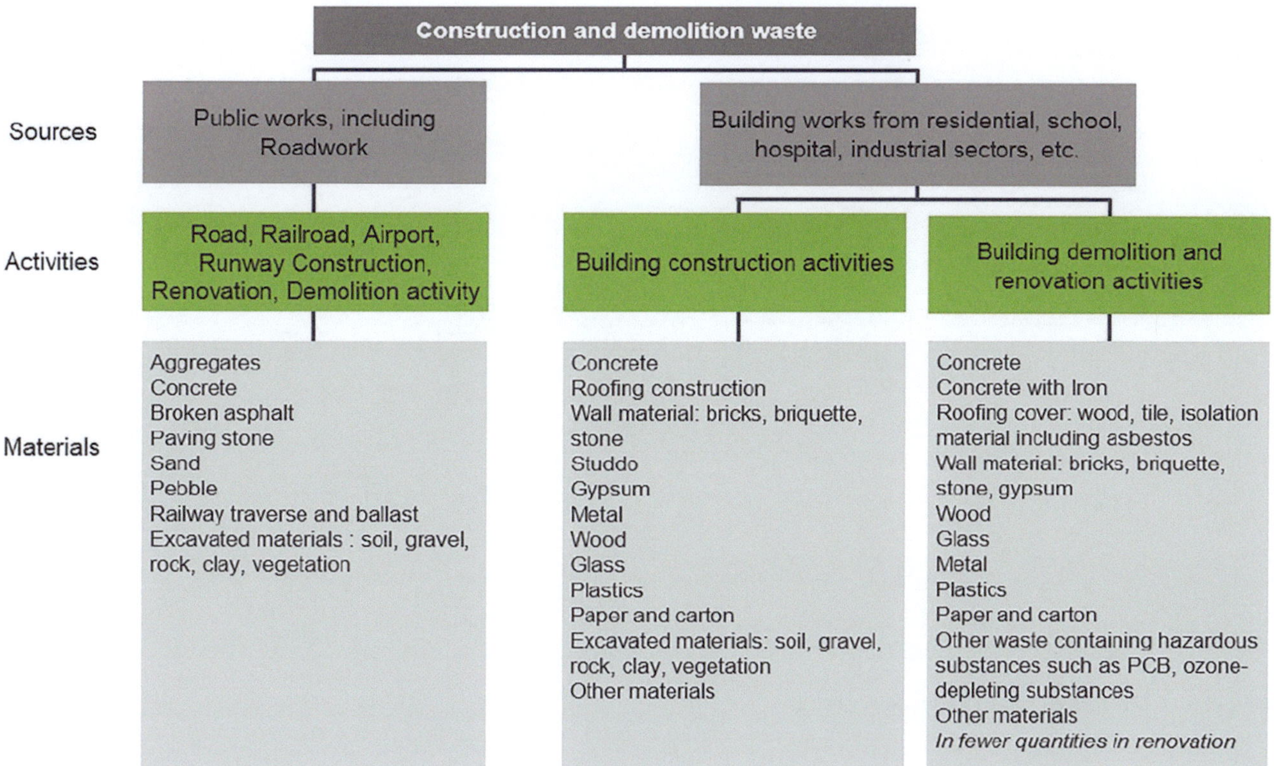

Fig. 9.8 Diversity of CDW in Europe (Deloitte 2017)

Table 9.3 Construction and demolition waste composition in Europe (BioIS 2011)

Waste category	%, min–max range
Concrete and Masonry	40–84
Concrete	12–40
Masonry	8–54
Asphalt	4–26
Others (mineral)	2–9
Wood	2–4
Metal	0.2–4
Gypsum	0.2–0.4
Plastics	0.1–2
Miscellaneous	2–36

Table 9.4 Major and minor components of CDW

Major components
• Cement concrete • Bricks • Cement plaster • Steel (from RCC, door/window frames, roofing support, etc.) • Rubble • Stone (marble, granite, sand stone) • Timber/wood (especially demolition of old buildings)
Minor components
• Conduits (iron, plastic) • Pipes (GI, iron, plastic) • Electrical fixtures (copper/aluminium wiring, wooden baton, wire insulation) • Panels (wooden, laminated) • Others (glazed tiles, glass panes)

Fig. 9.9 CDW formed basically by masonry bricks

Box 9.1—Concrete in CDW

Concrete is the most widely used building material in the construction industry. It is robust, durable, safe, and requires very little maintenance. Concrete is utilized in massive amounts for all types of buildings (from residential to office blocks) (◘ Fig. 9.10) *and in infrastructure projects (roads, bridges, dams,* etc.*), being a vital component of our society. In fact, it is impossible to understand the modern world without the existence of this construction material. Concrete, which is a man-made stone, can take virtually any shape and form because it is similar to a heavy liquid when produced and poured into a mold, being an extremely versatile building material. Its versatility and ready availability have assured that concrete is and will continue to be of increasing importance for all types of construction throughout the world. Therefore, concrete will also be the main component of the CDW stream.*

Mixing cement, aggregates, water, and admixtures and additions, if needed, make concrete. Generally, a mixture is approximately 10–15% cement, 60–80% aggregates (◘ Fig. 9.11) *and 15–20% water. In many concrete mixtures, entrained air can also sum up around 5–8%. Aggregates are quantitatively, with a large difference, the principal component of concrete, forming the skeleton of concrete.*

In general, there are different sources of concrete waste: fresh concrete from ready-mix trucks returned from site works (trucks are the most common methods of concrete transport), producing debris in the precast manufacturing facility (a precast concrete manufacturing plant produces waste during the manufacturing process of the component) as well as construction and demolition waste (◘ Fig. 9.12), *the latter being the most important source. Concrete can be 100% recycled in new concrete, using recycled aggregates* (see Sect. 9.3), *or in other applications. Recycled aggregate obtained from the demolition of concrete structures is typically utilized in unbound applications (i.e., road bases and subbases) and as aggregates for the production of new concrete.*

◘ **Fig. 9.10** The CN Tower in Toronto was built with concrete

◘ **Fig. 9.11** Coarse aggregates used in the production of concrete

Fig. 9.12 CDW constituted mainly by concrete

Wood originates in the demolition sector since it is ubiquitous in buildings and construction structures (Fig. 9.13) in many countries worldwide. For instance, a significant number of houses in the United States are wood framed. A barrier to growing the reuse market of wood is the increasing mechanization of demolition works since it makes salvage operations more difficult. The most common wood type originating in construction and demolition waste is classically dimensional lumber although wood-based products such as plywood, treated wood, and manufactured wood (i.e., composite panels) are also very common.

Asphalt materials are almost always produced in the road construction sector and are potentially 100% recyclable. Road-recovered materials are carried to an asphalt plant for sorting and batching to ensure that the mix incorporates an adequate ratio of bitumen to aggregate and an appropriate proportion of aggregate size and air voids. On the other hand, asphalt shingles (Fig. 9.14) have very good properties for implementation in many uses.

Gypsum plasterboard is one of the most recent materials incorporated into the CDW stream because it has been a common interior finish system since the 1950s. As a result, gypsum plasterboard is more present in new construction and remodeling waste (Fig. 9.15) than in demolition debris. Most gypsum plasterboard recovery is regarded as a contaminant when present in recovered construction and demolition waste. Gypsum plasterboard is one of the most challenging components for improving the recovery of mixed CDW, even though plasterboard itself is almost 100% recyclable.

Metals and plastics are present in minor quantities in the CDW stream compared to the principal construction and demolition waste materials commented on previously. There is a strong economic incentive to recover metals in the CDW stream (Fig. 9.16). The majority (approximately 90%) of metals recovered from CDW are generated in commercial demolition sites. Of this material, "up to 95% is steel and the remaining materials (approximately 5%) are nonferrous metals; these

Fig. 9.13 Wood generated in the demolition process (image courtesy of Demoliciones Anka)

nonferrous components are mostly aluminum (1–2%), stainless steel, and copper piping or wire" (Bustillo 2021). Ferrous metals (i.e., steel) are recovered from construction and demolition waste utilizing moderately inexpensive magnets. Metals are collected in bins and transported to the processing plant, either directly from the construction and demolition site or from sorting facilities. The treatment process separates the different metal types and removes any other materials (i.e., plastic and wood).

Regarding the plastics used in building and construction works, they present two basic types: packaging and durables (nonpackaging). The categorization of these materials is associated with the lifespan of the plastic product. There are many uses of plastic products in the building and construction sector, but the most common products are roofing membranes for roofing systems (Fig. 9.17), house wraps (Fig. 9.18), flexible cables (Fig. 9.19), piping (Fig. 9.20), and PVC windows and doors (Fig. 9.21). Nevertheless, few plastic components are recovered from the CDW sector.

The excavation materials (Fig. 9.5) are recovered where civil or site preparation works are undertaken. A great variety of materials can be generated as byproducts depending on the geology of the area. On the other hand, soil and sand originate from site preparation and excavation works linked to construction activities. Great quantities of fine materials are generated through these activities and, unless the materials can be reutilized on-site, they require further treatment and/or disposal.

9.2.2 Management of CDW

The management of construction and demolition waste is a predicament for humanity and the issues of managing this waste stream are global. Establishing an effective construction and demolition waste management system, that is, environmentally sound and economically feasible, is mandatory and requires comprehensive exploration and discussion. Because the price of raw materials for construction is continuously increasing,

Fig. 9.14 Postconsumer recycled asphalt shingles (PRAS) as feedstock material in hot mix asphalt (image courtesy of North Carolina Environmental Quality)

Fig. 9.15 Gypsum plasterboard waste

better management and reuse of CDW could become a cost-effective and reliable source of raw materials.

Compared with other forms of waste (i.e., household waste), "more than 90% of the construction and demolition waste may be utilized as a resource and a replacement of construction materials, especially for primary and natural raw materials" (Lauritzen 2019). Nevertheless, these applications depend on the quality and market for the materials as well as the environmental impact of the processes utilized in the transformation of CDW into new buildings and structures. In this sense, it is important to bear in mind that some elements in construction and demolition waste possess a high value although others can possess a lower value, but they can easily be reprocessed into new products.

Because recycling is considered the most environmentally friendly system for managing CDW, finding a market for recycled CDW materials is the most important stage for establishing a recycling program. At present, recycling in CDW plants is a reasonable option for the existing unsustainable landfill methods. This is because the technology for the separation and recovery of CDW is firmly established, easy to access, and as a rule inexpensive. As a consequence, best practices are necessary for the management of CDW across the complete construction value chain. "Systematic implementation of best practices could radically improve resource

Fig. 9.16 Metals separated from the CDW stream (image courtesy of Demoliciones Anka)

efficiency, reducing environmental impact by (a) reducing waste generation, (b) minimizing transport impacts, (c) maximizing re-use and recycling by improving the quality of secondary materials, and (d) optimizing the environmental performance of treatment methods" (Gálvez-Martos et al. 2018).

A waste management method hierarchy has been well established, comprizing four strategies: "waste reduction, reuse, recycling, and disposal" (Peng et al. 1997). In this hierarchy, the primary principles are minimizing resource consumption and preventing environmental pollution. Construction and demolition waste minimization involve the application of the 3Rs, that is, reduce, reuse, and recycling. They are listed in an order of preference, representing a hierarchy of environmental benefits. A reduction in the CDW sector means reducing the consumption and waste of building materials. It is commonly the most cost-effective possibility because it provides both raw material and waste disposal cost savings.

Reuse involves the direct reuse of CDW in its original form. Thus, materials from a demolition project can be utilized in new construction. Recycling involves reprocessing CDW into new products. It diverts waste from landfills but it requires labor, machinery, and energy to transport and treat materials. The amount of material that may be salvaged for reuse and recycling depends principally on the type of building to be demolished, time constraints, and the local recycling market. In general, recycling is the most common election in the CDW market. "The competitiveness of CDW recycling could be improved with several operations (i.e., raising the price of raw materials, through taxation, and setting end-of-waste criteria for certain CDW fractions)" (Pellegrino and Faleschini 2016). Table 9.5 shows an overview of the main fractions of construction and demolition waste, EU-27 estimates, their distribution, and the possibilities of recycling, reuse, and recovery.

After World War II, many of the regulations in developed nations encouraged CDW management. Most of them were promulgated to decrease environmental impacts. At present, it is common to carry out a life cycle analysis from the time the waste is generated until

9

▪ **Fig. 9.17** Polyethylene roofing membrane (image courtesy of KÖSTER)

▪ **Fig. 9.18** Nonwoven polyolefin used in house wraps (image courtesy of Barricade Building Products)

Fig. 9.19 Cables in the CDW stream (image courtesy of Demoliciones Anka)

Fig. 9.20 Plastics in piping (image courtesy of Carolina Bustillo)

Fig. 9.21 PVC windows

9

Table 9.5 Overview of the main fractions of construction and demolition waste, EU-27 estimates, their distribution, and the possibilities of recycling, reuse, and recovering (Lauritzen 2019)

CDW fraction	Type of material	Distribution minmax by weight %	Potential of recycling, reuse, recovering
Concrete, plain concrete, or reinforced concrete	Inert, ceramics	50–90	Recycling of crushed concrete, bound, and unbound applications
Masonry, bricks, and tiles	Inert, ceramics	35–50	Reuse of bricks recycling of crushed materials
Stones	Inert, ceramics	1–2	Reuse of granite blocks, recycling of crushed aggregates
Asphalt	Mixed inert and combustible	1–5	Recovering and recycling in new asphalt layers
Wood, timber, planks, doors, windows	Biodegradable, combustible	1–10	Reuse of doors, windows, timber and planks, production of chipboards, energy recovery
Scrap iron, steel, metals	Inert, metal	1–5	Recovering
Gypsum		0.2–0.4	Recovering
Glass	Inert	0.1–1.0	Recovering
Insulation materials, mineral wool, glass wool	Inert	1–10	Recovering
Paper, cardboard, plastics	Combustible		Incineration, energy
Ceramics and porcelains	Inert		Disposal, recycling possible
Hazardous substances			Disposal or special treatment

Table 9.6 Life cycle environmental burdens for one ton of CDW waste treated using different methods (Blengini and Garbarino 2010)

Treatment	Global warming potential, kg CO_2e/Mg	Primary energy MJ/Mg	Land use PDF m^2 a/Mg
Collection	6	100	0.15
Landfill	15	300	0.80
Recycling	2.5	45	0.18

the point that the material is recycled or dissipates into the environment to evaluate the environmental impacts of various methods of CDW management. This type of analysis (i.e., Hackenhaar et al. 2019) is a useful tool for conducting a systematic environmental impact assessment of different CDW management scenarios. As an example, Table 9.6 shows the life cycle environmental burdens for one ton of CDW treated using different methods. It can be concluded that recycling is the best treatment for CDW. Because any material incorporated into the building structure remains until removed during renovation or demolition, at which point it becomes demolition waste, the life-span values of different construction materials are essential in CDW management (Table 9.7).

Incentives have been incorporated to prevent the utilization of virgin products and to deter landfilling and incineration of CDW through strong systems of taxation and penalties. As a rule, adequate product management tends to lead to a better end product that can be utilized in a broader group of applications.

Construction and demolition waste is also classified based on its management requirements (a) it needs management at an approved facility (i.e., asphalt shingles, wallboard/drywall, plastic plumbing, insulation products, plaster) or (b) it can be used as clean fill (inert) (i.e., rock, aggregate, soil, bricks, concrete, asphalt pavement).

CDW management is basic and simple and the benefits of proper waste management are not solely environmental but also economic. In this context, managing CDW seeks to eliminate waste where possible, minimize waste where feasible, and reuse materials that might otherwise become waste. In Fig. 9.22, the upper figure illustrates a demolition project and a construction project without any synergy whereas the lower figure shows the exploitation of the recycled CDW materials from demolition as resources for the construction of a new building. Table 9.8 shows the principles of the business model of integrated CDW management compared with traditional CDW management, as illustrated in Fig. 9.22.

Several regulations impose the execution of a waste management plan to reduce the quantity of CDW generated. It must include information about the "characterization of the construction works, main waste streams, waste management framework, estimation of the quantities of each material, proposal for minimization, reuse, and recycling, and transport of the CDW" (Bustillo 2021). Effective management of building-related waste requires the coordinated action of governmental, business, and professional groups and their activities. Nevertheless, practices in the management and disposal of CDW are established by factors such

Table 9.7 Life-span values of different construction materials (EPA 2016)

Material	Source	Life span		
		Short	Typical	Long
Concrete	Buildings	50	75	100
	Roads and bridges	23	25	40
	Other structures	20	30	50
Lumber	Buildings	50	75	100
Railroad ties	Other structures	20	35	45
Plywood and veneers	Buildings	50	75	100
Wood paneling	Buildings	20	25	30
Drywall and plasters	Buildings	25	50	75
Steel	Buildings/roads and bridges	50	75	100
Brick	Buildings	50	75	100
Clay floor and wall tile	Buildings	15	20	25
Asphalt shingles	Buildings	20	25	30
Asphalt concrete	Buildings	20	25	30

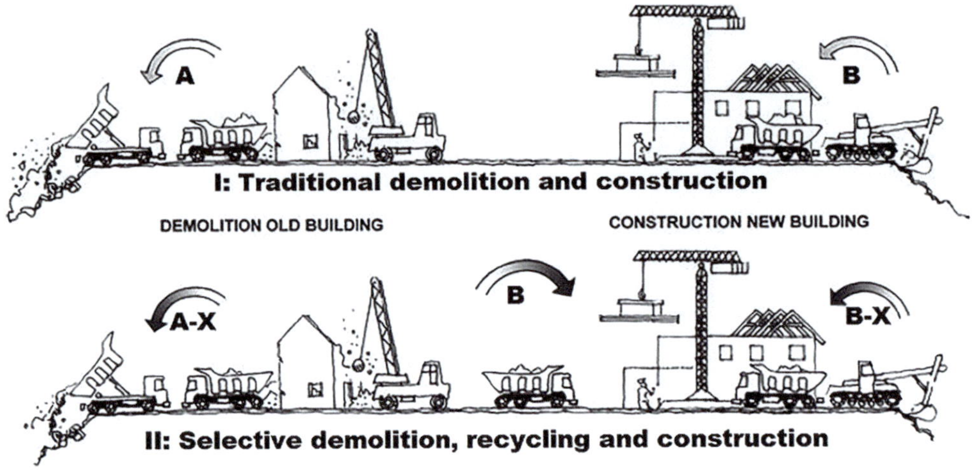

Fig. 9.22 Principle of the business model of integrated CDW resource management (Lauritzen 2019)

Table 9.8 Principles of the business model of integrated CDW management compared to traditional CDW management, as illustrated in Fig. 9.22 (Lauritzen 2019)

Type of CDW management	Cost/CO_2
I. Traditional CDW management, demolition, and construction	
Handling A tons CDW. transport, fees, disposal, etc.	I.A
Providence of B tons raw materials, production, transport, etc.	I.B
Total cost/load	I.A. + I.B.
II. Integrated CDW management, demolition, recycling, construction	
Handling A—X tons CDW, transport, fees, disposal, etc.	II.A–X
Providence of B—X tons raw materials, production, transport, etc.	II.A–X
Transformation, crushing, qualification of X tons recycled materials	X
Total cost/load	(IIA–X) + (II.B–X) + X
Preferences	
I.A + I.B > (II.A–X) + (II.B–X) + X → Integrated CDW management to be preferred	
I. A + I.B < (II.A–X) + (II.B–X) + X → Traditional CDW management to be preferred	

as "the availability of suitable disposal sites, economic conditions, societal priorities, availability of markets for recycling and reuse, transportation options, and the capabilities of local workforces and construction businesses to adapt demolition processes for management of wastes" (World Building Design Guide).

A key aspect of appropriate construction and demolition waste management is to keep materials separated on-site. If inert CDW is separated, recycling will be more effective and the quality of the different recycled materials will be higher (i.e., recycled aggregates). Figure 9.23 shows three possible construction and demolition waste pathways (the first pathway includes the separation of CDW on-site; the separation process depends mainly on the options offered at the site (i.e., space and labor) and on the costs and revenues of separated materials). Figures 9.24 (lead sheets) and 9.25 (roof tiles) show two examples of materials separated on-site. In this sense, "when waste management controls were firstly incorporated in the 1970s, illegal dumping of construction and demolition waste was prevalent" (UNEP 2015).

In addition to CDW reduction, recycling and efficient use of resources, the principal objective of integrated CDW management is to create and implement synergy. The following preconditions for integrated

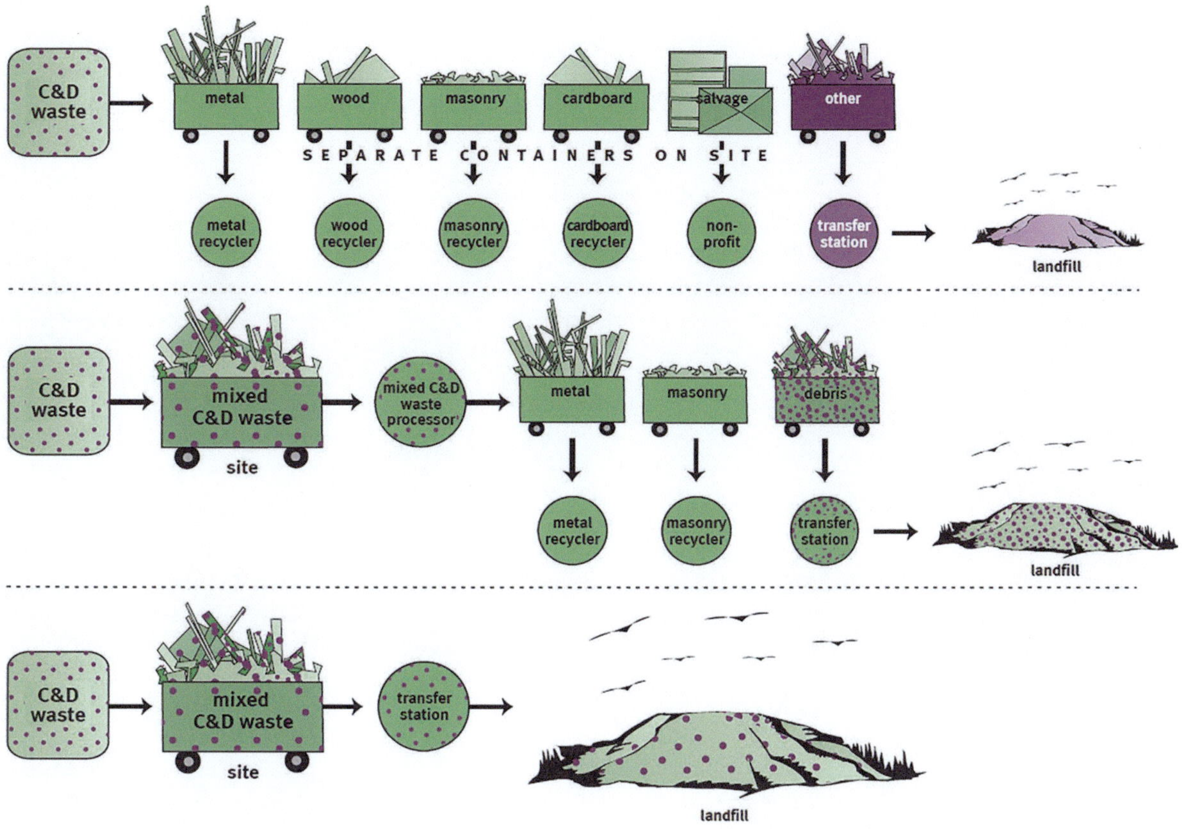

Fig. 9.23 Three construction and demolition pathways (NYC-DDC 2003)

CDW management must be met: "(a) basic understanding of the need for substituting natural resources with recycled waste materials, (b) dialogue based on mutual, equal understanding between the waste sector and the building and construction sector, (c) matching demolition contracts and production of CDW with new building and construction contract needs for resources, and (d) establishment of a professional and viable market for recycled CDW resources" (Lauritzen 2019).

Adequate regulations on recycling rates form a crucial step toward a more sustainable CDW management. Regulations can impose the execution of a waste management plant. For instance, as noted above, the revised waste framework directive no. 2008/98/EC that establishes the minimum recycling percentage for non-hazardous CDW at least 70% by weight until 2020 will increase the recycling rate in European countries. Construction and demolition waste can also undergo another material recovery, such as backfilling operations, in which waste substitutes for other materials. The Construction Resource Initiatives Council introduced a series of targets as a global call to action for the entire building industry under the international Mission 2030 initiative (Fig. 9.26). As an example of these practices, Fig. 9.27 illustrates the recovery and disposal of CDW (including excavation—CDEW) in the UK in 2014.

Current policies and standards influencing the management of CDW can be subdivided into five principal categories (BioIS 2011): "(a) waste framework policies: national policies or regulation on waste, usually at least transposing concepts, targets, and obligations set in the European Waste Framework Directive, (b) landfill regulations: stricter control of landfilling of certain types of waste represents a major driver toward better management of CDW, (c) CDW policies: when specific obligations regarding the management of CDW are not directly included in the waste framework policies, specific policy or legal documents may have been developed, specifically addressing the issue of C&D waste, (d) secondary raw material regulation and standards, i.e., standards on the quality of secondary materials from CDW, and (e) construction and demolition site regulations and standards: for instance, requirements for buildings specifically addressing CDW". Taking into account these factors, CDW can be classified based on "the type of work that gave rise, the type of material, how hazardous the components are, and the final treatment it can be undergone" (Margarido 2015) (Table 9.9).

Fig. 9.24 Lead sheets separated on-site (image courtesy of Demoliciones Anka)

9.2.3 Valorization Plants

After the demolition of a building structure (conventional or selective—sometimes termed deconstruction), CDW is sent to the valorization plant, in which the majority of components can be recycled. In the 1980s, as a consequence of the scarcity of landfills and the growing sensitization of the environmental impact and resource potential of construction and demolition waste, numerous plants for CDW valorization went into operation in many countries worldwide.

There are two types of valorization plants: stationary or fixed (Fig. 9.28) and mobile (Fig. 9.29). In both cases, it is very important that the plant location is close to the origin and final destination of the construction and demolition waste. Stationary plants, with recycling capacities ranging from 100 to 400 tons/h, commonly adopt higher-level technologies. They are typically provided with sorting equipment for the separation of unwanted fractions. This type of plant is suitable in high-density areas and capable of producing high-quality, efficient materials due to the production of various recycled products of different grading. Nevertheless, the initial investment for this type of plant can be in excess. Mobile plants, with recycling capacities up to 100 tons/h, can treat smaller quantities of CDW in temporary demolition worksites. They are very frequent and have the great advantage of a lower cost of operation and construction and demolition waste transport. Mobile plants deploy basic technologies and commonly produce low-grade recycling products through in-situ recycling. They are economically feasible from a quantity of 4000–7000 tons/site and can be moved to various locations. However, in this type of installation, limited cleaning facilities are present, generating a low-quality recycling product.

The stationary CDW valorization plant (▶ Box 9.2: Salmedina Valorization Plant) is an assembly of crushing, sieving, and washing machinery interconnected by conveyor belts for material movement. The machinery is permanently housed on steel/concrete platforms. The crushing units can incorporate dust control systems, noise control systems, magnetic separator devices (Fig. 9.30), and other additional devices based on requirements. The systems are either semiautomated or completely automated units. Containers with CDW are collected and transported to diversion facilities using trucks (Fig. 9.31). CDW is delivered to a licensed

Fig. 9.25 Roof tiles separated on-site (image courtesy of Demoliciones Anka)

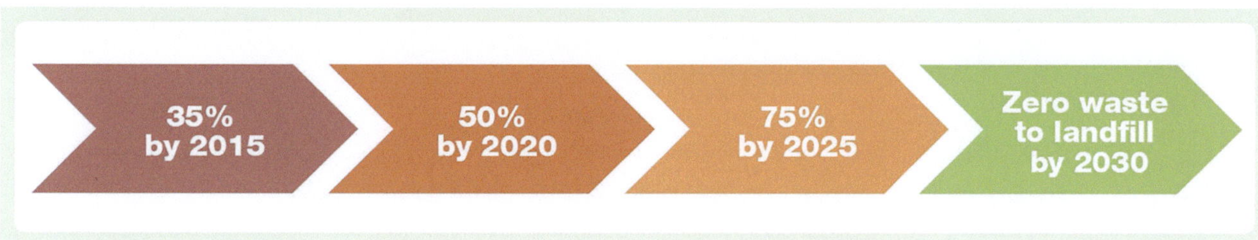

Fig. 9.26 Mission 2030 initiative (UNEP 2015)

CDW plant by truck and an initial inspection and acceptance of the load is made. The load Identification at CDW arrival is necessary for assuring that materials are appropriately handled.

At the entrance to the plant, the CDW must be weighed. Because separating the different fractions by a selective demolition process (selective demolitions can include the dismantling of elements classified as hazardous waste, electrical installations, etc.) can increase the cost of demolition, the CDW valorization plants must offer in their reception differentiated fees depending on the degree of contamination of the arriving CDW. Management plans describe materials that can be accepted. Loads with material that is not accepted can be turned away; in that case, the materials likely will be disposed of in a landfill.

To have an adequate size and weight for processing, large construction and demolition fragments are firstly break up using a hydraulic hammer or a hydraulic clamp. Some materials such as reinforcing bars of steel, large pieces of wood, concrete, and materials with sizes greater than one meter are picked before the sorting process may start. The lighter fractions can be effectively separated from the inert material using air separation devices (Fig. 9.32).

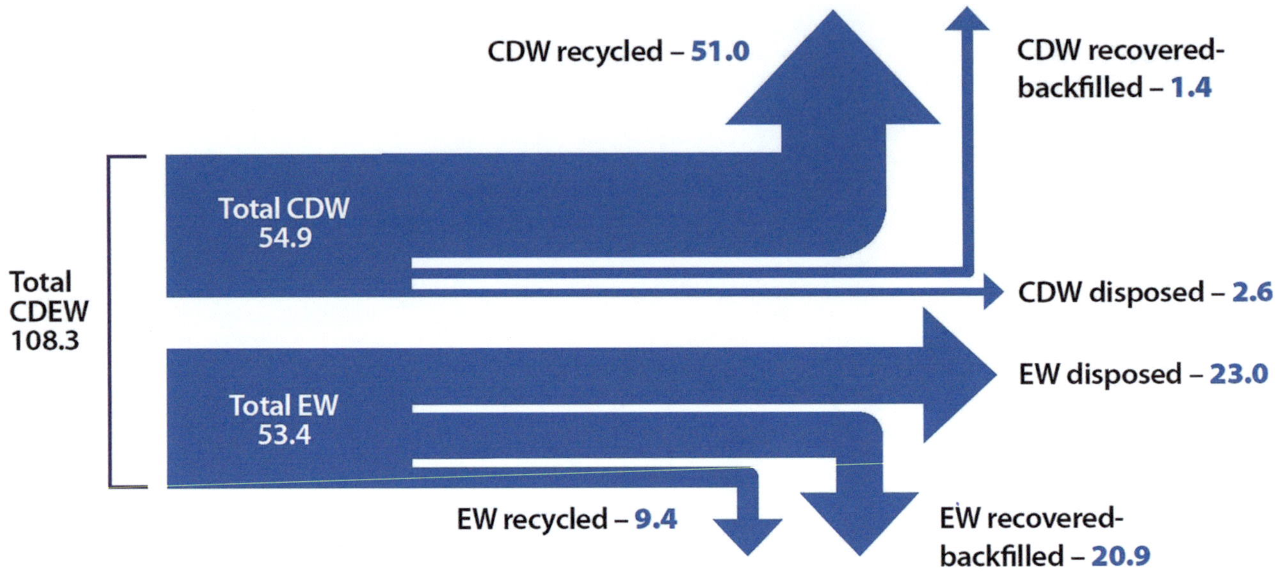

Fig. 9.27 Recovery and disposal of CDW (including excavation—CDEW) in the UK, 2014 (figures presented in million tons and exclude hazardous waste) (MPA 2019)

Table 9.9 Classification of CDW based on the type of work, type of material, dangerous nature of the material and destination (Margarido 2015)

Classification	Characteristics
Type of work	*Construction waste*: waste from new construction works for buildings and infrastructure
	Demolition waste: waste from demolition of buildings and infrastructure
	Repair work waste: waste from refurbishment and repair of buildings and infrastructure
Type of material	*Waste from structural concrete* (simple or reinforced)
	Waste from masonry
	Waste from asphalt material
	Waste from vegetation and unwanted soil
Dangerous nature (of this material)	*Inert waste*: this includes waste from concrete, soil, mortar, bricks, roofing tiles, masonry, etc.
	Noninert waste: this includes waste from different packaging, plastics, wood, metal, glass, etc.
	Hazardous waste: this includes waste from asphalt, used oil, tins of paint, and solvent asbestos, etc.
Final destination	*Reusable waste*: waste that can be reused directly on the work site or on another, consisting of clean bricks, concrete blocks, stone facades, roofing tiles and metal, vinyl coating, wooden cupboards, balconies, stairs, piping/electrical equipment carpets, clean insulation, and wooden/front beams
	Recyclable waste: waste that can be recycled, consisting of broken bricks, concrete or stone facades, ceramics and tiles, damaged or broken glass windows, wooden beams, trees, metal dividers, covering material, aluminum window, and door frames
	Nonrecyclable waste: waste that either due to its characteristics or because it is contaminated, cannot be recycled, particularly mixed waste that cannot be sorted, nonrecyclable materials, such as asphalt, linoleum flooring, dangerous waste including asbestos, treated wood waste (cladding or door frames), wood shavings, and wood contaminated by paint

Fig. 9.28 Salmedina stationary CDW recycling plant

Fig. 9.29 Mobile CDW recycling plant

Hand sorting is carried out by visual inspections and manual removal of unintended wastes by operators. It involves loading materials onto an inclined metal belt and passing across a manual sort line (Fig. 9.33) where materials are identified, grabbed, and deposited in vertical openings at each sorting station. Typically, 3 to 6 operators perform this operation, therefore labor costs may be relevant. Diverted materials are destined for incorporation into new products as recycled material (i.e., recycled aggregates). The production of recycled aggregates classically matches a crushing device (i.e., a jaw or impact crusher) as a primary crusher and other equipment such as a cone crusher for secondary or tertiary crushing. This results in well-graded and shaped aggregates. The circuit should also have magnetic separators for removing ferrous metals (i.e., steel). In other locations of the plant, additional magnetic separators can be utilized. On the

Fig. 9.30 Magnetic separator device

Fig. 9.31 Truck with a CDW container

Fig. 9.32 Air separation device

other hand, the plant can also be equipped with Eddy current separators to extract nonferrous metals. In some cases, air jigs and water can be utilized to efficiently stratify the recycled aggregates by density.

After sieving, the aggregates are stored. The storage of concrete recycled aggregates must be carefully handled to minimize breakage, moisture caused by the presence of organic materials and unhydrated cement, and stockpiling to reduce heterogeneity. A simplified scheme of a construction and demolition valorization plant is represented in Fig. 9.34.

Fig. 9.33 Sort or triage station

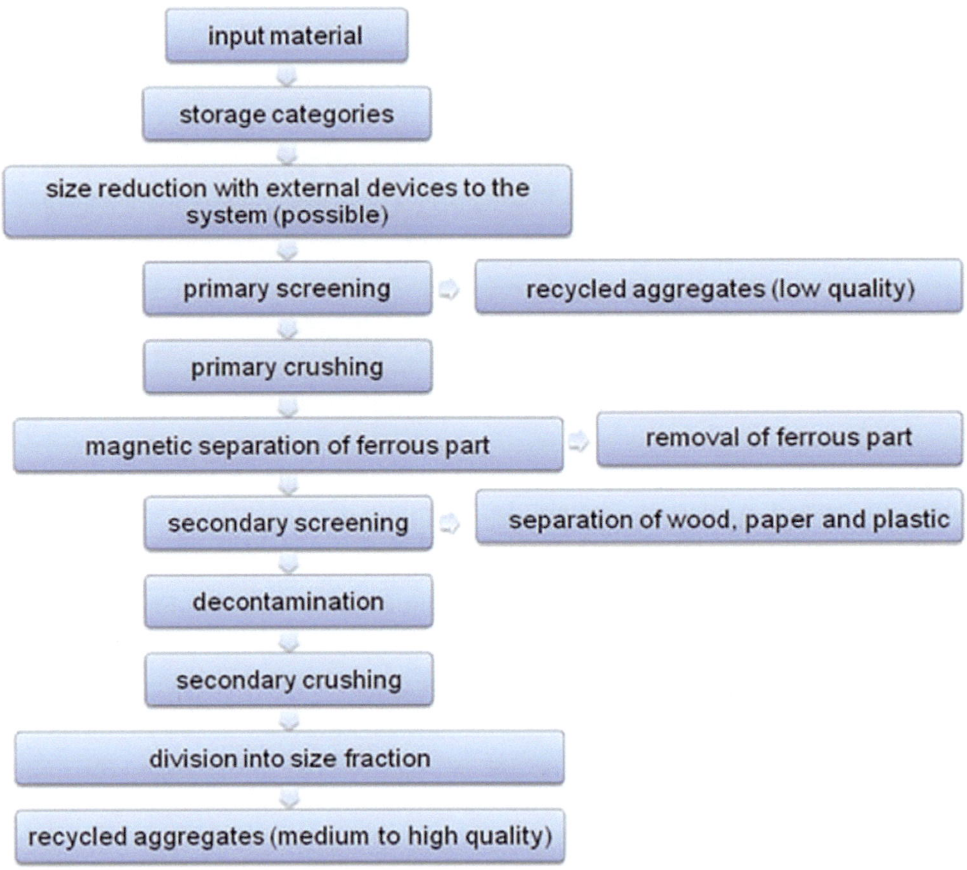

Fig. 9.34 Scheme of a construction and demolition waste valorization plant (Pellegrino and Faleschini 2016)

Box 9.2—Salmedina Valorization Plant (Courtesy of Salmedina TRI S.L.)

The Salmedina processing plant (▣ Fig. 9.28*) can treat 250 tons/h CDW concrete and 600 tons/h mixed CDW. It means approximately 7800 tons per day of CDW, with a recovery and recycling percentage of around 70% using both concrete and mixed debris plants. Environmental management of CDW at the plant begins with the right choice of materials to recycle and the treatments that they will receive according to their type. This result in new materials being reused, minimizing their environmental impact. There are three possible destinations in the plant when the truck arrives: concrete plant, mixed CDW plant, and purchase of material. In the first case, materials are made up almost entirely of concrete and undergo several processes (i.e. triage, crushing, and screening) to produce recycled concrete aggregate. In the second option, the trucks contain a lower percentage of concrete mixed with other CDW of various types. The purchase of material option is for trucks buying recycling concrete aggregate of different particle sizes depending on use; these aggregates are produced as a result of previous destinations (concrete plant or mixed CDW plant).*

The plant is divided into two-level platforms. The upper platform is used for the unloading of vehicles (▣ Fig. 9.35*) while the lower level incorporates the machinery and the area for stockpiling and dispatch of material. In the concrete plant, CDW arrives from building sites and is cleaned of other materials. This type of waste is subjected to a recycling process in three operational stages (a) the crushing stage to reduce the size of debris (*▣ Fig. 9.36*), (b) the sieving phase (*▣ Fig. 9.37*) to separate materials resulting from the previous stage, and (c) the stockpiling stage (*▣ Fig. 9.38*) in which the final aggregates are stored in an area intended for this purpose.*

In the mixed CDW plant, CDW is processed to extract materials such as plastics, paper, cardboard, or wood that are mixed with the concrete, separating and removing them both manually and mechanically. Once the materials have been separated, the procedure is similar to the concrete plant, that is, the material is crushed and screened.

▣ **Fig. 9.35** Upper platform for unloading the trucks (image courtesy of Salmedina TRI S.L.)

Fig. 9.36 Crushing process (image courtesy of Salmedina TRI S.L.)

Fig. 9.37 Sieving process (image courtesy of Salmedina TRI S.L.)

Fig. 9.38 Stockpiling stage (image courtesy of Salmedina TRI S.L.)

9.3 Recycled Aggregates

As previously mentioned, UEPG defines recycled aggregates (RA) as "reprocessed materials previously used in construction". Some examples include recycled concrete from construction and demolition waste and reclaimed aggregates from asphalt pavements. This definition can be considered universal since all countries worldwide take it into account, apart from small differences on some occasions. The utilization of recycled aggregates is based on environmental reasons because, in technical terms and in a generic way, recycled aggregates present a worse quality than natural aggregates. In comparison to natural aggregates, "concrete recycled aggregates are weaker, more deformable, more porous and lighter and have larger water absorption; because of their composition, concrete recycled aggregates are also rougher than natural aggregates" (Pacheco and de Brito 2021). For instance, for geotechnical applications, the utilization of RA is not common and natural aggregates are favored.

The concept of the use of recycled aggregates from demolished concrete structures "was introduced into practice at the time of World War II in Europe" (CCANZ 2011). While recycling is cited as the best way to manage CDW, there are still many disadvantages to utilizing recycled aggregates in construction. The following are the most important: "(a) lack of confidence of clients and contractors, (b) uncertainty as to its environmental benefits, (c) lack of standards and specifications that concrete producers can take into account, (d) low quality of the final product, owing to lack of knowledge and/or interest of CDW recycling plant owners, (e) distance between construction and demolition sites and recycling plants and (f) lack of a consistent supply of good quality RA that can satisfy existing demand" (Silva et al. 2014).

Many specifications classify recycled aggregates based on their composition and each has its designation. Four principal types of materials generated from most construction and demolition waste and road renovations can be highlighted (a) crushed concrete, (b) crushed masonry, (c) mixed demolition debris and (d) road planning. After crushing and beneficiation, aggregate is included in one of the following categories: (a) recycled concrete aggregate (RCA) (Fig. 9.39), (b) recycled masonry aggregates (RMA), (c) mixed recycled aggregates (MRA), (d) reclaimed asphalt pavement (RAP), and (e) construction and demolition recycled aggregates (CDRA. Many contaminants present in recycled aggregates from the demolition of previous structures can seriously deteriorate the concrete strength produced with them. Some examples of these contaminants are gypsum, metals, plastic, rubber, soil or, wood.

RA from crushed concrete is generally utilized in the production of concrete and high-performance pavements whereas those including components generated by crushing masonry, RMA and MRA, can be used in the construction of certain types of pavements. This is because the strength and size requirements of recycled

Fig. 9.39 Coarse and fine recycled concrete aggregate (image courtesy of Salmedina TRI S.L.)

aggregates are less demanding and thus safety is less important than those in structural concrete. The materials produced with aggregates from construction and demolition waste constituted basically by concrete have been shown to have better properties in comparison with materials manufactured employing RA constituted principally of masonry. High-quality RA is currently produced in construction and demolition waste recycling plants, which are very similar to the plants producing crushed natural aggregates.

According to the Construction and Demolition Recycling Association (CDRA), recycled concrete aggregate "(a) is high quality, meeting or exceeding all applicable specifications, (b) can be an acceptable source of aggregate into new concrete, (c) is currently being used in concrete and asphalt products with better performance over comparable virgin aggregates, (d) provides for superior compaction and constructability, (e) weighs 10–15% less than comparable virgin quarry products, (f) offers a way to reduce landfill waste streams, and (g) means minimization of environmental impacts."

Several countries have established standards for the use of recycled aggregates. They manage recycled aggregates by dividing them into high-, medium-, and low-quality aggregates. The classification is generally based on tests such as water absorption, density, chloride content, impurity content, soundness, and Los Angeles abrasion. Once the type of aggregate is established, the recycled aggregates replacement ratio as well as the purpose of utilization are defined. In summary, some properties of recycled aggregates (i.e., shape of aggregate, particle size distribution, water absorption, strength, and presence of contaminants) are essential to be assessed prior to their utilization in many applications, specifically the production of concrete.

9.3.1 Properties of Recycled Aggregates

The overall quality of recycled aggregates depends on many factors: type of construction and demolition waste, methods of treatment, and a long etcetera. For instance, as far as crushing is concerned, an impact crusher generates better-recycled aggregates than a jaw crusher. In this sense, some countries have specific processing techniques to obtain recycled aggregates of similar quality to natural aggregates, usually through the elimination of the mortar adhered to the recycled aggregate particles.

An understanding of how natural aggregates change after being utilized in concrete production can improve the capacity to describe why recycled aggregates can perform differently when utilized in the production of new concrete than natural aggregates. Although the properties to be considered depend on the destination of the recycled aggregate, they can be considered the following as the most important: (a) adhered mortar, (b) particle size distribution (gradation), (c) shape, (d) density, (e) absorption, (f) impurities/contaminants, and (g) chemical composition. This is because these properties influence the two main recycled aggregate applications: concrete production and road bases and subbases. ◘ Table 9.10 shows the comparison between the principal properties of natural aggregates and recycled aggregates (Cardoso et al. 2016).

- Adhered Mortar

The most distinguishing characteristic of recycled aggregates is that they are formed by natural aggregates and adhered mortar (◘ Fig. 9.40), unlike natural aggregates. The properties of recycled aggregate are dependent on the amount of mortar content in the CDW aggregate. The old clinging mortar basically contains fine aggregate, hydrated and unhydrated cement particles. Unless specific processing methods are carried out to eliminate the mortar adhered to the recycled aggregate particles, these will always have, to a greater or lesser extent, this component in the recycled aggregate composition. In fact, adhered mortar is the cause of the main differences that exist between the properties

◘ Table 9.10 Comparison between some properties of RA and NA (Cardoso et al. 2016)

Property	NA	RA	RA *versus* NA
Size	Depends on the crushing process. Breakage may occur for softer rocks. Fines may be generated by friction between rough grains	Depends on the crushing process. Breakage and fines generation are expected during compaction, disaggregating the grains	Higher breakage is expected for RA
Shape and roughness	Depends on the type of rock and its cleavage plans	Depends on the original material and crushing process. Less adhered old mortar helps to create smoother surfaces	Depends on the crushing process and type of materials. May be more difficult to attain homogeneity for RA
Particle density and porosity	Depends on the type of rock	Depends on size and materials and their proportions present in the particles	Larger density expected for NA. Larger porosity expected for RA, but not homogeneous
Particle crushing strength	Depends on size and type of rock	Depends on size and materials and their proportions present in the particles	Smaller strength is expected for RA, but is not homogeneous. Smaller strength for masonry fragments than for concrete particles. Lower resistance is expected when adhered cement paste is present
Particle stiffness	Depends on size and on the type of rock	Depends on size and on the materials and their proportions present in the particles	High stiffness is expected for RCA aggregates of structural concrete. Smaller stiffness is expected masonry particles than for concrete fragments
Water retention properties	Depends on the type of rock and amount of accessible pores	Depends on the materials and their proportions present in the particles	Larger water absorption is expected for RA due to the presence of adhered cement paste but the value decreases significantly for bituminous, glass or vitrified masonry
Frost resistance, soundness and toughness	Depends on the type of rock	Depends on the materials and their proportions present in the particles	Toughness depends mainly on the number of cracks and defects in the grains. They are present in both types of aggregates but more in large fragments of rocks. Freeze–thaw resistance and soundness depend on the accessible porosity and size of pores

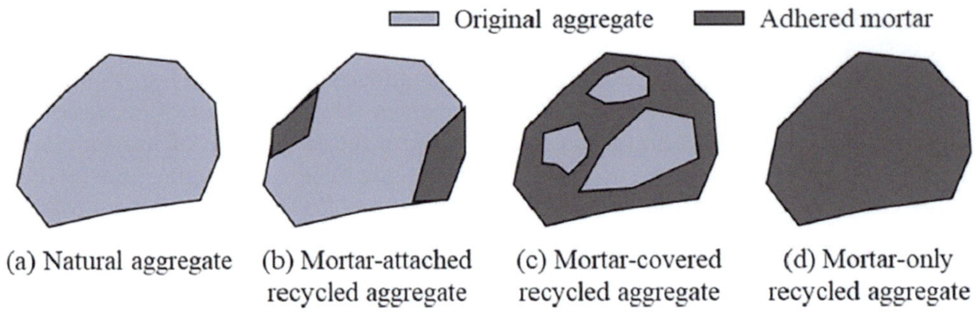

Fig. 9.40 Types of recycled aggregates depending on adhered mortar (Kim 2022)

of natural aggregates and those of recycled aggregates: lower density, greater absorption, etc. There is always a reciprocal relationship between adhered mortar and the quality of recycled aggregates and the negative effect of the mortar presence on a certain property of recycled concrete aggregate is proportional to its quantity.

9

The quantity of adhered mortar on the recycled aggregate surface depends principally on the crushing process, aggregate size, and strength of the original concrete and is often proportional to water absorption and inversely proportional to density. "Recycled aggregates obtained from the concrete having lower strength have the higher quantity of adhered cement paste for the given size of aggregate" (Rao et al. 2019). The volume of the residual adhered mortar in recycled aggregates ranges from 25 to 60% depending on the size of the aggregate. Nassar and Soroushian (2012) reported that "around 20% of cement paste is found attached to the surface of RA for particle size range from 20 to 30 mm" while Poon et al. (2004) reported that "RA extracted from crushing of waste concrete consists of 65–70% natural coarse and fine aggregate and 35–30% of cement paste by volume".

All these properties generate a recycled aggregate product of lower quality, which, in turn, has a negative impact on the materials manufactured with said type of aggregate. The principal factor controlling the quantity of adhered mortar in the recycled aggregate is the processing method because the fines generated in the crushing process come basically from the adhered mortar. As previously cited, an impact crusher generates a large quantity of fines, producing a lower content of adhered mortar and a higher quality aggregate. It can be stated that the amount of adhered mortar can decrease with an increase in the size of the aggregates. However, increasing the number of crushing processes increases the production costs of aggregates and optimization is necessary to obtain high-quality CDW aggregates with minimal production costs. In this sense, apart from increasing the mechanical processing steps, other methods are proposed to remove the mortar content and improve the quality of recycled aggregate. These are "thermal treatment, mechanical treatment, thermal–mechanical treatment, acid soaking, chemical–mechanical treatment and microwave-assisted treatment" (Akbarnezhad et al. 2013). In summary, continuous efforts have been developed to enhance the quality of recycled aggregates by removing or strengthening the adhered mortar.

- Particle Size Distribution (Gradation)

The size distributions of coarse and fine construction and demolition waste aggregates differ from the corresponding natural aggregate gradations. Recycled aggregates are poorly graded due to their poor particle size distribution. Well-graded recycled aggregate that abides by ASTM standards could be produced without much difficulty. It can be stated that recycled coarse aggregate can be utilized in concrete production from the consideration of grading of aggregates.

The particle size distribution of recycled aggregates largely depends on the processing techniques used for crushing the construction and demolition waste. As a rule, the most suitable crushing sequence is the one that includes first a jaw crusher to reduce CDW size down to 40 mm and subsequently an impact crusher or a cone crusher to reduce CDW particles to 20 mm or less. Nevertheless, from an economical point of view, a single crushing process can be more effective. The composition of the CDW is also important because the presence of large quantities of ceramic products generates more quantities of fines than if the CDW is principally concrete. However, small variations in the CDW composition do not have a significant influence on the gradation of the recycled aggregate. Chen et al. (2003) found "similar grade size distributions in the fine and coarse fractions of two types of CDW aggregates, which are slightly different in their compositions but processed by similar crushers with the same maximum size". The percentage of recycled coarse aggregate generated in the crushing process ranges from 70 to 90% of the total recycled aggregate produced.

The fineness modulus of the recycled aggregate can change according to the number of crushing processes, but the differences obtained are related to the composition of the CDW. Moreover, the quality of the concrete

Table 9.11 Fineness modulus of the recycled and natural aggregates (de Brito and Saikia 2013)

Recycled aggregate	Natural aggregate
2.61–2.68	2.95 (fine)
6.39–6.69	6.48 (coarse)
6.35 (coarse); 2.74 (fine)	2.78 (fine)
2.38 (fine)	2.38 (fine)
3.55 (fine); 6.40 (coarse)	3.04 (fine); 6.04 (coarse)
6.68 (coarse)	6.78 (coarse)
6.75 (coarse); 3.10 (fine)	3.28 (fine)

is the most important factor affecting the fineness module. CDW formed by high-strength concrete originates RA with a better fineness module than RA originated from low-strength concrete, keeping the same crushing process. The fineness modulus of recycled and natural aggregates reported in various studies is shown in Table 9.11. Due to the typical production process of concrete, concrete recycled aggregate is not as optimized as that utilized for producing natural aggregate (regarding crushing and sieving), "the grading of CRA may not be as suitable for concrete as that of NA and the voids content of CRA tends to be larger, with detrimental influence on the mechanical and durability properties of concrete produced using recycled aggregates" (Larrard and Colina 2019).

- Shape

There are no great differences between the shape of natural aggregates and that of recycled aggregates (Chen et al. 2003) since shape index values are very similar in both types of aggregate. However, finer recycled aggregates have a more angular shape. Regarding the shape of coarser aggregate particles, some studies have found important differences between natural aggregates and recycled aggregates. This is attributed to the fact that the adhered mortar tends to round the more angular particles and reducing the percentage of slabs.

In terms of geometry, concrete recycled aggregates are flakier and more elongated, which is principally caused by "(a) the nature of part of their constituents (i.e., ceramic tiles and bricks), which tend to fragment into elongated shapes, and (b) many CDW plants use primary crushing only, typically with a jaw crusher; this will also result in more elongated particles" (Pacheco and de Brito 2021). For NCA and RCA, "the content of elongated and flaky particles is 4.8 and 6.2%, respectively, and the difference between these two values is not large" (Xiao 2018), indicating that the shape of RCA is similar to that of NCA.

If RCA only undergoes a primary crushing stage, the recycled aggregates present higher shape and flakiness indexes than those of natural aggregates. Nevertheless, it should also be noted that there are cases in which natural aggregates typically present flatter and sharper shapes than recycled aggregates, mainly if these are from crushed slates or other natural stones with distinct cleavage planes. Table 9.12 shows some shape parameters for recycled and natural aggregates reported in different studies. In general, "recycled coarse aggregates demonstrate better indices when compared to natural aggregates" (Rao et al. 2019).

- Density

Table 9.12 Shape parameters of CDW and natural aggregate (de Brito and Saikia 2013)

Parameters	Recycled aggregates	Natural aggregates
Shape index (%)	28	25
Shape coefficient	10–20 mm: 0.363	10–20 mm: 0.364
	5–12 mm: 0.444	5–12 mm: 0.576
Elongation index	10–20 mm: 6	10–20 mm: 15
	5–12 mm: 8	5–12 mm: 19
Flakiness index (%)	4–20 mm: 7	8–20 mm: 7
		4–12 mm: 14
Flakiness index (%)	4–12 mm: 9	4–12 mm: 25
	10–25 mm: 7	10–25 mm: 11
Flakiness index (%)	10 mm: 10.44–17.82	–
	20 mm: 5.70–12.96	
Shape index (%)	22.3	14.0–18.3
Shape index (%)	24.3	11.1
Flakiness index (%)	10	9

Density is one of the most important physical properties of aggregates because it defines how densely the material is packed in a given volume. In addition to water absorption, the density of an aggregate is a crucial characteristic. The density of recycled aggregates is lower than that of natural aggregates, reflecting the lower density of some CDW components (i.e., ceramic products or mortar adhered to the surface of the natural aggregate particles). The variation in density is dependent on the specific aggregate in question. The recycled aggregate density generally ranges between 2100 and 2400 kg/m^3, which is approximately 10% lower than that of NA. Gonzalez-Fonteboa and Martinez-Abella (2008) affirm that "the density of recycled aggregate sizes ranging from 4–12 mm and 10–25 mm were 11.4 and 10.1% lower respectively than that of corresponding natural aggregate".

The main factors that influence the density of the recycled aggregate are the size of the fraction considered, processing methods, presence of impurities, and quality of the original concrete present in the CDW. Regarding the first factor, the density of recycled aggregate changes with the particle size distribution, with high densities in coarser fractions for the same CDW composition (the presence of low-density materials decreases the density of the recycled aggregate). Mortar, with a density lower than 2 g/cm^3, is another component present in recycled aggregates that can definitely influence the density of the final product. Nevertheless, its amount is controlled by the production process of recycled aggregates, as is the case for other impurities (i.e., wood, gypsum, and glass).

The processing methods also control the density of the recycled aggregate, with higher density values where impact crushers are used. The reason is that this type of crusher most effectively removes the adhered mortar in the aggregate particles. The successive crushing process can increase the density of the recycled aggregates due to the reduction of the adhered mortar, which is the principal cause of the decrease in the average density of the recycled aggregate. Apart from decreasing the quantity of adhered mortar, two or more crushing steps classically lead to rounder and less sharp particles.

Regarding the quality of the concrete present in the CDW, the type of natural aggregate used in its production has an influence on the density of concrete and the density of the recycled aggregate obtained from this CDW. The density values of RMA are commonly lower than those of RCA due to their very high porosity levels. Variables such as the water/cement relationship in the concrete do not seem to have a decisive influence on the density of the recycled aggregates. Nevertheless "the RA that results from crushing high-strength concrete will present higher oven-dried density, lower water absorption and higher resistance to fragmentation when compared to the RA coming from low-strength concrete" (Cardoso et al. 2016). When mixing recycled concrete aggregates and recycled masonry aggregates, to produce mixed recycled aggregates, the density of the resulting aggregates increases with the recycled concrete aggregate amount, regardless of the strength of the source material, and decreases with the recycled masonry aggregate content.

- Water Absorption

The ability to absorb water is one of the main characteristics of recycled aggregates (▶ Box 9.3: Water Absorption of Recycled Aggregates).

Box 9.3—Water Absorption of Recycled Aggregates

Water absorption capacity is one of the major differences between natural and recycled aggregates. This is because the water absorption capacity of RA is higher than that of natural aggregates due to the porosity of adhered mortar in recycled aggregates. The water absorption of RA is 3–12% higher than that of NA for the coarse and fine fractions. The water absorption values of fine recycled aggregates are still higher than those of coarse recycled aggregates. For this reason, some legislation limits the utilization of recycled aggregate with a water absorption capacity greater than 7–10% to be utilized in structural concrete. It is considered that five factors control the highest value of absorption in recycled aggregates: the amount of adhered mortar content, particle size, density (◘ Fig. 9.41*), processing method, and strength of parent concrete when RCA is used.* ◘ Table 9.13 *shows the mean 24 h water absorption of three types of recycled aggregate: concrete waste, masonry and concrete waste, and unsorted CDW waste.*

The water absorption and adhered mortar on the surface of the aggregates are directly proportional to each other. Thus, the higher the content of adhered mortar in recycled aggregates, the higher the water absorption. Regarding the particle size, the fine fractions of recycled aggregates always have greater water absorption (for the same original concrete) than the coarse fractions. This is because a higher percentage of the mortar is concentrated in the finer fractions. The impact of the high-water absorption of recycled fine aggregates on some properties of the final concrete (i.e., shrinkage and gas permeability) has been highlighted.

The density of recycled aggregates also influences the water absorption capacity of recycled aggregates: the lower the density is, the higher the water absorption. According to

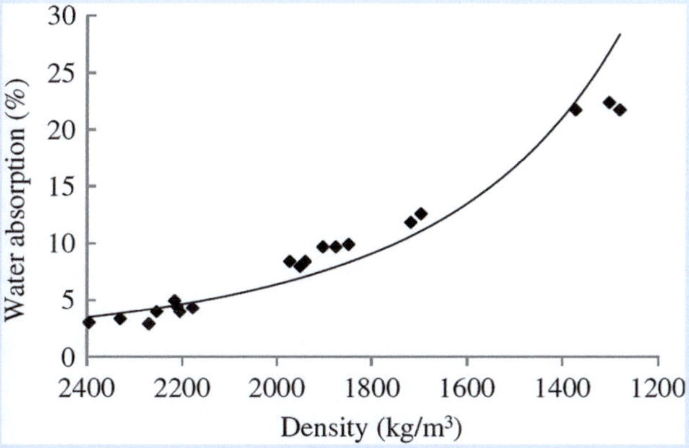

Fig. 9.41 Water absorption as a function of the density of recycled concrete aggregates (Rao et al. 2019)

Table 9.13 Mean 24 h water absorption of three types of recycled aggregate: concrete waste, masonry and concrete waste and unsorted CDW waste

Type of RA	Fraction	24-h water absorption (%)
Concrete waste	FRA CRA	9.5 4.9
Masonry and concrete waste	FRA CRA	9.3 7.2
CDW waste	FRA CRA	8.0 5.0

FRA = fine recycled aggregate; CRA = coarse recycled aggregate (Pacheco and de Brito 2021)

Zega et al. (2010), *"the water absorption of recycled aggregate depends on the type of natural aggregate used in the original concrete; higher density of natural aggregate used in original concrete, higher the adhered mortar content in recycled aggregate and hence higher was the water absorption of recycled aggregate".*

In recycled aggregates, the content of materials such as concrete, brick, and asphalt can change according to the processing method. Thus, "the water adsorption of RA that has not been subjected to effective sorting or contamination removal techniques will naturally increase due to the presence of highly porous materials (i.e., brick, tiles, wood and soil)" (Silva et al. 2014). *Moreover, CDW subjected to a secondary crushing process, for instance with an impact crusher, will result in recycled aggregates with less adhered mortar.*

As the density, the water absorption of recycled aggregate is also a function of the strength of parent concrete. Water absorption increased with the increase in the strength of the concrete from which the recycled aggregate was derived. This is because the high quantity of old mortar content adhered to recycled aggregates obtained from high-strength concrete. It is important to remember that, "in low-strength concrete, which has relatively weak bonding strength between mortar paste and original aggregate compared to high-strength concrete, most of the adhered mortar is separated from the original aggregate during the crushing process" (Kim 2022).

- Mechanical Properties

As a rule, recycled aggregate is found to be weaker than natural aggregate against mechanical resistance (i.e., Los Angeles abrasion test). The high value of Los Angeles abrasion test in recycled aggregates, which denote low resistance to abrasion processes, is caused by the weaker mortar adhered to the recycled aggregates and the weaker bond between adhered mortar and aggregate in recycled aggregates. Therefore, the abrasion loss is directly proportional to the adhered mortar content. The value of mechanical resistance depends, as in the case of density or water absorption, on "the strength of concrete from which the recycled aggregate derived, the quantity of adhered mortar and the quality of the aggregate used in the production of the original concrete; in addition, the method of crushing also influences these properties" (Rao et al. 2019).

The particle size of recycled aggregates affects their Los Angeles abrasion value. Gonzalez-Fonteboa and Martınez-Abella (2007) reported "lower abrasion values for coarse aggregates than that for fine aggregates". For instance, Hansen and Narud (1983) found abrasion values of 22.4 and 41.4 for the same type of construction and demolition waste aggregates with size ranges of 16–32 mm and 4–8 mm, respectively. This is because the fine fractions include the greatest amount of adhered mortar.

For a given strength of concrete from which the recycled aggregate is derived, "the resistance against mechanical actions decreased with the reduction in maximum size of aggregate; this can be attributed to the higher surface area of smaller size aggregates facilitating higher amount of mortar adhered" (Padmini et al. 2009). Hansen and Narud (1983) reported that "Los Angeles abrasion test values for recycled aggregates obtained by crushing 40 MPa strength concrete had lower abrasion than those obtained by crushing 16 MPa strength concrete".

Regarding the crushing value of recycled aggregates, the crushing values of recycled coarse aggregates are approximately double those of natural aggregates. For this reason, the higher the aggregate crushing value is, the lower the strength of the aggregate. Regarding the crushing methods, the increase in the crushing levels decreases the amount of adhered mortar, so the Los

Angeles abrasion test values will decrease as the number of crushing steps increases.

- Contaminants

As commented previously, many contaminants (i.e., gypsum, metals, plastic, rubber, soil or wood) that can be found in RA can severely degrade the quality of concrete made with them. Bituminous materials have a general effect of reducing the strength of concrete. Glass is removed from buildings before demolition and it tends not to be present in construction and demolition waste and recycled aggregates. This presorting is necessary due to their similar density to rocks and bricks, which makes it difficult to separate glass from other heavyweight materials. Before crushing, organic materials are difficult to separate from construction and demolition waste. Soil and clay particles often stick to stone and brick components. When they are not removed, their presence can adversely influence the properties of concrete. In some countries (i.e., The Netherlands), construction and demolition waste containing gypsum is regarded as contaminated. These contaminated CDW must be extensively washed before they can be utilized for producing recycled aggregates.

In summary, the main issues generated by the presence of contaminants in concrete manufactured with recycled aggregates are the following: "(a) glass enables the alkali-aggregate reaction; the separation of glass is difficult because its density is similar to that of the aggregate; (b) gypsum can cause sulfate attack; (c) wood and paper are unstable to wet-dry and freeze–thaw cycles, producing surface spalling; (d) paints entrap air in concrete; (e) clayey soils increase the demand for water and cause greater shrinkage; (f) brick particles can worsen freeze–thaw behavior and, in the case of refractory bricks, cause expansive phenomena; in this sense, a small amount of brick in the recycled aggregate (i.e., 5%) does not noticeably affect the strength of the concrete, but it does substantially worsen its durability, and (g) the presence of metals can produce surface stains due to oxidation, especially when the chloride content is high; zinc or aluminum presence can originate problems due to the evolution of hydrogen in fresh concrete and fractures, caused by expansion, in hardened concrete" (Sánchez de Juan 2004). The maximum contents of these components are usually controlled worldwide by different specifications. For instance, Table 9.14 shows the recommendations of the RILEM for recycled concrete aggregates for concrete. They were the first international recommendations for the application of recycled aggregates in concrete manufacturing.

- Chemical Composition

Organic matter and salts (chlorides, sulfates, and others) from several origins can contaminate recycled aggregates and their detrimental effects depend on the substance. They can range from changes in the setting and hardening of concrete and long-term adverse chemical reactions to reinforcement corrosion. For this reason, the chemical properties of recycled aggregates are important aspects of the study of aggregates since the durability of RA depends on their chemical composition. Ascertaining the chemical compositions of recycled aggregates are essential since the history and

Table 9.14 Mandatory requirements for recycled coarse aggregates for concrete (RILEM 1994)

Mandatory requirements	RCAC	RCAC	RCAC
	Type I	Type II	Type III
Min. dry particle density (kg m^{-3})	1500	2000	2400
Max. water absorption (% m/m)	20	10	3
Max. content of material with SSD < 2200 kg m^{-3} (% m/m)	–	10	10
Max. content of material with SSD < 1800 kg m^{-3} (% m/m)	10	1	1
Max. content of material with SSD < 1000 kg m^{-3} (% m/m and % v/v)	1	0.5	0.5
Max. content of foreign materials (metals, glass, soft material, bitumen) (% m/m)	5	1	1
Max. content of metals (% m/m)	1	1	1
Max. content of organic material (% m/m)	1	0.5	0.5
Max. content of filler (<0.063 mm) (% m/m)	3	2	2
Max. content of sand (<4 mm) (% m/m)	5	5	5
Max. content of sulfate (% m/m)	1	1	1

Type I aggregates which are implicitly understood to originate primarily from masonry rubble
Type II aggregates which are implicitly understood to originate primarily from concrete rubble
Type III aggregates which are implicitly understood to consist of a blend of recycled aggregates and natural aggregates

Table 9.15 Chemical characterization of different recycled aggregates (Jiménez 2013)

	RCA—siliceous (%)	RCA—limestone (%)	RMA and RMCA (%)
SiO_2	45–60	4–5	40–50
AI_2O_3	15–20	1–2	6–8
Fe_2O_3	2–5	1–2	2–4
CaO	5–7	52–54	20–28
MgO	0.5–1.5	0.2–0.8	0–1

RCA: recycled concrete aggregates; RMA: recycled mixed aggregates and RMCA: recycled mixed ceramic aggregates

properties of the original components of construction and demolition waste are unlikely to be known. Since CDW has been exposed to a vast range of environments and conditions, the chemical composition of recycled aggregates must be established with the aim of define limitations to be imposed that will result in good-quality aggregates. For example, the quantity of sulfur compounds in road aggregates is controlled to assure their chemical stability. Table 9.15 shows the chemical characterization of different recycled aggregates.

The utilization of RA requires a specific approach for guaranteeing a sufficiently low concentration of chloride or sulfate ions. Water-soluble sulfates in recycled aggregates are potentially reactive and can generate expansive reactions. Gypsum occurs in finely dispersed form and it is originated principally from plasterwork. It has a negative effect due to its solubility, low hardness, and low density. On the other hand, it has been suggested that recycled concrete aggregates can have higher sulfate content than natural aggregates because of sulfates from the cement of the adhered mortar.

Chloride content is another chemical property of great importance because an excessive concentration of chloride ions may lead to quick corrosion of reinforcement steel. It is important to define strict limits for the chloride content in recycled aggregates. The excessive chloride content can be eliminated by immersion of the aggregates in water. This is because these components are not linked to the cementitious microstructure and, consequently, they are easy to remove from recycled concrete aggregates. Regarding the presence of alkalis (usually from cement) and reactive silica, they can often produce expansive alkali-silica reactions. For this reason, it is necessary to control the alkali content of the concrete components.

9.3.2 Applications

The main application of recycled concrete aggregates has been its use in pavement subbase materials although it is common their utilization in concrete and asphalt paving layers and as general fill and embankment material.

- Recycled Aggregate Concrete (RAC)

The fresh and hardened properties of recycled aggregate concrete, that is, concrete produced using recycled aggregates (fine or coarse or both) obtained from CDW either as a partial or 100% replacement of conventional natural aggregates, "depend on the strength of original concrete from which the recycled aggregate derived, the w/c ratio, the size of recycled aggregate, amount of recycled aggregate and moisture condition of recycled aggregate, etc." (Rao et al. 2019).

In the manufacturing of RAC, it is possible to substitute only the coarse fraction of natural aggregates with recycled aggregates without significantly affecting the properties of concrete. In this sense, many studies have been carried out utilizing different percentages of coarse recycled aggregates for substituting natural aggregates. In this substitution, it is specifically recommended that recycled aggregates should come from crushed concrete due to their higher quality. Regarding the maximum admissible amount and quality permitted in structural concrete mixes, the international codes, guidelines, and recommendations change notably from country to country. The amount of the adhered mortar attached to the natural aggregates is also a significant feature to be studied for designing RAC because the adhered mortar influences the performance of RAC, basically the strength characteristics.

The use of recycled aggregates notoriously influences the fresh concrete properties, specifically its workability. This is because the porosity of recycled aggregates is higher than that of natural aggregates and the workability of the recycled aggregate concrete mix is lower for the same water content than that using natural aggregates (it has been clearly established that RA absorb more water than natural aggregates). The high absorption characteristic of RA makes it difficult to control the concrete consistency. The quick absorption of the recycled aggregates directly affects the water demand for concrete. If only limited quantities of coarse recycled aggregates are utilized (i.e., 20–30%), the influence of the substitution is almost negligible. In other cases, an extrawater demand must be taken into account for the production of recycled concrete, in which several alternatives are applied: "(a) direct increase of the mixing water, (b) use of chemical admixtures (plasticizers or superplasticizers) and (c) pre-saturation of the recycled aggregate" (Agrela et al. 2013). For instance, it has been observed that the utilization of superplasticizers leads to a considerable increase in workability.

The incorporation of recycled fine aggregates "up to 20% has been observed to have no major effect on the workability of the RAC" (Kisku et al. 2017). Further increases in the recycled fine aggregate content led to a clear decrease in workability. Recycled aggregate concrete exhibits higher slump loss compared to conventional concrete. The moisture state of recycled aggregates is also a major parameter affecting the workability of RAC. For its part, the density of fresh RAC is slightly lower than that of the mixture including natural aggregates because recycled aggregate densities are lower than those of natural aggregates. The existence of lower density adhered mortar particles attached to the aggregate is the principal factor for lowering the density of concrete due to the addition of recycled aggregates. Finally, other factors such as the surface texture and angularity of recycled aggregates also have a considerable influence on the workability of concrete.

Regarding the hardened concrete properties of RAC, the recycled aggregate utilization in the production of concrete originates a loss of compressive strength, which is higher if mixed aggregates are employed, and increases further with the percentage of aggregate replacement. The influence of the recycled aggregate substitution ratio on the compressive strength of RAC is shown in ◘ Fig. 9.42. The strength of the parent concrete is another important feature in the production of RAC. The compressive strength of recycled aggregate concrete increases with an increase in the strength of the parent concrete. Recycled aggregate concrete obtaining using higher-grade concrete shows significantly less difference in compressive strength than the compressive strength of concrete manufactured using natural aggregates. The compressive strength of RAC can be enhanced using different methods such as the use of admixtures, increasing the cement content, and using plasticizers or superplasticizers.

When the recycled aggregate substitution ratio increases, or when specific performances are needed (i.e., high-strength concrete or durability in severe exposure conditions), the utilization of recycled aggregates can cause a loss of performance, both in mechanical and durability terms. The lower properties of recycled aggregates (i.e., the lower density and mechanical strength) make RAC less durable in terms of carbonation, chloride penetration, permeability, and freezing/thawing resistance. The poor durability performance of RAC is associated with the inferior quality of recycled aggregates "due to the presence of numerous cracks and fissures and pores present inside the aggregate, making it more susceptible towards permeation" (Behera et al. 2014). As an example, ◘ Fig. 9.43 shows the permeability behavior of RAC with time in different studies. The durability of RAC can be enhanced using different mixing approaches and/or incorporating different quantities of mineral admixtures.

- Road Construction

The use of RA from CDW in the construction of subbases and foundations for roads (◘ Fig. 9.44) is a very good management idea to increase the recycling rates and creating a market opportunity for RA. Since the 1970s, recycled concrete aggregates have been utilized in the construction of hundreds of highways in the United States and worldwide. These projects have included the

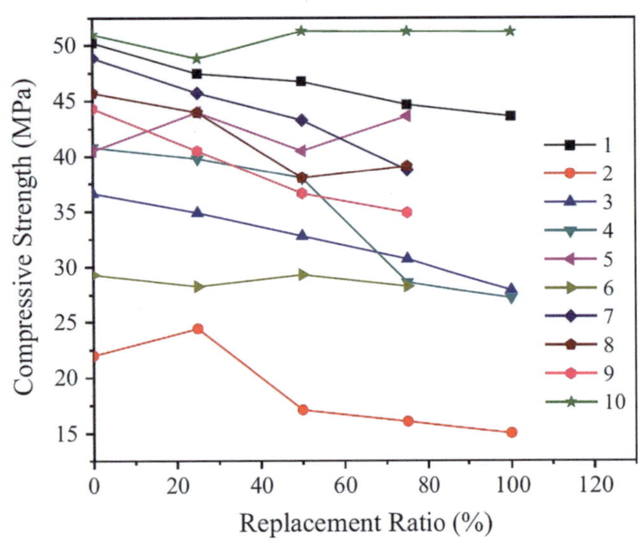

◘ Fig. 9.42 Compressive strength of RAC for different experimental studies (Kisku et al. 2017)

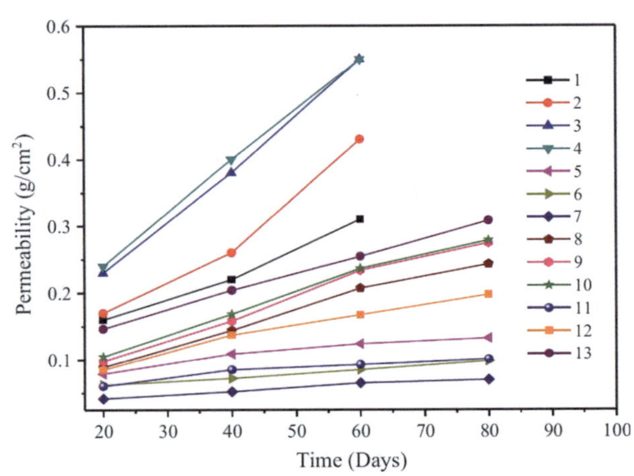

◘ Fig. 9.43 Permeability behavior of RAC with time for different experimental studies (Kisku et al. 2017)

■ Fig. 9.44 Use of recycled construction and demolition waste aggregate for road course surfacing (Herrador et al. 2012)

utilization of recycled concrete aggregates for pavement fill, foundation, subbase, and surface layers.

The most common utilization of recycled aggregates in many countries is in road construction. This is because recycled aggregates usually meet all requirements for their utilization in this application. Reliable and strong base layers are built by using recycled aggregates. Taking into account the different pavement construction applications, recycled aggregates are principally utilized in unbound applications (i.e., in subbase and base layers) and less in hydraulically bound and bituminous bound mixtures. The recycled materials "are not susceptible to any weathering or appreciable physical or chemical changes over their life spans, regardless of their application, and recycled aggregates do not produce water solutions that can damage concrete structures and cement-treated pavement layers or contaminate soils and water currents" (Jiménez 2013).

Unbound Pavements

As unbound roadbed materials, unclassified recycled aggregates, recycled mixed concrete aggregates, and materials from the prescreening process in recycling plants are utilized for roadbed materials for roads with medium and low average daily traffic volumes. This utilization produces a low value-added product although consumes large amounts of recycled materials with low embodied energy. ■ Figure 9.45 shows the use of RCA in unbound aggregate shoulder surface applications.

Recycled aggregates utilized for road bases or foundations must be resistant to heavy loads. Consequently, "requirements regarding grading and compression should be met; aggregates used for road base and foundation layers also become part of the subsoil, so environmental aspects should be paid attention to as well" (Lauritzen 2019). In this sense, "the recycling processes of the CDW and the correct classification and selection of the raw waste materials is of great importance in the quality of the constructed layers; minor additional caution should be used when their use is foreseen in foundation or base layers" (Pourkhorshidi et al. 2020). Recycled aggregates have also been used as "cement treated materials in recent years and this application is one of the best ways to recycle these materials because it does not have a complex treatment process in recycling plants and gives high added value to the RA" (Lauritzen 2019).

Fig. 9.45 RCA aggregate shoulder (50% blend with natural aggregate) on US 34 east of Fairfield, Iowa (Snyder et al. 2018)

9

Bituminous Mixtures

The use of recycled aggregates in bituminous mixtures is an attractive approach for constructing new pavements sustainably since their incorporation generates many advantages (i.e., preservation of natural resources, lower energy consumption, lower harmful gas emissions, and cost benefits). Despite these advantages, the inferior performance of RCA incorporated in bituminous mixes restricts its field utilization. Currently, the use of recycled aggregates in bituminous mixtures is mandatory in highway engineering, "although their use is still one step behind the application to other construction materials such as concrete" (Giménez-Meijide and Pérez 2013). Lower strength and higher porosity of recycled concrete aggregates produce important negative effects on the different properties of asphalt mixes.

Cement mortar adhered to the surface of the recycled concrete aggregate particles increases the internal friction between aggregate particles. As a consequence, it is generated an increase in the mixing time required to achieve a 100% coating, complicating the compaction processes. Further recycled concrete aggregate incorporation decreases the density of the asphalt mix. Thus, the increase of the optimum bitumen content of the asphalt mixes is due to the higher absorptive nature of recycled concrete aggregates; it can result many times in uneconomical construction.

The utilization of several treatment methods to enhance the performance of recycled concrete aggregated included in asphalt mixes is one of the most broadly employed improvement techniques. These treatment methods can be applied to recycled concrete aggregates and mix itself. The most important issue linked to some of these treatment methods is their high resource and power consumption nature, which can many times jeopardize the economic and environmental benefits of replacing natural aggregates with recycled aggregates.

9.4 Questions

Short Questions

- Define construction and demolition waste (CDW)
- List the main components of CDW
- What are the objectives of the management of CDW?
- Explain the aim of a valorization plant for CDW treatment
- What is the more recycled CDW component? Explain the reasons
- Definition of recycled aggregate
- Enumerate the main types of recycled aggregates based on their composition
- List the most important properties of recycled aggregates

- Explain the significance of adhered mortar in recycled aggregates
- Comment the importance of water absorption in recycled aggregates

Long Questions

- Describe the processes that take place in a construction and demolition valorization plant
- Explain the principal applications of recycled aggregates

References

Agrela F, Alaejos P, de Juan MS (2013) Properties of concrete with recycled aggregates. In: Pacheco-Torgal F, Tam VWY, Labrincha JA, Ding Y, de Brito J (eds) Handbook of recycled concrete and demolition waste. Woodhead Publishing Ltd., Cambridge, pp 304–329

Akbarnezhad A, Ong K, Tam C, Zhang M (2013) Effects of the parent concrete properties and crushing procedure on the properties of coarse recycled concrete aggregates. J Mater Civ Eng 25:1795–1802

Behera M, Bhattacharyya SK, Minocha AK, Deoliya R, Maiti S (2014) Recycled aggregate from C&D waste and its use in concrete—a breakthrough towards sustainability in construction sector: a review. Constr Build Mater 68:501–516

BioIS (2011) Management of C&D waste. European Commission DG ENV Final Report Task 2, 240 p

Blengini GA, Garbarino E (2010) Resources and waste management in Turin (Italy): the role of recycled aggregates in the sustainable supply mix. J Clean Prod 18:1021–1030

Bustillo M (2021) Construction materials—geology, production and applications. Springer, Cham, 603 p

Cardoso R, Silva RV, de Brito J, Dhir R (2016) Use of recycled aggregates from construction and demolition waste in geotechnical applications: a literature review. Waste Manag 49:131–145

CCANZ (2011) Best practice guide for the use of "recycled aggregates in new concrete". Cement and Concrete Association of New Zealand, Technical report TR 14

Chen HJ, Yen T, Chen KH (2003) Use of building rubbles as recycled aggregates. Cem Concr Res 33:125–132

De Brito J, Saikia N (2013) Recycled aggregate in concrete. In: Use of Industrial, construction and demolition waste. Springer-Verlag, London, 445 p

Deloitte (2017) Study on resource efficient use of mixed wastes, improving management of construction and demolition waste—final report. Prepared for the European Commission, DG ENV., 213 p

EPA (2016) Construction and demolition debris generation in the United States, 2014. United States Environmental Protection Agency, 23 p

EPA (2019) Advancing sustainable materials management: 2017 fact sheet—assessing trends in material generation, recycling, composting, combustion with energy recovery and landfilling in the United States. United States Environmental Protection Agency, 22 p

European Commission (2016) EU construction and demolition waste management protocol. European Commission, Directorate-General for Internal market, Industry, Entrepreneurship and SMEs, 52 p

Gálvez-Martos JL, Styles D, Schoenberger H, Zeschmar-Lahl B (2018) Construction and demolition waste best management practice in Europe. Resour Conserv Recycl 136:166–178

Giménez-Meijide B, Pérez I (2013) Recycled aggregates (RAs) for asphalt materials. In: Pacheco-Torgal F, Tam VWY, Labrincha JA, Ding Y, de Brito J (eds) Handbook of recycled concrete and demolition waste. Woodhead Publishing Ltd., Cambridge, pp 378–393

Gonzalez-Fonteboa B, Martinez-Abella F (2007) Shear strength of recycled concrete beams.Construction and Building Materials, 21:887–893

Gonzalez-Fonteboa B, Martinez-Abella F (2008) Concrete with aggregates from demolition waste and silica fume materials and mechanical properties. Build Environ 43:429–437

Hansen TC, Narud H (1983) Strength of recycled concrete made from crushed concrete coarse aggregate. Concr Int 5:79–83

Hackenhaar IC, Waskow RP, Tubino R, Passuello A (2019) Life cycle assessment applied to construction and demolition waste treatment: proposal of a Brazilian scenario. In: IOP Conference Series: Earth and Environmental Science 323, 10 p

Herrador R, Pérez P, Garach L, Ordóñez J (2012) The use of recycled construction and demolition waste aggregate for road course surfacing. J Transp Eng 138:182–190

Jiménez JR (2013) Recycled aggregates (RAs) for roads. In: Pacheco-Torgal F, Tam VWY, Labrincha JA, Ding Y, de Brito J (eds) Handbook of recycled concrete and demolition waste. Woodhead Publishing Ltd., Cambridge, pp 351–377

Kim J (2022) Influence of quality of recycled aggregates on the mechanical properties of recycled aggregate concretes: an overview. Constr Build Mater 328:127071

Kisku N, Joshi H, Ansari M, Panda SK, Nayak S, Dutta SC (2017) A critical review and assessment for usage of recycled aggregate as sustainable construction material. Constr Build Mater 131:721–740

Larrard F, Colina H (2019) Concrete recycling—research and practice. Routledge, Taylor & Francis Group, p 636

Lauritzen EK (2019) Construction, demolition and disaster waste managemet—an integrated and sustainable approach. CRC Press, Taylor & Francis Group, 337 p

Margarido F (2015) Environmental impact and life cycle evaluation of materials. In: Goncalves MC, Margarido F (eds) Materials for construction and civil engineering. Springer International Publishing, pp 799–835

MPA (2019) From waste to resource—a UK mineral products industry success story. Mineral Products Association, London, p 6

Nassar R, Soroushian P (2012) Strength and durability of recycled aggregate concrete containing milled glass as partial replacement for cement. Constr Build Mater 29:368–377

NYC-DDC (2003) Construction and demolition waste manual. NYC Department of Design & Construction, 28 p

Pacheco-Torgal F, Vivian T, Labincha JA, Ding Y, de Brito J (2013) Handbook of recycled concrete and demolition waste. Woodhead Publishing, 672 p

Pacheco J, de Brito J (2021) Recycled aggregates produced from construction and demolition waste for structural concrete: constituents, properties and production. Materials 14:18

Padmini AK, Ramamurthy K, Mathews MS (2009) Influence of parent concrete on the properties of recycled aggregate concrete. Constr Build Mater 23:829–836

Pellegrino C, Faleschini F (2016) Sustainability improvements in concrete industry—use of recycled materials for structural concrete production. Springer, Cham, 175 p

Peng CL, Scorpio DE, Kibert CJ (1997) Strategies for successful construction and demolition waste recycling operations. Constr Manag Econ 15:49–58

Poon CS, Shui ZH, Lam L, Fok H, Kou SC (2004) Influence of moisture states of natural and recycled aggregates on the slump and compressive strength of concrete. Cem Concr Res 34:31–36

Pourkhorshidi S, Sangiorgi C, Torreggiani D, Tassinari P (2020) Using recycled aggregates from construction and demolition waste in unbound layers of pavements. Sustainability 12:9386

Rao MC, Bhattacharyya SK, Barai SV (2019). Systematic approach of characterisation and behaviour of recycled aggregate concrete. Springer, 318 p

RILEM (1994) Specifications for concrete with recycled aggregates. Mater Struct 27:557–559

Sánchez de Juan M (2004) Estudio sobre la utilización de árido reciclado para la fabricación de hormigón estructural. Thesis, ETSI Caminos, Canales y Puertos, Madrid, UPM, 514 p

Silva RV, Brito J, Dhir RK (2014) Properties and composition of recycled aggregates from construction and demolition waste suitable for concrete production. Constr Build Mater 65:201–217

Simões dos Reis G, Quattrone M, Monteiro W, Grigore B, Hoffmann C (2021) Current applications of recycled aggregates from construction and demolition: a review. Materials 14:1700

Snyder MB, Cavalline TL, Fick G, Taylor P, Gross J (2018) Recycling concrete pavement materials: a practitioner's reference guide. national concrete pavement technology center. Iowa State University, Institute for Transportation, 92 p

Strufe N (2005) Eliminating hazardous materials from demolition waste. Proceedings of the institution of civil engineers, paper 13762. Eng Sustain 158:25–30

UNEP (2015) Global waste management outlook. United Nations Environment Programme, 2015, 346 p

Xiao J (2018) Recycled aggregate concrete structures. Springer-Verlag GmbH Germany, 632 p

Zega CJ, Villagran-Zaccardi YA, Di Maio AA (2010) Effect of coarse aggregate type on the properties of recycled coarse aggregates. Mater Struct 43:195–202

9